Textile Pollution

Textile Pollution

By
Dr. D.S. Malik
and
Dr. Pawan Kr. Bharti
Department of Zoology and Environmental Science,
Gurukula Kangri University, Hardwar – 249 404 (UK)

2010
DAYA PUBLISHING HOUSE
Delhi - 110 035

ISBN 9789351242086

Published by : **Daya Publishing House**
A Division of
Astral International Pvt. Ltd.
– ISO 9001:2008 Certified Company –
4760-61/23, Ansari Road, Darya Ganj,
New Delhi-110 002
Ph. 011-43549197, 23278134
E-mail: info@astralint.com
Website: www.astralint.com

Laser Typesetting : **Classic Computer Services**
Delhi - 110 035

Printed at : **Chawla Offset Printers**
Delhi - 110 052

PRINTED IN INDIA

Preface

Environmental pollution by heavy metals such as cadmium, nickel, zinc, lead, copper etc., is accelerated dramatically during the last two decades due to mining, use of agricultural fertilizers, pesticides, traffic emissions, smelting, manufacturing, municipal wastes, industrial effluents and chemicals. Industrial effluents are the waste products include liquid, solid and gaseous waste resulting from industrial processing operation. Liquid waste adversely affect the wastes supply and sewage of towns and cities while a high quantity of water is required for different industrial processes, only a small fraction of it is incorporated in their products, the rest finds its way into the wastewater course. This greatly contributes to the pollution problems to surface water, groundwater, agriculture soil and sediments.

The textile industry provide high economic sources to generate the income and job opportunities to the all strata of human society but the textile industry also play a vital role to deteriorate the surrounding groundwater quality in various manners. The textile effluents have consisting high concentrations of heavy metals, organic pollutants and toxic colours, which may alter the surface water quality of the surrounding environment. Toxic pollutants may percolate down via soil profile and reach in groundwater and agriculture soil through irrigation by wastewater of industrial zone.

The purpose of writing this book is to arrange the available matter and research findings in a comprehensive manner for easy understanding to the research scholars. The book is designed primarily for research scholars providing the fundamental research on the accumulation pathway of heavy metals from industrial effluents to different trophic levels of existing environment in relation to environmental pollution standards and health hazards.

We are indebted to Prof. A.K. Chopra, Registrar, Gurukula Kangri University, Hardwar; Prof. D.R. Khanna, F.O. GKV, Hardwar, Dr. R. Bhutiani and Mr. Nitin Kamboj, Department of Zoology and Environmental Science, Gurukula Kangri University, Hardwar, for their kind suggestions and motivations in the preparation of manuscripts.

The authors are thankful to all persons who have helped directly or indirectly in the completion of this book. The authors express their sincere thanks to all staff members of Daya Publishing House, Delhi for bringing out this book in time.

Finally, the authors deeply acknowledge the patience, support and understanding of their family during the preparation of the manuscript. Suggestions, constructive criticism for improvement of the text will be highly appreciated and acknowledged.

Dr. D.S. Malik
E-mail: malikdsgkv@gmail.com

Dr. P.K. Bharti
E-mail: gurupawanbharti@rediffmail.com

Contents

Chapter 1

Introduction

In the last two decades, the rapid growth of industrialization and urbanization has created negative impacts on the environment. The industrial, municipal and agricultural wastes containing pesticides, insecticides, fertilizers residues, organic pollutants and heavy metals in their effluents have been polluted surface and groundwater. In India, the industrial effluents have contributed a major source of pollution. In the modern industrial era, rapidly growing of industrialization and urbanization has posed adverse impacts on the environment and its all components. Air, soil and water everything on earth surface may changed its quality due to the increasing pollution. Also the underground water resources may affect after a long time by continuous pollution condition.

In modern industrialization period, the most of water resources have affected enormously by seepage, leaching and intermixing of industrial effluents in maximum metropolitan cities and industrial townships. The effluents have consisting high concentrations of heavy metals, organic pollutants and toxic colours, may alter the water quality of the region and ultimately causes the health hazards among livestock and human being.

Water is one of the prime necessities of life. We can hardly live for a few days without water. Water has become an essential commodity for the development of industries and agriculture (Kudesia, 1992). Water is one of the widely distributed and abundant substance found in the nature it covers about 75 per cent of the earth surface therefore earth is sometimes called a "water planet" water is in sea, river, ocean, pond, streams and even in the atmosphere in the form of humidity. Most of the earth surface water is in the sea, the ocean contain about excessive quantities of salts. Fresh water is in lakes, ponds, rivers and steams. Fresh water is also available in the forms of rain, snow, dew etc. The Earth's atmosphere contains 0.02 to 4 per cent water by volume, depending on the location. In addition to providing sources for precipitation, atmospheric water vapors intercept some of the ultraviolet radiation and intercept heat loss from the earth sand redirect part of it to the Earth.

The available natural freshwater resources today are threatened by hazard of pollution particularly rivers are greatly polluted due to release of untreated effluents and waste material from agriculture practices and industries located around water bodies. The poor living conditions of people in settlement around water bodies or rivers, non-availability of treatment from urban areas and negligence of industries for treatment of effluent before release of natural water bodies are the major reason of pollution of Indian rivers and other water bodies like ponds, lakes, etc.

Drinking water are obtained from all sorts of sources, some good, some not well, some bad and some outright dangerous. These are reflected in the health, vitality and longevity of people. Since one gulp water and it is taken straight into the body its cleanliness is vitally important. It should be free from pathogens, it should not contain excessive amount of salts and toxic elements. Scientifically, distilled water is the cleanest, but it is not good drinking water. Water is the mostly used for industrial, municipal and agricultural purpose. The quantity and quality of available water are very important for the purpose of industries. Each industry has its own water requirements and sometimes adequate supply of water may be very suitable for one industry but the same may be dangerous for other. It is therefore, extremely important to take into account the use of water to be carried out and its suitability based on the result of chemical analysis.

1.1 Pollution Aspects

As we know our life depends upon water and man needs water for his drinking and other domestic and industrial processes. With the fast increasing in our industrial civilization, the demand for water is increasing tremendously day-by-day. Although water pollution is an age old problem but in this modern age, the problems like population increase, sewage disposal, industrial waste, radioactive waste etc. have polluted our water resources so much so that about 70 per cent rivers and streams not only in India but of all the countries contain polluted waters (Kudesia, 1992). The industrial effluents and trade waste of pulp and paper, distillery, fertilizer, electroplating, asbestos, silt, alcohol, detergents, steel, tannery, textile, cane sugar, oils, pesticides and herbicides, radioactive wastes etc., play a significant contribution in pollution of water.

In India, the industrial effluents have contributed a major source of pollution. Different industries of the country contribute about 16 per cent of the total wastewater generated. The treatment facilities have been installed in less than one third of the polluting industries (Gopal, 1994). Most of effluent treatment plants (ETP) have not performed physical and chemical treatment processes satisfactory due to economic reasons on commercial scales. Water may pollute by many anthropogenic activities like discharge of household sewage, disposal of various industrial wastes, run off form agricultural fields, acid rain and many other man-made infrastructures and activities. Different types of industries exploit the huge amount of freshwater and discharge wastewater in the form of effluents. These effluents discharged directly into any freshwater body, *viz.* pond, lake, stream or any seasonal drain.

Textile industry is one of the oldest type of industry in India, manufacturing different types of clothes. There are about 850 textile mills established in big cities of India, manufacturing approximately 11000 million meters cloth. Textile industries can be classified into three categories: cotton, woolen and synthetic fibres depending upon the used raw materials. The cotton textile industries are one of the oldest industries in our country having more than 1000 mills mainly located in Mumbai, Surat, Ahmedabad, Coimbatore, Delhi and Kanpur. The woolen mills in the country are mostly located mainly in Amritsar, Bikaner, Ludhiana and Panipat. Synthetic fibre, a comparatively new entry, has become an integral part of the textile industry and is being used to manufacture purely synthetic fabric or is mixed with cotton or wool to manufacture blended fabrics in mills throughout the country (Groff, 1993).

Textile industries in general consume large volume of water of high purity. Consequently, these units discharge large quantities of effluent that normally exhibit polluting characteristics. According to review of available literature the quantity of effluent discharged and quantity of consumed water by woolen units and composite cotton textile industries, always remain high. In the cotton industry, the solid waste generated during processing fibres and spinning amounts to be about 5 per cent of raw cotton weight. In woolen industry, about 4 per cent of the raw wool weight comes out as solid waste prior to and during spinning. Other solid waste generated in the textile industry includes rags, felts, waste piece of cloth and yarns, packaging materials, ash from coal-fired boiler.

As far as the textile mills are concerned boiler feed water should be as soft as possible and should contain least amount of nitrate and organic matter in order to prevent encrustations or corrosion of boilerplates. For the purpose of dyeing water should be free from iron and should posses little hardness only. The consumption of water in dyeing and printing mills is 3,45,000 liters/day.

1.2 Textile Industries

Consumer demand for textiles is ever increasing in the domestic and international market. More cotton textile processing industries are set up at a rapid rate in many Indian cities, the problem of treating wastewater generated at these industries are compounded. Small-scale units cannot afford to have their own treatment facilities because of non-availability of land and skilled manpower to operate it. A common effluent treatment plant (CETP), for the treatment of combined wastewater from a cluster of small scale units is generally designed based on primary and secondary waste treatment system, which involves screening, degreeting, primary clarifier and aerobic biological system.

In the preparation of these raw materials for spinning, weaving and finishing, liquid wastes are produced, the volume and strength of which extend over a wide range, depending not only on the type of fibre being processed but also on the source of raw product, the ultimate use, availability of process water, control of manufacture and other variables.

The textile mills not only add to air pollution by ejecting chimney gases into the atmosphere but also the considerable water pollution by discharging their liquid effluents into various receiving bodies like public sewer, inland surface water like ponds, rivers, *Nallas* or on irrigation land. In fact, major pollution load from the textile mills can be attributed to the liquid effluents from the various wet processing operation like scouring, bleaching, mercerizing, dyeing, etc. amount of composite effluents discharged from the textile mills varies from 1 to 10 million liters per day depending upon quantity of the cloth produced and various manufacturing processes employed. Selection of treatment methods for these wastewaters will be governed by their characters, the policy of regulatory body covering the waters of receiving stream, economics of treatment and present day knowledge of method of purifications.

Textile industries consume substantial volume of water and chemicals in wet processing of textile. Approximately, 40,000 different dyes and pigments are used in the industry and over 7×10^5 tons of these dyes are produced annually worldwide (Zollinger, 1987). It was estimated that 10-15 per cent of the dyes used in textile processing were lost in the effluent during the dying processes (Vaidge and Datye, 1982). These dyes include several structural varieties of dyes, such as acidic, reactive basic, disperse, azo, diazo, anthraquinone-based and metal complex dyes. The only one thing in common among them is the ability to absorb light in the visible region (Wong, 2000). The effluents from dye industries contain high amount of cation, anion, organic pollutants in the form of colour and heavy metals.

Dying is the process of applying to the fabric stock yarn or fabric. The dying is carried out with six different dye classes, vat, direct, developed, napthol, sulphur and aniline black. If aniline and sulphur dye are used, dye wastewater may contain colours, which are extremely difficult to be removed by common waste treatment methods. The water requirement for textile industry is very large for their various processes. The use of water varies from mill to mill and depends on source of water, availability and quantity and quality of fabric produced. To produce one meter of cloth 12-65 liters of water is required, as textile industry requires large volume of water, it releases large amount of wastewater after processing. The main source of wastewaters of textile industry *viz.* preparation of yarn-slashing process waste, textile processing- desizing, kiering, slashing, bleaching, mercerizing, dying and printing waste, washing waste in each step, other waste streams-like cooling water, boiler, house keeping, spills, leaks and sanitary wastes. The quantity of waste discharged from each mill process is very small as compared to the volume of waste generated from the washing and rinsing of cloth. The volume of wastes is highly variable from one mill to another. The pollution load in wastewater posed by textile industry is very high and affects water quality and other environmental components in many ways. These wastes are coloured due to use of dyes and dying aid agents in the different textile processing of material.

1.3 Dyes

Synthetic dyes are major contaminants in textile wastewater and have to be removed before discharging into water bodies or on land. The presence of very small amounts of synthetic dyes in water (less than 1 mg/l for some dyes) is highly visible and affects the aesthetic qualities, water transparency, and gas solubility, in lakes, rivers and other water bodies. The removal of synthetic dyes from wastewater is often more important than the removal of the soluble colourless organic substances, which usually contributes the major fraction of the biochemical oxygen demand. Methods for the removal of the BOD from most effluents are fairly well developed, synthetic dyes, however, are more difficult to be degraded because of their synthetic origin and mainly complex aromatic molecular structures. Such structures are often constructed intentionally to resist fading on exposure to sweat, soap water, light or oxidizing agents (Mckay, 1979) and this renders them more stable and less amenable to biodegradation (Seshadri *et al.*, 1994).

Interest in the pollution potential of textile dyes comes from their possible toxicity and carcinogenicity. This is because many dyes are synthesized from known carcinogens, such as benzidine or their aromatic compounds. Some azo dyes are reported to be reduced in sediments and aquatic environment to form toxic, mutagenic or even carcinogenic aromatic amines among different trophic organisms.

Dye wastewater discharged from textile and dyestuff industries have to be treated due to their impact on the water bodies and growing public concern over their toxicity and carcinogenicity in particular. Dyes usually have synthetic origin and complex aromatic molecular structure. According to their dissociation in an aqueous solution, dye can be classified as (i) Anionic acid, direct and reactive dye (ii) Cationic, basic dye and (iii) Non ionic, dispersive dyes. Dyes such as acid, basics and direct are all water soluble but dispersive dyes have low solubility and colloidal dispersion properties, thus they are agglomerations. Physically, dyes are mostly used in the powder, liquid and pearl or granulate form. Chemically, dyes may be categories as Nitro dyes (napthol yellow), Nitroso dyes (napthol green), Azo dyes (synthetic), Di and triphenol methane dyes, Vat dyes, Anthraquinonoid dyes, Quinoline dyes, Sulphur dyes. Broadly dyes may be classified as direct dyes and developed dyes. Six main classes of dyes for cotton fabric namely vat, direct, developed, napthol, sulphur and aniline black. Many different and complicated molecular structures of dyes make dye water difficult to be treated by conventional biological and physico-chemical processes. About 6,34,900 metric tons of dyes are produce worldwide each year and nearly 10-15 per cent of them are discharged through effluents. Generally conventional biological treatment processes have certain difficulties in degrading those dues causing high COD and colours in the wastewater. A significant portion about 40-70 per cent of COD and colour in textile effluents, are still remaining, and is required an advanced treatment. Studies using physical and chemical processes to further reduce COD and colour in textile wastewater have been intensively reported (Zhang *et al.*, 1994). Developing countries like India and other Asian countries discharge the effluents to the surface water without any treatment or sometimes little treatment due to technological and economical limitations. Colours affect the nature of water, inhibit sunlight penetration and reduce the photosynthetic action. Some of the dyes cause rapid depletion of dissolved oxygen affecting aquatic life adversely. Some of the dyes, especially Benzidine based dyes are toxic and carcinogenic.

Thus, uses of dyes contaminated water without any treatment may cause adverse effect on the health of human beings, domestic animals, wildlife and aquatic environment. So, it is very much essential to treat dye-contaminated wastewater before discharging in to the surface water and develop simple and economical method to treat the stream suitable for developing countries. The conventional methods of removal of dyes using alum, ferric chloride, activated carbon, lime etc. are not economical. Adsorption is a physico-chemical treatment process, which has gained a lot of attraction for the removal of dye from wastewater, since it produces a high quantity treated effluent. The most widely used adsorbent is activated carbon for industrial wastewaters application. But the cost of the activated carbon is high and it increases with its quality. Agricultural solid waste such as coir, pitch, banana pitch, coconut husk, biogas slurry, rice husk, bagasse and paddy straw, saw dust etc. have been investigated for the removal of dye from wastewater.

1.4 Heavy Metals

Heavy metals are chemical elements with a specific gravity that is at least 5 times the specific gravity of water. The specific gravity of water is 1 at 4°C (39°F). Simply stated, specific gravity is a measure of density of a given amount of a solid substance when it is compared to an equal amount of water. Some well-known toxic metallic elements with a specific gravity that is 5 or more times that of water are arsenic, 5.7; cadmium, 8.65; iron, 7.9; lead, 11.34; and mercury, 13.546. In small quantities, certain heavy metals are nutritionally essential for a healthy life. Some of these are referred to as the trace elements (e.g., iron, copper, manganese, and zinc). These elements or some form of them are commonly found naturally in foodstuffs, in fruits and vegetables and in commercially available multivitamin products. Heavy metals are also common in industrial applications such as in the manufacture of pesticides, batteries, alloys, electroplated metal parts, textile dyes, steel and so forth. Twentieth century industrialization has led to an increased exposure to heavy metals for everyone. Certain people suffer a lot due to toxic exposure because of where they live, their occupation, or their lifestyle. Others may be genetically predisposed to heavy metal accumulation due to poor detoxification systems in their bodies.

Heavy metals are usually present in trace amounts in naturally waters but many of them are toxic even at very low concentration. Although many of the metals are essential components of the biological system yet some of these are potentially toxic. Metals such as arsenic, lead, cadmium, nickel, mercury, chromium, cobalt, zinc and selenium are highly toxic even in minor quantity. Increasing quantity of heavy metals in aquatic resources is currently an area of greater concern especially since a large number of industries are discharging their metal containing effluents in to fresh water with out any adequate treatment (Canter, 1987).

Heavy metals become toxic when they are not metabolized by the body and accumulate in the soft tissues. Heavy metals may enter the human body through food, water, air, or absorption through the skin when they come in contact with humans in agriculture and in production of different industrial products, pharmaceutical, biotechnological or residential settings. Industrial exposure accounts for a common route of exposure for adults and children. Ingestion is the most common route of exposure in children. Children may develop toxic levels from the normal hand-to-mouth activity of small children who come in contact with contaminated soil or by actually eating objects that are not food (dirt or paint chips).

Most of the industries discharged their waste directly (without any treatment) into the stream, lakes and oceans as well as in the open land and that contaminate the groundwater. There are number of pollutants like fertilizers, pesticides, heavy metals, which seriously affect human like by entering into the system directly or indirectly through food material. Pollution of environment with heavy metals is a serious problem. Besides causing specific toxicity symptoms, these metals may also contribute to global warming by destroying the atmosphere ozone layer like atmosphere methane, nitrous oxide and sulphur dioxide because of potentially harmful effects on human and animal health, few toxic metals (lead, cadmium, mercury, arsenic, and chromium). It is major concern to ecologist or researcher, because air or water pollution from point and non-point sources may contribute significantly to the load of metals on natural ecosystem (Mani *et al.*, 2005). Environmental contamination and exposure to heavy metals such as mercury, cadmium and lead is a serious growing problem throughout the world. Human exposure to heavy metals has risen dramatically in the last 50 years as a result of an exponential increase in the use of heavy metals in industrial processes and products.

The most common heavy metal pollutants are As, Cd, Cr, Cu, Ni, Pb, and Hg. There are different types of sources of metal pollutants; such as, point sources (localized pollution). The second types of pollutant sources are non-point sources, where pollutants come from dispersed (and often difficult to identify) sources. There are only a few examples of localized metal pollution, like the natural weathering of ore bodies and the little metal particles coming from coal burning power plants via smokestacks in air, water and soils around the factory. The most common metal pollution in freshwater comes from mining operations. They usually use an acid mine drainage system to release heavy metals from ores, because metals are very soluble in an acid solution. After the drainage process, they disperse the acid solution in the groundwater, containing high levels of metals (www.lenntech.com).

Heavy metals are introduced into aquatic system from various anthropogenic sources. Heavy metals constitute a special group of contaminants of aquatic systems and deserve special attention. Since natural degradation processes do not remove metals, they may become enriched in sediments over time. Metal contamination of sediments is an issue of growing concerns worldwide (NRC, 1989). Both natural processes and anthropogenic activities are responsible for introducing metals in to the aquatic system (Nriagu, 1989). Many contaminants discharged into surface waters rapidly become associated with the particulate matter and incorporated in sediments. Metals in aquatic systems become part of the water-sediment system and their distribution is controlled by a dynamic set of physico-chemical interactions and equilibria, largely governed by pH and type of legends and chelating agents, oxidation state of the mineral components and the redox conditions of the system. Metal contaminated sediments may release heavy metals back to the overlying water column and, thus, pose risk to aquatic life and ecosystems. Due to their particle reactivity, heavy metals tend to accumulate in sediment as a result may persist in the environment long after their primary sources have been removed (Forstner and Wittman, 1981).

In the recent years there has been a growing concern with environmental protection. This can be achieved either by decreasing the afflux of pollutants to the environment or by their removal from contaminated media. The former is a feasible choice only for pollutants of anthropogenic origin, whereas, the latter is unavoidable for those of natural origin (Gomez-Serrano *et al.*, 1998). The presence of heavy metals in the aquatic environment has been of great concern to scientists and engineers because of their increased discharge, toxic nature, and other adverse effects on receiving waters.

1.5 Groundwater

Groundwater is used for domestic supply, industries, and agriculture in most part of the world as it is a replenishable resource and has inherent advantage over surface water. There has been a tremendous increase in the demand of fresh water due to growth in population, advanced agriculture practices and industrial usages (Bachmat, 1994). In the last two decades, the rapid growth of industrialization and urbanization has created negative impacts on the environment. Due to industrial, municipal and agricultural wastes containing pesticides, insecticides, fertilizers residues and heavy metals with water; groundwater has been polluted by leaching process of pollutants present in effluents (Sastry and Rathee, 1999).

In Haryana, about 90 per cent of the population of rural areas utilizes groundwater for drinking, washing and other daily uses. But groundwater in Haryana is very saline and large variation in salinity can be observed over a short distance. Water table is also very high. Therefore, scarcity of drinking water is very big problem faced by population of this state. Panipat district of Haryana is well known for the textile industries. A large number of small-scale industries mainly textile printing and textile dying have been established in Panipat city. These industries discharge their effluent openly into pond or other water bodies. The seepage of effluents posed serious threat to the groundwater resources (Groff, 1993). So, groundwater must be managed and protected carefully for the most beneficial use in the future.

Clean groundwater is needed for more than drinking purposes. Agriculture depends highly on groundwater for irrigation; as a large number of groundwater pumps were used for crop irrigation in India. Poor or contaminated groundwater could jeopardize crops and threaten the health of livestock. Groundwater is connected to surface water in the hydrological cycle, and some aquifers actually feed area springs and rivers. Sometimes contamination occurs naturally by mixing the sallow aquifers with other contaminated sources of water table. Saline water from deeper aquifers may reach aquifers that provide water for human beings, livestock and others agricultural activities. Some groundwater may have naturally high levels of nitrates, metals, iron, sulfate, or chloride, all of which provides an odd odor, color, or taste in water. Groundwater contamination has become a major problem in recent decades due to exploring industrialization. Groundwater, as a wide resource for irrigation in most part of country, may contaminate the soil quality by repeated irrigation of agricultural fields. Bed quality of groundwater used in irrigation may alter the characteristics of agriculture soils of the area.

1.6 Agricultural Soil and Sediment

Agricultural soils require sufficient irrigation for high production and irrigation water with poor quality can result in a buildup of soil salts and high soil pH. The chief minerals in irrigation water are chloride, sulfate, bicarbonate, sodium, calcium, and magnesium. These minerals contained in the irrigation water can build up in the soil and cause problems. Sodium, bicarbonate, and chloride are the three minerals that contribute most to soil salinity and alkalinity. More frequent irrigations become necessary to keep up with increasing soil salinity levels. Plants are most sensitive to saline soils during germination. Once established, they have the ability to tolerate higher soil salinity levels. Soil acidity affects a plant's ability to absorb nutrients.

The sediments existing as the bottom of the water column play a major role in the pollution scheme of an aquatic system by heavy metals (Forstner, 1985). They reflect the current quality of the water system and can be used in detecting the presence of contaminants that do not remain soluble after discharge into surface water. As a result of complex physical, chemical and biological processes a major fraction of heavy metals (contributed naturally as well as through various anthropogenic activities) is found to be associated with bottom sediments (Baruah *et al.*, 1996). Bed sediments in surface water systems thus act both as a sink and source of metals. Metal accumulation in sediments provides a record of the spatial and temporal history of pollution (Martin and Whitfield, 1983). Hence, sediment monitoring can provide important information on various pollution events.

Discharges of textile waste on land and used for agricultural purpose have also affected the crop and soil productivity due to high dissolved salts. The suspended solids, which are present in the waste, may carry out clogging of soil pores. The sodium, which is present in the waste, may lead to hardening of texture of soil, thereby preventing penetration of roots in soil.

Textile effluent pollution may also alter the characteristics of sediment and soil of industrial area directly or indirectly. Composite effluents from textile industries and dye houses of Panipat industrial area normally discharged openly into a common effluent drain without any adequate treatment and commences in a large pond near Binjhole village near to industrial area, which may change the quality of bottom sediments of drain, pond sediment quality and also the physico-chemical characteristics of surrounding agricultural soil.

Dyes may be harmful for living organisms because these contain toxic heavy metals. Labour and handling workers may affect by toxic metal with contact directly to soft organs, wounds or by inhalation. Heavy metals present in the dyes, may affect a community by consumption of contaminated water. These are harmful aspects of textile industries regarding the loss of properties and human health hazards, while the rest negative impacts of textile industries are on environment and its all components, living and non-livings, such as water, soil, agriculture plants, cattle, human beings. So, the present study was aimed for an assessment of groundwater and soil quality affected by textile industrial effluents.

Underground water is one of the major water resources for domestic, agricultural and industrial uses. Pollutants are being added to the groundwater and soil system through various human activities and rapid growth of industrialization, which affects the human health directly or indirectly. This study would provide baseline data on status and management of industrial effluent pollution as well as its impact on groundwater potability and agricultural uses and suggest remedial and preventive measures in future for remediation studies of pollutants removal from groundwater and agricultural soil system.

Chapter 2

Overview of Textile Industries in Panipat

Geographically, Haryana is situated between 27° 37' to 30° 35' North latitude and between 74° 28' to 77° 36' longitude on the globe. Uttar Pradesh borders Haryana on the East, Punjab on the West, Himachal Pradesh on the North, and Delhi and Rajasthan on the South direction. In the geographical conditions, Haryana is agriculture based green state of the India. The state of Haryana came in 1966, when it was curved out of Punjab. Haryana ranks as one of India's most prosperous states among all other states.

Demographically, the population of Haryana, according to the 2001 census, is 2,10,83,000, with 1,13,28,000 males and 97,55,000 females. The population density is 477 people/sq km. Hindus are about 80 per cent of the population, Sikhs 16 per cent, Muslims 4 per cent and Christians 0.10 per cent (wikipidia.org, 2006).

2.1 History

The Shivalik range rises in the North and the ancient Aravalli hills slope in the South. Haryana displays four main geographical features:

The Shivalik Hills

Altitude varying between 900 to 2300 meters. These hills are the source of the rivers like Ghagger, Tangri and Markanda, parts of Panchkula, Ambala and Yamunanagar districts are in these hills.

The Ghagger-Yamuna Plain

This is divided into higher and lower plain. The higher one is called 'Banger' and the lower one called 'Khadar'. The alluvium plain is made up of sand, clay, silt and hard calcareous balls like gravel, locally called 'Kanker'.

A Semi-Desert Sandy Plain

These areas border the state of Rajasthan and include the districts of Sirsa and parts of Fatehbad, Hissar, Bhiwani and Mahendargarh district.

Aravalli Hills

This mountain range is 20 million year old, older than the Himalayan range. It is dry area with uneven, irregular landscape (www.indiatravelogue.com).

The river Yamuna flows along the Eastern boundary of the state. The ancient and some say mythical Saraswati River may flow throw Haryana but there are no traces of it evident today. The river Ghagger is a seasonal river, which flows from the outer Himalayas and transverses the district of Panchkula, Ambala and Hissar before it flows into Rajasthan and is absorbed by that arid land.

2.2 Background of Panipat City

Panipat city (District Panipat) situated 90 KM from Delhi (National Highway No.-1) on *'Sher Shah Suri Marg'* has a significant importance in the history. Panipat city is famous in India by the name of "City of Weaver".

Panipat District has significant placed in International Market for Handloom Production. Darri, Carpet Mat, Table Cover, Bed sheet, Bed Cover, Curtain etc. are export to Canada, Japan, Germany and Australia. In addition to the above Panipat city is the biggest centre of "Shoddy Yarn" in the World. Blankets prepared through Handloom and Powerloom are sent to soldier. The sub division Samalkha of this district is famous for Foundry of Agriculture instruments. In this way, this district, which is continuously developing on the base of industry, has unlimited employmentness. Not only from Haryana but Businessman, Engineer, unemployed worker artist Weavers and labour from other state of India visits here in search of employment and settled here permanently.

Panipat is situated on the bank of river Yamuna, it lies between 29' 09' 50" and 29' 50' 32" North latitude and 76° 31' 15" and 77° 12' 45" East longitude and its height is 255 meters above sea level. Here the three historical battles of Panipat were fought. Karnal, Sonipat, Jind, Kaithal and Uttar Pradesh bordered this district. Population of Panipat city is 261874 approximately. Panipat town is located on the national highway no.1 about 90 km toward north of Delhi has a population of about 0.27 million. Average population density of the urban area of Panipat city is 25,278 per sq. km. (Figures 2.1).

Climatic Conditions

Topographically, Ground is flat but gently slopes towards south-east. Water level is 15 m below G.L. (Ground Level). Soil is almost sandy and alkaline in nature.

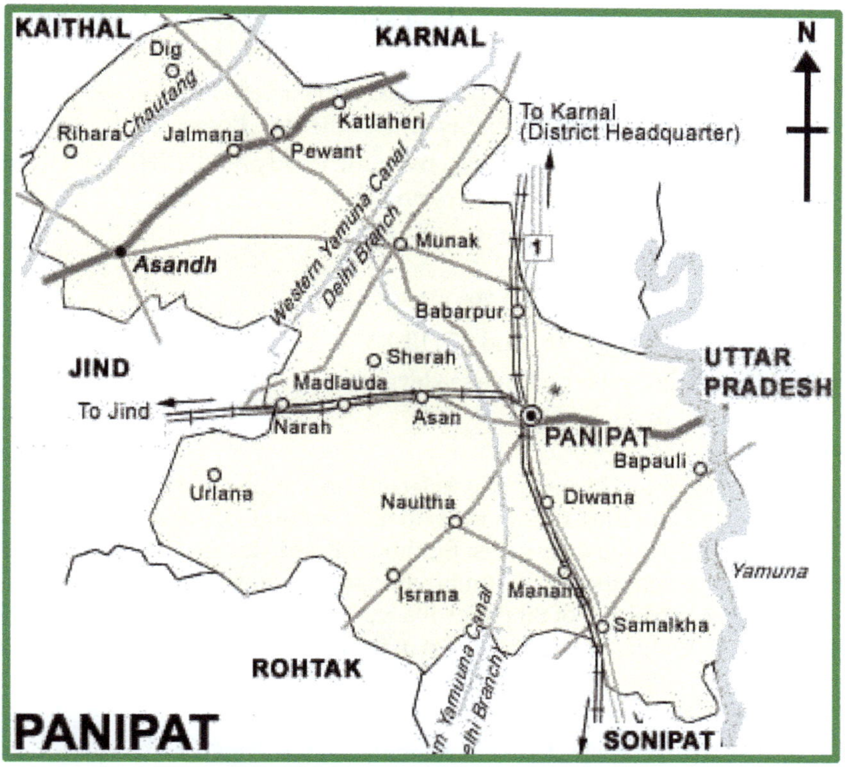

Figure 2.1: Territorial Position of Panipat District of Haryana
(*Source*: haryanaonline.com)

The area of town is 10.8 sq. kms. Regarding the atmosphere and temperature, in summer, warmer and in winter the temperature goes up to minimum 4.4 °C in December, January. However, in summer the temperature goes up to 42.4 °C in June (www.haryanaonline.com).

2.3 Status of Textile Industries in Area

Panipat is a famous city of handloom business in India and abroad also. Industrial scenario and status of Panipat city and salient features of the industries in the Panipat are given below (Verma and Kharat, 2000):

- Town has three major projects under public sector–IOC Refinery, NFL and Thermal Power Station.
- Biggest Centre in the Country for producing Shoddy Yarn and low priced blankets.
- Biggest Centre in the Country engaged in Export of Cotton Durries, Made-ups, Throws and Mats and larger consumer of rags for re-processing.
- Panipat Town is a very old supplier of Barrack Blankets to the Armed Forces.

- ☆ Samalkha, a smaller town near Panipat, is known for the foundry work and supply of Agriculture machinery to nearby U.P. area.
- ☆ A very high per day wage for the labour vis-à-vis national average–a worker earns Rs.70/- to 120/- per day on Handloom/ Power loom.
- ☆ A town has infrastructures such as Rail, Road and ICDS (Inland Container Depots) best suited to Industry and Export.

Geographical Position of Site

- ☆ Latitude: 29° 09' 50" and 29° 11′ 50" N
- ☆ Longitude: 76° 31' 15" and 77° 12' 45" E
- ☆ Altitude: 255 meters above sea level.

Study Area

Panipat district of Haryana was selected on the basis of existence of a large number of textile industries, those results in huge amount of effluents. Panipat city is well known for the textile industries. Binjhol village is located in the south-west of Panipat city. A large number of dye units have come up along the both side of the Jatal road in very close proximity of Binjhole village. Due to the effluent being discharge by textile industries, water used to dry down in the *Nala* itself but with the dramatic rise in the dying units, huge quantities of the effluents are transported through this *Nala* that ultimately joins the village pond.

A pond (12 hac), attached with a effluent drain is situated near this village and collected complete treated and non-treated effluents from the textile industries. As farmers of the area complained that the main source of the groundwater pollution due to seepage from the village pond that is being regularly fed by the effluents from the dyeing industries. Samples were collected from Binjhole village pond, textile effluents, drain and irrigated agricultural fields. Control site was selected in non-industrial area, 10 km. far from experimental sampling site.

Satellite imageries are also attached here as Figures 2.3a and b taken by GoogleEarth (2006), which clearly capture the industrial area as well as affected pond also. Yamuna river may be easily identified by these imageries. Effluent drain Binjhole village and CETP can be identified in Google imageries.

Panipat district of Haryana was selected on the basis of existence of a large number of textile industries, those results in huge amount of effluents. Panipat city is well known for the textile industries. Many dye units have come up along the both side of a common seasonal drain (*Nala*) nearby Binjhole village is located in the south-west of Panipat city. Due to the effluent being discharge by textile industries, water used to dry down in the *Nala* itself but with the dramatic rise in the dying units (Figure 2.2), huge quantities of the effluents are transported through this *Nala* that ultimately joins the village pond.

A pond (12 hac), attached with a seasonal drain is situated near Binjhole village and collected complete treated and non-treated effluents from the textile industries

Figure 2.2(a-b): Showing Sampling Points and Sites in the Study Area at Panipat Region

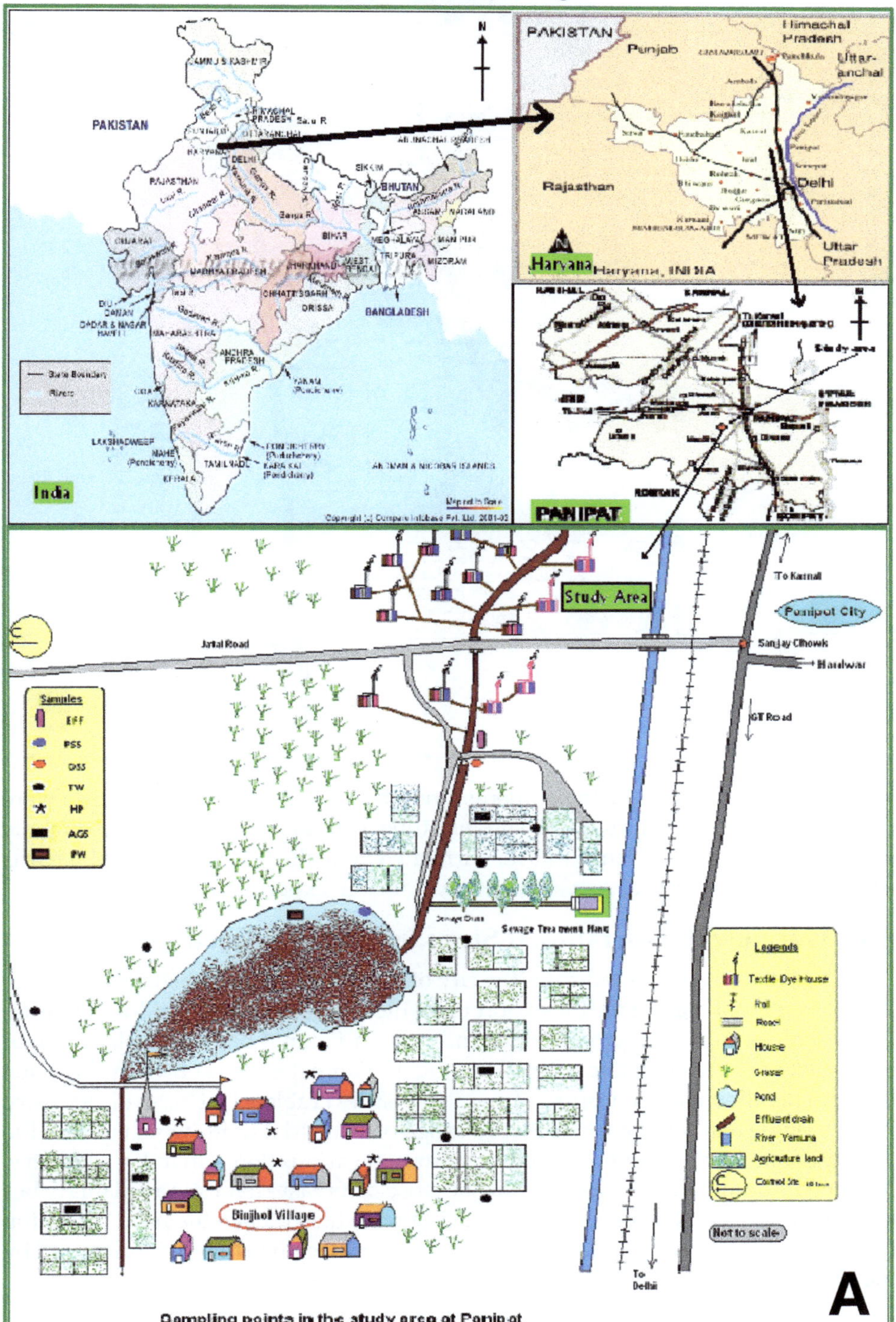

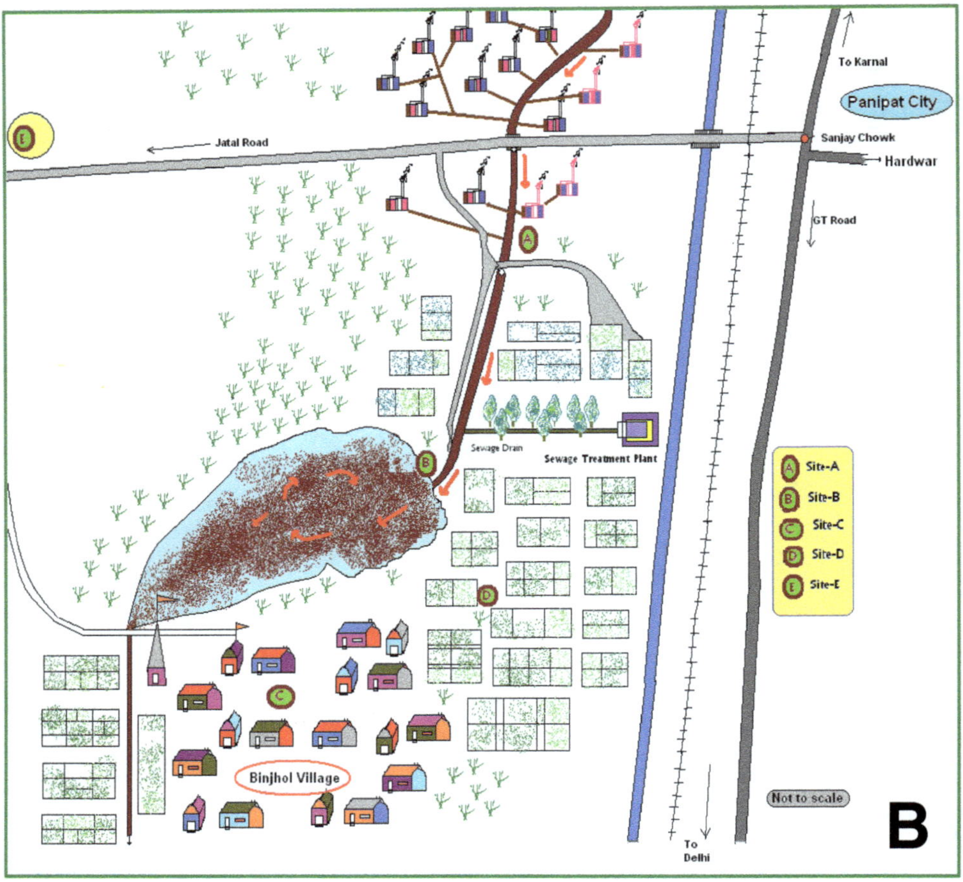

and being regularly fed by the effluents from the dyeing industries. Samples were collected from different sampling sites *viz.* tube wells and hand pumps of Binjhole village, pond, textile effluents drain and irrigated agricultural fields. Control site was selected in non-industrial area about 10 km. far from experimental sampling site.

2.4 Overview of Operational Steps in Textile Industries at Panipat

Panipat as the name indicates the city of textiles. It has established its name in the international market for the export of textile and handlooms products. During three decades rapid growth of textile industry in Panipat has taken place due to consistent very high demand of textile products. Panipat is also known by the name of city of handlooms. It comprises of many industries (Table 2.1). The textile industry is not a single entity but encompasses a range of industrial units, which use a wide variety of natural and synthetic fibres top produce various fibres. In India, cotton has retained its dominant position not only because of its easy domestic availability but also due to the climatic condition in the country which dictate the need for cotton based wear. The growth of the textile industry in any region is greatly influenced by the climatic condition and availability of raw materials.

Figures 2.3 (a-b): Satellite Imageries Showing Industrial Area and Affected Pond

A

Panipat dye houses
Pond
Pond
Image © 2006 DigitalGlobe
© 2006 Europa Technologies
Google
Pointer 29°21'57.31" N 76°56'23.23" E elev 764 ft
Streaming 100%
Eye alt 3410 ft

B

Woolen Industries

Till the 1950s the woolen industry was not developed in India, as demand for woolen garments is seasonal – during the winter season only. Even this demand is noticed only in the northern parts of the country, where winters are much severed. The climate of the country is warm for most part of the year and therefore heavy woolen garments are not preferred. The limited market for this industry is also due to the fact that woolen cloth is expensive and therefore most people do not buy it.

From 16 units in the organized sector in 1950 to 102 units in the late seventies, there are now 621 units in the India. They are mainly concentrated in Punjab and Haryana. Panipat state is ranked seventh in woolen textile industry. Besides the organized sector there are a wide range of decentralized traditional handicraft centers (Figure 2.4).

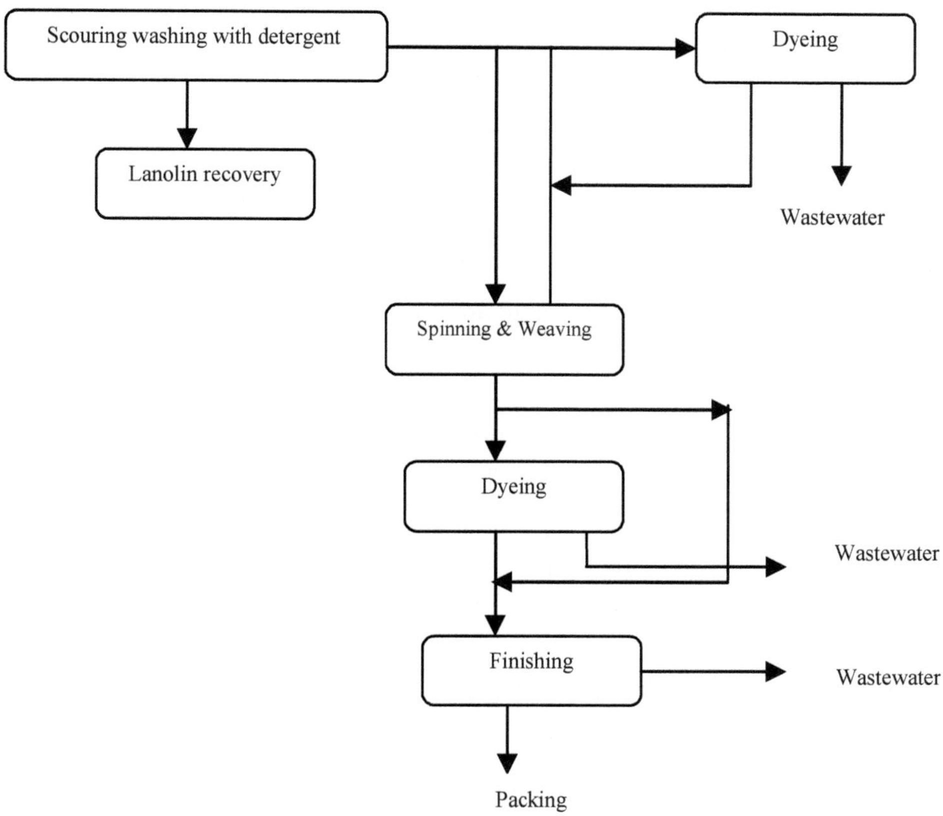

Figure 2.4: Textile Processing Procedures in Woolen Industry

The textile industries can be broadly classified in two groups, namely, cotton industry and woolen industry. The industries based on cotton and woolen fibres also include synthetic fibre blends. For study and development of standards, the textile industries have been classified in following two groups:

Composite Textile Industry

The composite textile industries employ all the process operation starting from cleaning of fibres to dyeing, printing and finishing of fabrics.

Textile Processing Industry

The processing industries are those units, which take grey fabrics as raw material and process it to obtain finished fabrics.

There are various methods of dyeing and many types of dyes may be used for each method. Various dyeing methods have been developed depending on the specific requirements of quality, type of fabric and other practical parameters.

The common methods of dyeing are as follows:

(*a*) Batch Method (Piece Dyeing)
- (*i*) Jigger,
- (*ii*) Winch,
- (*iii*) Beam,
- (*iv*) Jet, and
- (*v*) Padding

(*b*) Continuous Method

Dyeing Application

Based on the application to the fibre stock, yarn or fabric dyeing may be described as yarn dyeing or fabric dyeing.

Yarn Dyeing

Yarn dyeing section handles goods in the form of yarn wound as suitable package, loose fibres of cotton, viscose polyester etc, are also dyed depending upon the requirements. The package are suitably mounted and entered into the required machinery, initial scouring followed by light bleaching as required is given using standard procedures and chemicals. The lot is dyed by circulating the dye liquor through the packages. The dyed lot is subsequently finished and washed as usual.

Khaki Dyeing

For khaki dyeing the fabric is first impregnated with the solution of a mixture of ferrous sulphate and basic chromium sulphate and dried subsequently. It is then passed through a strong solution of caustic soda to precipitate iron and chromium salts as hydroxides. After proper ageing in a steaming chamber, drab khaki shade is developed. Depending upon the quantity of salts and iron used, various shades of khaki can be produced. The fabric is subsequently washed in an open soaper to remove unfixed salts and excess alkali.

Printing

Printing is similar to dyeing except that various colours are applied according to the desired pattern. In order to prevent spreading of the colour, using various printing gums and similar products thickens the dyestuff used for printing. The print paste consists of the dyestuffs, suitable auxiliaries and thickener. Fixation of the print is affected by streaming, dry heating etc. as required for the particular class of dyestuff applied. After fixation of the print, the fabric is given a thorough wash in an open soaper or a winch to remove unfixed dyestuffs and thickening materials.

Finishing

The objective of this is to impart various finishes to the fabrics. Softening treatment is given or stiff finishes are imparted depending upon the requirements, Resin finishing improving the wrinkled recovery is also given to selected varieties of cloth. Compressive shrinking on a rubber belt sanforizer imparts required degree to dimensional stability.

Other Processes

Beside the above mentioned wet processing steps, there are other plants and processes which through not directly involved in the manufacture of cloth, employed by industries for improving the quality of water and for the generation of stream required for various processes. Normally, ion exchange process is used for treatment of water. Chemicals used for regeneration of ion exchange resins are sodium chloride and sometimes an acid and alkali. The effluent coming out from regeneration and rinsing operations contain considerable amount of salinity and sometimes week acid and alkali in case of demineralization process.

Through the proportion of the effluent from this process as compared to the total effluent from the industry are marginal, its high salinity can influence the composition effluent when it is to be discharged on open irrigation land.

Woolen Industries

A brief description about the generation of wastewater and water consumption is given in Table 2.2. The woolen industries may which employ some or all of the following processes are, Scouring, Stock Dyeing, Carding, Fulling, Washing, Carbonising, Dyeing, Bleaching and Brightening (Verma and Kharat, 2000).

Scouring

The wool fleece, prior to treatment has both natural and acquired impurities, which must be removed. The natural wool impurities consist of wool grease, a source of lanolin and suint, a source of potassium. Scouring is used to remove these impurities and is performed using either detergents or solvents. Detergent scouring uses three to five bowls, each with a capacity of 4500 to 13500 litres.

The waste from this type of scouring has an alkaline pH and high BOD concentrations, which contributes 55 to 75 per cent of total BOD load. The solvent scouring process is not widely used in this country; however, it reduces the high BOD strength generated in detergents scouring by recovery of the wool grease and suint. Detergent washing is required after using solvents such as benzene, carbon tetrachloride, ethyl alcohol and methyl alcohol.

Dyeing

Dyeing is performed in either pressure of open vessels. The amount of dye required varies with the type of dye and process being used and the grade of wool to be processed. Stock dyeing is used more often than piece dyeing because of the increasing popularity of multi coloured fabrics. Stock dyeing imparts colour to the woolen fibre prior to weaving operation.

Carding

The carding process (mixing and oiling) does not contribute directly to the pollution load, however, the excess oil is washed out of the fabric in a subsequent washing. In recent years, a non-ionic emulsifier has been substituted for the oil in order to reduce the BOD of the washing operation.

Fulling

The fulling operation follows carding and imparts a felt quality to the wool. Fulling is accomplished by either an alkali or an acid procedure. Soap or synthetic detergent, soda ash, and sequestering agents are the process chemicals used in alkali fulling, whereas sulphuric acid, hydrogen peroxide and minor amounts of metallic catalyst are used in acid fulling.

Washing

The washing process follows fulling to remove the excess process chemicals added in both the carding and fulling operations.

Due to the large amount of chemicals sticking to the wool fabric during above processes, the washing step contributes significant BOD load. The procedures consist of two soaping with an intermediate rolling operation after which the fabric is rinsed with both cold and hot water.

Carbonizing

Carbonizing is a final attempt to remove the vegetable matter from the fabric. This is a sulphuric acid treatment in which the fabric is impregnated with 4 to 6 per cent sulphuric acid solution and oven dried at temperature above 140°C. The fabric is then rinsed and neutralized with a solution of sodium carbonate.

Bleaching

Bleaching with sulphur dioxide, hydrogen peroxide or optical brightener may be performed if white or light coloured fabric is desired.

The textile industries in Panipat are broadly classified into three categories and the water consumption also varies to one another. The industries can broadly be categorized on the basis of capital investment, production, employment generation and the items they are producing (Table 2.2). The industries are growing tremendously in the district.

Table 2.1: Types of Textile Industries at Panipat

Industry	*Numbers*	*Employment (No. of Persons)*	*Turnover (Crores)*
Handloom	>40000	50000	1600
Powerlooms	>15000	28000	76.20
Shoddy and woolen industry	314	6900	568.90
Carpet industry	150	5000	100
Open and spinning	280	–	325 Crores
Barrack yarn spinning units	15	–	28
Carpet woolen yarns	60	–	175
Processing and printing	525	–	–

Source: Verma and Kharat (2000).

Table 2.2: Broadly Classification of Industries and Wastewater Generation in Panipat Industrial Area

Textile Industry	*Water Usage (m^3/1000kg of cloth)*	*Wastewater Generation (m^3/1000kg of cloth)*
	Cotton Textile Industry	
Unit Process		
Desizing	57-65.5	49-55
Keiring	600	245
Bleaching	187.5	155
Mercerizing	244	140
Dyeing	164	134
Printing	261	165
Washing	800	430
	Synthetic Textile Industry	
Unit Process		
Scour	32-82	30-80
Scour and Dye	22-62	20-60
Salt Bath	22-42	20-40
Fine Scour	22-103	20-100
Special Finishing	5-70	5-70
	Woolen Textile Industry	
Unit Process		
Scouring	38-40	35-38
Stock Dyeing	23-25	22-24
Washing	336-340	320-325
Fulling	45-48	43-45
Neutralization	125-132	120-125
Bleaching	3-5	2-5

Source: Verma and Kharat (2000).

2.5 Sampling Programme and Sampling Sites

Keeping in mind to assess the effect of textile industrial effluent on the quality of the groundwater and soil characteristics, the following routine sampling and analysis programme was devised:

Collection of samples	:	Monthly [Between 20th–22nd of each month]
Time of sample collection	:	Between 9.00 A.M. to 11.00 A.M.
On site analysis	:	Temperature, pH, DO and free CO_2
Analysis within 24 hours	:	COD and physico-chemical parameters
Analysis within 15 days	:	Heavy metals
Calculation of data	:	After completion of analytical work

Textile industrial effluent, Groundwater and soil samples were collected monthly from different sampling stations for the study of the status of groundwater and agricultural soils, *i.e.*

(A) Effluent Drain on Jatal Road, (EXP)
(B) Effluent Drain End with Pond, (EXP)
(C) Binjhole Village, (EXP)
(D) Agricultural fields, (EXP) and
(E) Control 10 Km Jatal Road, (C)

The samples were collected from the selected sampling sites during the month of April 2005 to March 2007 in morning hours (from 9.00 am to 11.00 am) as follows:

(A) Experimental Sampling Site

(*i*) Water Samples

1. Effluent from textiles outlet [EF]
2. Tube wells/Hand pumps of the area [TW/HP]
3. Surface water from pond [SW/PW]

(*ii*) Soil Samples

1. Effluent drain sediment [EFS]
2. Pond bottom sediment [PDS]
3. Agricultural soil [AGS]

(B) Control Sampling Site

Water and soil samples were collected from two sampling points of control, located 10 km distance in non industrial area along with Jatal Road at Panipat as follows:

1. Groundwater from tube well of 80 ft [GWC]
2. Agriculture soil [AGSC]

Parameters have Studied

(*a*) Physico-chemical Parameters of Textile Effluents, Surface Water and Groundwater

Temperature, pH, Total solids, Total suspended solids, Total dissolved solids, Electrical conductivity, Dissolved oxygen, Free-CO_2, Biochemical oxygen demand (BOD), Chemical oxygen demand (COD), Total alkalinity, Calcium (Ca^{++}), Magnesium (Mg^{++}), Potassium (K^+), Sodium (Na^+) Chloride (Cl^-)

(*b*) Physico-chemical Parameters of Agriculture Soils, Effluent Sediment and Pond Sediment

Temperature, Water holding capacity, Bulk density, Soil moisture, pH, Electrical conductivity, Chloride, Potassium (K^+), Sodium (Na^+)

(*c*) Heavy Metals in Textile Effluent, Surface Water, Groundwater, Effluent Sediment Soil, Pond Sediment Soil and Agriculture Soil

Cadmium (Cd), Copper (Cu), Iron (Fe), Manganese (Mn), Nickel (Ni), Lead (Pb), and Zinc (Zn).

Chemicals and Glasswares

The chemicals, used in the present investigation were procured by E. Merck India, S.D fine chemicals and B.D.H., India of analytical grade. All the glassware used was of corning grade, manufactured by Borosil India Ltd. All the glassware and Jeri canes were cleaned or washed with soda and chromic acid and rinsed with double distilled water 2-3 times.

Instruments

Following instruments were used during the study:

Sl.No.	*Name of Instruments*	*Manufacturers*
1.	BOD incubator	Tasco
2.	Spectrophotometer	Besto/Elico
3.	Electronic balance	Roy Electronic
4.	Conductivity meter	Tasco Chipset
5.	COD Assembly	-
6.	Atomic absorption spectrophotometer	ECIL
7.	Hot plate	Sunvic
8.	Hot air oven	Modern Scientifics
9.	Mantle heater	A.J. Enterprises
10.	pH meter	Hanna/Tasco
11.	Refrigerator	LG
12.	Distilled water assembly	-
13.	Computer system	IBM/HP

2.6 Sampling Methods

Grab sampling was generally applied during the sampling. A grab sample is an ordinary sample, which is taken from a particular place representing the whole water quality. This type of sample is valid only when it is certain that the water quality is not changing in a short time and effluent discharges, if any, are fairy regular. It is advisable to collect and analyze the grab samples separately at various timings, if the water quality is known to change with time. In such cases the schedule of timings shall be dependent on the frequency of discharge or change in water quality. This approach is used in studies to understand diurnal changes in pH, free CO_2, alkalinity, dissolved oxygen, etc. by collecting a 2 or 3 hour sample round the clock. In any kind of water, at least for dissolved oxygen, several grab samples spread over the whole day or even the night shall be important in limnological study, but in the case of groundwater, there is not any more diurnal changes in general water quality parameters, so a single grab sample is enough to represent the quality status of groundwater.

The sampling of water/soil from stream banks/shallow groundwater/ pond/ drain sediment can be easily carried out using a wide mouthed polythene bottle but a sampler has to be used if the studies are to be undertaken at various strata of water. The sampling of sediments from the bottom of natural water bodies, stabilizing ponds or effluent drains is invariably carried out using special samplers. A large number of water and sediment samplers have been developed over the year with varying degree of sophistication. Most common of them and those, which are readily available in India are Van Dron water sampler, simple laboratory made water sampler, shallow water sampler and Eckmen Dredge sediment sampler.

Eckmen Dredge sediment sampler is a most common type of sampler used for sampling the soil for physical, chemical and biological analysis. It is made up of a brass or steel box provided at the bottom with two jaws hinged together and to the box. Two hinged lids are provided at the top of the box. It has a fixed sectional area when the jaws are in open position, to sample the soil from a fixed area. The box is lowered in water with jaws in open position. When box reaches the bottom, a messenger is sent with the rope which closes the jaws with the bottom sediment enclosed in the box. The box is then hauled up with the upper lids closing automatically under the pressure of water.

The textile effluents, water and groundwater samples were analyzed by standard methods (Wetzel, 1975; Trivedi and Goel, 1984; APHA, 1998 and Khanna and Bhutiani, 2004).

Soil samples were analyzed by standard methods (Trivedi and Goel, 1984; Kalra *et al.*, 1991; Saxena, 1994 and Hesse, 1994).

The trace heavy metals were analyzed in the laboratory by Atomic absorption spectrophotometer *AAS 4129 model* (Singh *et al.*, 1987; APHA, 1998 and ECIL Methods Manual, 2004).

2.7 Some Snaps of Textile Pollution and Assessment

Figure 2.5: Died Yarns in Dye Houses after Dyeing

Figure 2.6: Different Type of Dyes Used in Dye Houses of Textile Industries

Figure 2.7: A View of Textile Effluent Flowing in the Drain

Figure 2.8: Effluent Discharging from Textile Industrial Area

Figure 2.9: A View of Common Effluent Drain for all Dye Units

Figure 2.10: A View of Confluence Point of City Sewage and Textile Effluents

Figure 2.11: Confluence Point of Textile Effluent Drain and Village Pond

Figure 2.12: A View of Pond are Encroached by Villages

Figure 2.13: Hand Pump for Groundwater Sampling in Binjhole Village (Panipat)

Figure 2.14: Soil Sample Collection from Agricultural Fields near Binjhole (Panipat)

Figure 2.15: AAS in the Laboratory of Department of Zoology and Environment Science

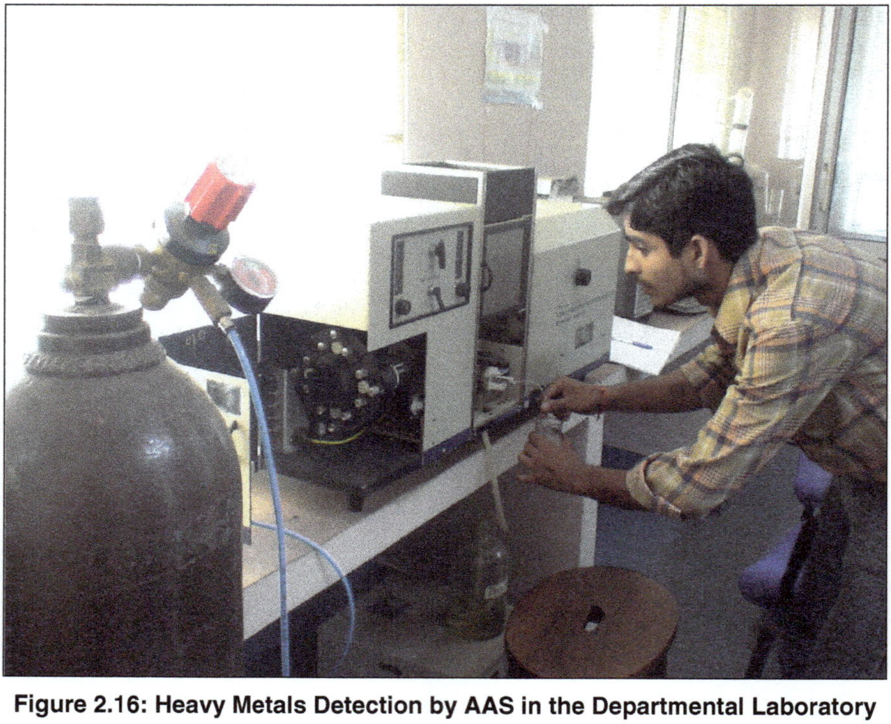

Figure 2.16: Heavy Metals Detection by AAS in the Departmental Laboratory

The sampling was done during April 2005 to March 2007 in morning period (from 9.00 am to 11.00 am) from the selected sampling points of experimental sites as well as control site.

The samples were taken in BOD bottles and plastic Jeri cane for physical, chemical and heavy metals analysis. Following materials were used in sampling process.

(1) Sampling Bottles

300ml capacity BOD bottles were used for collecting the samples. These were washed with chromic acid and then rinse well with tap water followed by rinsing with distilled water. Then these were oven dried at 160°–180° for 3 hours.

(2) Pipettes

Pipettes were washed in same manner as in above. All the pipettes used were of borosil glass certified grade pipettes used were of 1ml capacity (micro pipettes), 5ml capacity and 10ml capacity.

(3) Reagents and Glassware

Some reagents were taken for on spot analysis and sample preservation. Glassware used in study *viz.* conical flask, beaker, burettes etc. were of Borosil glass certified grade. All the glassware were washed with chromic acid and then rinsed well with tap water and then with distilled water.

(4) Jeri Canes

Jeri cane used for carrying the samples from sampling sites to laboratories. Jeri cane made of plastic and washed well with water and then rinsed well with distilled water.

(5) Chemical Bottles

Some small chemical bottles were used for transporting some necessary chemical urgently required during the sampling for immediately analysis for selected parameters.

(6) Bag

A bag of large carrying capacity was taken during the sampling for the collection of all water, soil and sediment samples.

(7) Khurpi

A small paddle (*Khurpi*) was used for digging the soils from agricultural fields and relevant areas related to study.

(8) Polythene Bag

Poly bags were used for the collection of agricultural soils and sediment samples.

(9) Plastic Cane

Many small and big Jeri canes were used for the collection of textile industrial effluent sample, surface water and groundwater samples from different sampling spots for the study.

(10) Thermometer

Mercury thermometer was used for on spot measurement of the soil and water samples.

(11) pH Meter

A handy pen pH meter was used for on spot evaluation of the pH of surface water, groundwater, soil and sediment samples.

(12) Eckmann Dredge Sampler

This sampler was used for the collection of sediment samples from pond sediments and also effluent drain sediment.

2.8 Sample Collection, Preservation, Dilution and Digestion

Collection of Water and Effluents Samples

Due to the discharging of textile effluents in to the pond directly without any adequate treatment by the dye houses, surface water may pollute (Malik *et al.*, 2006). After percolation of pollutants through the soil/sediment profile may affect the groundwater quality of the region. Effluents, surface water and groundwater samples were collected from effluent drain, pond and hand pumps/tube wells respectively. Samples collected in Borosil BOD bottles and some Jeri canes for laboratory experiments.

Some parameters were determined immediately on sampling sites, for rest parameters and heavy metals; samples were stored in refrigerator at 4°C. Heavy metals may be analyzed in the 6 months period from the preservation date.

For determination of the parameters like conductivity, total solids, pH (hydrogen ion concentration), COD, free CO_2, total alkalinity, total hardness Ca/Mg, chlorides, lead, manganese, zinc, calcium and magnesium samples were taken from surface water (Pond) and groundwater from Hand pumps and Tube wells by keeping and opening Jeri canes below the source of water. Caps of cans were closed tightly after filling up of cane.

For dissolved oxygen/biochemical oxygen demand, samples were collected in sterilized 300 ml capacity BOD bottles. The bottles were filled completely with sample water up to the rim and stopper was placed only avoiding any kind of air bubble inside it. DO was immediately fixed by adding 2 ml each of alkaline potassium iodide and manganese sulphate ($MnSO_4$) at sampling spot. While the sample for BOD was incubated for 5 days in BOD incubator at 20°C in the laboratory.

Water samples were acidified in the field with concentrated HNO_3 (5 ml/l of water sample, to reduce the pH of the sample, pH > 2.0) for the total metal estimation. The total metal content was determined by digesting 200 ml of sample with a mixture of concentrated HNO_3 and $HClO_4$ acid (10 ml + 2 ml) (Singh *et al.*, 2005). The digested samples were filtered through Whatman filter No. 42 and finally volume were made 10 ml with 0.1 N HNO_3 and analyzed for heavy metals using AAS (Model ECIL-4129).

Collection of Soil and Sediment Samples

In recent two decades, industrialization creates some negative impacts on environment and its all biotic and abiotic components. Industrial effluents contaminate the groundwater of the industrial region, which may further affect the agricultural soil quality due to the repeated irrigation by contaminated water tapped from groundwater resources with the help of bore wells. Affected soil samples from agriculture fields and effluent drain bed sediments were collected and analyzed for the physico-chemical parameters to assess the quality, fertility, nature of soil and effect of textile pollution on soil surface and texture.

Soil sample selection depends on the nature of analysis. Composite sampling is the best method for soil analysis. Nutritional study requires sample collected at the depth of 10-15 cm. borer samples are used for this purpose. Soil samples were normally collected in polythene bags and dispatched or transported to laboratory as early as possible. Since some parameters like pH, soil moisture, were analyzed immediately. For rest parameters soil samples were dried below 40°C (Millar and Turk, 2002). Soil samples were collected monthly during the period April 2005 to March 2007 for analytical purpose. Samples have been collected separately from each area, which showed some difference in soil colours, texture and others parameters. During collection of samples, stony cities were avoided; sampling was done after the dry sample of soil. The surface material is removed by the help of scrubber. Approximately 500gm soil sample were collected from a depth of 15 cm for each site. These soil samples were taken and placed in a polythene bags. The stones and organic matter, like roots and pieces of leaf were removed from the samples. For the analysis of physico-chemical parameters, the 1:5 ratio soil suspensions are required. After it all the procedures of analysis are same to water analysis. Digestion of the same suspension is required for the total metal estimation same as in water analysis. Sediment and soil samples were collected in the polythene bags and transported to laboratory as early as possible. Some parameters should be determined immediately. For other parameters, the soil samples were dried and stored. For drying, temperature for soil is critical, as many constituents change with change in temperature. High temperature drying generally affects the exchangeable characteristics of clay and organic colloids.

Soon after drying, stones and other similar objects are picked up and the soil was grounded in a mortar to break up the aggregates or jumps, taking care not to break actual soil particles. The soil was passed through a 2 mm sieve. The mesh size allows all the nutritionally important factors to pass through. Brass sieve should be avoided where copper and zinc are to be estimated stainless steel or nylon sieves are preferable with smaller mesh sieve of 0.5 mm.

Total metal content in soil and sediment was determined by digesting 0.5 g of soil/sediment sample from each site with a mixture of concentrated HNO_3 and $HClO_4$ (10 ml + 2 ml). The digested samples were filtered through Whatman filter No. 42 and finally volume were made 10 ml with 0.1 N HNO_3 and analyzed for heavy metals using AAS (Model ECIL-4129).

Sampling Stations

To monitor the groundwater and soil quality in textile industrial territory, it is necessary to collect some random samples from a specific location known as sampling station. For comparison of polluted environment a control site was selected to compare the status of water/soil quality in affected area. So firstly, at study point in the industrial area of Panipat, following sampling sites were selected for sampling of composite textile effluent, pond water, groundwater from tube wells and hand pumps and soil/ sediment samples.

For the study of the status of groundwater and agricultural soils, groundwater and soil samples were collected monthly from different sampling stations. All the sampling points are showing in the map given in Figure 2.2b.

Experimental Sampling Sites

(A) Effluent Drain on Jatal Road, (Site-A)

Around 35 textile dyeing units are located on Jatal road in Panipat industrial area. All the units discharge their effluent as wastewater collectively in to a common drain. The effluent from that drain may easily represent the overall composite characteristics of textile effluents. Some parameters like temperature, pH, DO, free CO_2 were analyzed on site, while for the analysis of other parameter effluent was collected in BOD bottles and Jericanes.

Figure 2.17: Showing Sampling Site A

(B) Effluent Drain End with Pond, (Site-B)

The common effluent drain flows towards a large pond at the distance of about 2 km. There is a confluence point of drain with pond, where pond receives all the wastewater carried out by drain. The samples were also collected from the pond (Figure 2.12) to assess the surface water quality of pond and for sediment quality also.

Figure 2.18: Showing Sampling Site B

(C) Binjhole Village, (Site-C)

Binjhole village is situated in south-east direction from Panipat city. It is important for study because this is very close to Site-B (pond) affected by dye wastewater. Local people, cattle and agriculture of villagers completely depends upon the groundwater. To assess the groundwater quality of Binjhole, it was selected as Site-C. Groundwater

Figure 2.19: Showing Sampling Site C

samples were collected form hand pumps, tube wells of different depths *viz.* 40 ft, 80 ft and more than 120 ft.

(D) Agricultural Fields, (Site-D)

Agricultural fields near Binjhole village were selected for the assessment of soil quality and the soil samples were collected in poly bags for further analysis.

Figure 2.20: Showing Sampling Site D

Control Sampling Site

(E) Control Site (Site-E)

A non-industrial area was selected as control site for present study. It is 10 km far away from textile units on Jatal road. Control samples of groundwater as well as soil were collected from this site.

Nomenclature of Samples

Samples were collected at different sites for the collection of water and soil samples from the above sites. During the sampling, the nomination (for labeling and identification) of water and soil samples collected at various sites was as follows:

(i) Water Samples

1. Effluent from textiles outlet [EF].
2. Tube wells of the area [TW80, TW120]
3. Hand pumps of the area [HP40, HP80]

Figure 2.21: Showing Sampling Site E

4. Surface water from pond [SW/PW]
5. Groundwater from Control [GWC]

(ii) Soil Samples

1. Effluent drain sediment [EFS]
2. Pond bottom sediment [PDS]
3. Agricultural soil of irrigated field [AGS]
4. Soil from control [SC]

2.9 Analytical Methods

Physico-chemical parameter were analyzed for:

1. Groundwater,
2. Surface water,
3. Textile effluents,
4. Agriculture soil,
5. Effluent drain sediment,
6. Pond sediment.

The physico-chemical and heavy metal analysis of water sample were made of the entire collected samples following the method outlined in a standard method for examination of water and wastewater APHA (1998) Trivadi and Goel (1984) and

Mathur (1993).

Temperature[For 1, 2, 3, 4, 5 and 6]

Temperature measurements are usually made with mercury filled centigrade thermometer. The thermometer should be of small thermal capacity to attain equilibrium rapidly and must be graduateted upto an accuracy of 0.1 to 0.2°C. The accuracy of the thermometer mat vary depending upon the type of analysis. To minimize the errors, it is always essential to calibrate the thermometer with another thermometer of known accuracy. While taking the reading, the scale of the thermometer should be immersed in the water or soil suspension upto the level of mercury in the capillary column. The temperature of textile industrial effluent and water samples was recorded by dipping the thermometer into water, while for the temperature measurement of soil and sediment thermometer was dipped into soil suspension in distilled water.

Total Solids (mg/l)[For 1, 2, and 3]

Total solids or total residue is the term applied to material left in a beaker after evaporating a well mixed sample and subsequently evaporating and dried it in an oven.

Procedure

(*i*) Taken a 100 ml washed and dried beaker and weight.

(*ii*) Evaporated 100ml of unfiltered samples in the beaker, on water bath than dried it in an oven at 105 ± 1°C.

(*iii*) Taken final weight of the beaker after drying the sample.

Calculation

$$\text{Total Solids (mg/l)} = \frac{A - B}{V} \times 10^6$$

where,

A: Final weight of beaker in gm.

B: Initial weight of beaker in gm, and

V: Volume of sample.

Total Dissolved Solids (mg/l)[For 1, 2, and 3]

Total dissolved solids of filterable residue are those solids which left after evaporated of the filtered sample.

Procedure

(*i*) Taken a 100 ml washed and dried beaker and weight, immediately before used.

(*ii*) Placed filter paper Whatman no.42 in funnel.

(*iii*) Poured 100 ml of well-mixed sample and filtered through filter paper.

(*iv*) Collected the filtrate in the 100 ml weighed beaker.

(*v*) Evaporated the sample after complete filtration and subsequently dry in an oven at 105°C ±1 for 4 to 6 hrs.

(*vi*) Cooled the beaker and weight.

Calculation

$$\text{TDS (mg/l)} = \frac{A - B}{V} \times 10^6$$

where,

A: Final weight of beaker in gm.

B: Initial weight of beaker in gm, and

V: Volume of sample.

Total Suspended Solids (mg/l)[For 1, 2, and 3]

Total suspended solids are the retained material on Whatman no. 42 filter paper after filtration TSS was determined by taking difference between the total solids and total dissolved solids.

Calculation

$$\text{TSS, mg/l} = \text{TS} - \text{TDS}$$

pH (Hydrogen Ion Concentration)[For 1, 2, and 3]

pH is the Hydrogen ion concentration in the given water sample. pH equals to negativity $\log_{10}$ of H^+ concentration. It can be measured by colorimetric methods using various indicators or paper strips. However, the use of colorimetric methods is less convenient and less accurate. For accurate measurement of pH, electrometric methods are used employing the hydrogen ion sensitive electrodes. Generally, the pH value is:

$$\text{pH} = -\log_{10} (H^+)$$

The pH was measured by any one of these methods:

(*i*) By using pH strips.

(*ii*) By using electronic pH meter.

(*iii*) By handy digital pH meter.

Electrical Conductivity (mho/cm)[For 1, 2 and 3]

Electric conductance is the ability of a substance to conduct the electric current. In water, it is the property caused by the presence of various ionic species.

It was generally measured with the help of a conductivity meter having a conductance cell containing electrodes of platinum coated with Pt black or carbon. These electrodes are mounted rigidly and placed parallel at a fixed distance. Conductance, when measured between the electrodes having a surface area of 1 cm^3 and placed at a distance of 1 cm, is called electrical conductivity and is the property of water sample, rather then the measuring system. The term specific conductance is also used in place of electrical conductivity, but is an obsolete term. The unit of conductivity measurement is Siemens (S) per cm. The older unit mho per cm is now rarely used. The conductivity of most waters is generally low so the unit µ S/ cm shall be much appropriate.

A simple whetstone bridge circuit was used to measure electric conductivity by null method. The unit of specific conductivity is mhos/cm or millimhos/cm (mhos/cm × 10^3) or micromhos/cm (mhos/cm × 10^6). The most common unit of electric conductivity being followed is dS/m (deci Siemen/m) which is equal to the older unit of millimhos/cm. Electric conductivity was measured by digital conductivity meter.

As the ionization of the solutes depends on the temperature, conventionally the results are reported at 25°C.

Procedure

1. Conductance depends on the area of the metallic electrodes and the distance between them. The factor used to convert the observed conductance into conductivity is called as the cell 'constant'. Cell constant of the 'conductance cell' is usually supplied by the manufacturer.
2. Note the temperature of the sample and find out the factor from conversion table to convert the values at 25°C.

Table 2.3: Conversion Table for Measurement of Electric Conductivity

°C	Factor	°C	Factor	°C	Factor
32	0.89	22	1.06	12	1.30
31	0.90	21	1.08	11	1.33
30	0.92	20	1.10	10	1.36
29	0.93	19	1.12	9	1.39
28	0.95	18	1.14	8	1.42
27	0.97	17	1.16	7	1.46
26	0.98	16	1.19	6	1.50
25	1.00	15	1.21	5	1.54
24	1.02	14	1.24	4	1.58
23	1.04	13	1.27	3	1.62

Calculation

Conductivity = Observed conductance × cell constant × temperature factor at 25°C

Also the conductivity has a relation with total. It may be calculated (in mho/cm) by dividing the total dissolved solids by 640 (Mor *et al.*, 2003).

Free Carbon Dioxide (mg/l)For 1, 2 and 3

Determined free CO_2 by titration of the sample using a strong alkali to pH 8.3. At this pH all the free CO_2 is converted in to bicarbonates.

Reagents

(A) *Sodium hydroxide (0.05N)*: Prepare 1.0 N NaOH by dissolving 40 gm of NaOH in CO_2 free distilled water diluted 50 ml of 1.0 N NaOH to 1 liter in CO_2 free distilled water.

(B) *Phenolphthalein indicator*: Dissolve 0.5 gm of phenolphthalein in 50 ml of 95 per cent ethanol and 50 ml of distilled water.

Procedure

(*i*) 100 ml of sample was taken in a conical flask. Add few drops of phenolphthalein indicator.

(*ii*) If the colour turns pink, free CO_2 declared to absent or if the sample remained colourless, titrate it against 0.05 N NaOH.

(*iii*) At the end point a pink colour will appear.

Calculation

$$\text{Carbon Dioxide (mg/l)} = \frac{(\text{ml} \times \text{N}) \text{ of NaOH} \times 1000 \times 44}{\text{ml of sample}}$$

Dissolved Oxygen (mg/l): For 1, 2 and 3

Dissolved oxygen is the measure of oxygen concentration present in a given water sample. The concentration of oxygen reflects whether the process undergoing is aerobic or anaerobic.

Dissolved oxygen was measured by the wrinkle's iodometric method. The manganous sulphate reacts with the alkali (KOH and NaOH) to form a white precipitate of manganous hydroxide which in the presence of oxygen, get oxidized to a brown colour compound. In the strong acidic medium manganic ions are reduced by iodide ions which get converted into iodine equivalent to the original concentration of oxygen in the sample. The iodine can be titrated against thiosulphate using starch as an indicator for DO determination.

Reagents

(A) *Sodium thio sulphate (0.025N)*: Dissolved 24.82 gm of $Na_2S_2O_3$. SH_2O in boiled distilled water and make up the volume to 1 liter. Add 0.4 g pallet of NaOH a stabilizer this is 0.1N stock solution. Diluted it to 4 times with boiled distilled water to prepare 0.025 N solution.

(B) *Alkaline potassium iodine solution*: Dissolve 100 gm of KOH and 50 gm of KI in 200 ml. of boiled distilled water.

(C) *Manganous sulphate solution*: Dissolved 100 gm of $MnSO_4$. $4H_2O$ in 200 ml. of boiled distilled water and filter.

(D) *Starch solution*: Dissolved 1 gm of starch in 100 ml. of warm distilled water and add a few drops of formaldehyde solution.

(E) *Sulphuric acid*: Conc. H_2SO_4

Procedure

300 ml. volume of the sample was taken in BOD bottle. Add 2 ml. of each $MnSO_4$ and alkaline KI solution by pipette. Stopper and mixed through let the precipitate settle, titrated the contents against, sodium thio-sulphate solution, using starch as an indicator. At change end point, initial dark blue colour changes into colourless.

Calculation

$$\text{Dissolved Oxygen (mg/l)} = \frac{(\text{ml} \times \text{N}) \text{ of titrant} \times 8 \times 1000}{V_2\,(V_1 - V/N)}$$

V_1: Volume of sample bottle after placing the stopper.

V_2: Volume of the part of the contents titrated.

V: Volume of $MnSO_4$ and KI added.

Biochemical Oxygen Demand (mg/l)[For 1, 2 and 3]

Biochemical oxygen demand is the measurement of the degradable organic material present in a water sample and can be defined as the amount of oxygen required by the microorganisms in stabilizing the biologically degradable organic matter under an aerobic. The principle of the method involves measuring the difference of the oxygen concentration between the sample and after incubated it for 5 days at 20°C.

Apparatus and Reagents

(A) BOD bottles

(B) BOD incubator (at 20°C)

(C) *Phosphate Buffer*: Dissolve each 2.1g H_2SO_4, 5.43g KH_2PO_4, 8.35g $Na_2HPO_4.7H_2O$ and 0.42g NH_4Cl in distilled water to prepare 250 ml of solution.

(D) *Magnesium Sulphate*: Dissolve 8.25g $MgSO_4$ in distilled water to prepare 100 ml of solution.

(E) *Calcium Chloride*: Dissolve 2.75g of anhydrous $CaCl_2$ in distilled water to prepare 100 ml of solution.

(F) *Ferric Chloride*: Dissolve 0.25 $FeCl_3.6H_2O$ in distilled water to prepare 1 liter of solution.

(G) *Sodium Sulphite Solution*: Dissolve 1.57g Na_2SO_4 in 100 ml distilled water and dilute to 1000 ml.

Procedure

Prepare dilution water in a glass container by bubbling compressed air in distilled water for about 30 minutes. Add 1 ml each of phosphate buffer, magnesium sulphate, calcium, calcium chloride and ferric chloride solutions for each liter of dilution water and mix thoroughly. Dilution factor was 20. Neutralize the sample to pH around 7.0. Prepare dilutions in a large glass bottle, mix the content thoroughly. Fill 2 sets of the BOD bottle. Keep one set of the bottles in BOD incubator at 20°C for 5 days, and determine the DO content in another set immediately. DO in the sample bottle immediately after the completion of 5 days incubation period. Similarly for blank, take 2 BOD bottles, for dilution water. In one, determine the DO content and the other inculcate with the sample to determine DO after 5 days.

Calculation

$$\text{BOD (mg/l)} = (D_0 - D_5) \times \text{Dilution factor}$$

where,

D_0: Initial DO in the sample

D_5: DO after 5 days.

Chemical Oxygen Demand (mg/l)[For 1, 2 and 3]

Chemical oxygen demand is the measurement of oxygen consumed during the oxidation of the oxidizable organic matter by a strong oxidizing agent. Potassium dichromate used in the presence of sulphuric sulphate in determination of COD. The sample is refluxed with $K_2Cr_2O_7$ and H_2SO_4 in presence of mercuric sulphate to neutralize the effect of chlorides and silver sulphate. The excess of potassium dichromate is nitrated against of $K_2Cr_2O_7$ used is proportional to the oxidization organic matter in the sample.

(A) *Potassium Dichromate Solution (0.25N)*: Dissolve 6.13 g of $K_2Cr_2O_7$ in distilled water to make 500 ml of solution.

(B) *Ferrous Ammonium Sulphate (0.1N)*: 39.2g of Fe $(NH_4)_2(SO_4).6H_2O$ was dissolved in water adding 20 ml conc. H_2SO_4 to make 1 liter of solution.

(C) *Ferrion Indicator*: 1.48g of 1-10, phenolphthalein and 0.69g of ferrous sulphate was dissolved in distilled water to make 100 ml of solution.

(D) Sulphuric Acid (Sp.Gr.1.83)

(E) Mercuric Sulphate

(F) Silver Sulphate

Procedure

20 ml of sample was taken in a COD flask. 10 ml of 0.25N $K_2Cr_2O_7$ was added in sample. Add a pinch of $AgSO_4$ and $HgSO_4$. Add 30 ml of sulphuric acid. Reflux at least for 2 hour on a remove the flask, cool and add distilled water to make the final volume to about 140 ml. Add 2-3 drops of ferrion indicator. Mix thoroughly and

titrate with 0.1 N Fe $(NH_4)_2(SO_4)_2$ solution. Run a blank with distilled water using same quantity of the chemicals.

Calculation

$$\text{COD (mg/l)} = \frac{(b-a) \times \text{N of FAS} \times 1000 \times 8}{\text{ml of sample}}$$

where,

A: ml of titrant with sample

b: ml of titrant with blank

Alkalinity (mg/l)[For 1, 2 and 3]

Alkalinity is the measure of the capacity of the water to neutralize a strong acid. The alkalinity in the waters is generally imparted be the salts of carbonates, bicarbonates, phosphates, nitrates, borates, silicates, etc. together with the hydroxyl ions in free state. However, most of the waters are rich in carbonates and bicarbonates with little concentration of other alkalinity imparting ions. Total alkalinity, carbonates and bicarbonates can be estimated by titrating the sample with a strong acid (HCl or H_2SO_4), first to pH 8.3 using phenolphthalein indicator and then further to pH between 4.2 to 5.4 with methyl orange or mixed indicator. In first indicator, the value is called as phenolphthalein alkalinity (PA) and in second case, it is total alkalinity (TA). Values of carbonates, bicarbonates and hydroxyl ions can be computed from these two types of alkalinities.

Reagents

(A) *Hydrochloric acid, 0.1 N*: Dilute 12 N concentrated HCl to 12 times to prepare 1.0 N HCl. Dilute it further to make 0.1 N HCl.

(B) *Methyl orange indicator*: Dissolved 0.5 gm of methyl orange in 100 ml of distilled water.

(C) *Phenolphthalein indicator*: Dissolved 0.5 gm of phenolphthalein in 50 ml of 95 per cent ethanol add 50 ml distilled water.

(D) *Sodium carbonate*: Dissolved 5.300 gm of Na_2CO_3, previously dried at 250°C for about 4 hours, in distilled water to prepare 1 liter of solution.

Procedure

A sample of 50 ml was taken in a comical flask and adds 2 drops of phenolphthalein add now 2 or 3 drops of methyl orange indicator is added to the above sample and titrate against 0.1N HCl until a yellow colour changes to pink.

Calculation

$$\text{Total Alkalinity (mg/l)} = \frac{V - N \times 1000 \times 50}{\text{ml of sample}}$$

where,

V: Volume of total HCl used with methyl orange,

N: Normality of HCl

Chloride (mg/l)For 1, 2, 3, 4, 5 and 6

Chloride gain access in many way *viz.* solvent power of water dissolve salts from top soil, spray form ocean, seas invading fresh waters during tides, underground formation and sewage, human excreta etc.

Silver nitrate reacts with chloride to form very slightly soluble white precipitate of AgCl. At the end point when all the chlorides get precipitated, free silver ions react with chromate to form silver chromate of redish brown colour.

Reagents

(A) *Potassium chromate indicator*: Dissolved 50 gm of potassium chromate in a small quantity of distilled water. Add $AgNo_3$ solution to produce a light red precipitate allows standing night and filter makes up 1.0 liter.

(B) *Silver Nitrate solution*: (N/35.5): Dissolved 4.78 gm of silver nitrate in 1.0 liter of distilled water standard with standard sodium chloride solution.

Procedure

Taken 100 ml of sample in a conical flask add 1-2 drops of potassium chromate solution. Titrated the contents against 0.02 N $AgNO_3$ until a persistent red tinge appears. Note the amount of titrant used.

Calculation

$$\text{Chloride (mg/l)} = \frac{\text{N} \times \text{ml of AgNO}_3 \text{ used} \times 1000 \times 35.5}{\text{ml of sample}}$$

(*Note*: For analysis in soils or sediments, 1:2.5 soil suspensions are urgently required. Take 20 gm of soil and add 50 ml distilled water to it. Stir for about one hour at regular intervals and then filter for chloride determination like in the case of water.)

Calcium (mg/l)For 1, 2 and 3

Calcium is one of the most abundant substances of the natural water. Being present in high quantities in the rocks, it is leached from there to contaminate the water. The quantities in natural waters generally very from 10 to 100 mg/l depending upon the types of the rocks.

Reagents

(A) *EDTA solution 0.01 N*: Dissolved 3.723 gm of disodium salt of EDTA in distilled water to prepare 1 liter of solution store in polythene or pyrex bottle.

(B) *Sodium hydroxide (1N)*: Dissolved 40 gm of NaOH in distilled water and dilute to 1 liter.

(C) *Muroxide indicator*: Mix 0.2gm of ammonium perpurate with 100 gm of NaCl and grind it.

Procedure

Taken 50 ml sample in conical flask. If the sample is having high alkalinity use smaller volume diluted to 50 ml. Add 2 ml of NaOH solution in the sample Add 100-200 mg of murexide indicator a pink colour develops titrate against EDTA solution until the pink colour change to purple.

Calculation

$$\text{Calcium (mg/l)} = \frac{\text{Volume of EDTA} \times 400.8}{\text{ml of sample}}$$

Magnesium (mg/l)For 1, 2 and 3

Calcium and magnesium form a complex of wine red colour with Eriochrome black-T at pH 10.0. The EDTA has got a stronger affinity for Ca^{++} and Mg^{++}, the former complex is broken down and a new complex of blue colour is formed. The value of Mg^{++} can be obtained by subtracting the value of calcium from the total of Ca^{++} + Mg^{++}.

Reagents

(A) *EDTA solution, 0.01M*: Dissolve 3.723 g of disodium salt of EDTA in distilled water to prepare 1 liter of solution store in polythene or pyrex bottle.

(B) *Buffer solution*: (*a*) Dissolved 16.9 g. ammonium chloride in 143 ml of concentrated ammonium hydroxide (NH_4OH); (*b*) Dissolved 1.179g of disodium EDTA and 0.780 g of $MgSO_4 . 7H_2O$ in 50 ml distilled water. Mix both (*a*) and (*b*) solution and dilute to 250 with distilled water.

(C) *Eriochrome Black-T indicator*: mixed 0.40g of Eriochrome Black-T with 100g NaCl (A.R.) and grind.

Procedure

1. Find out the volume of EDTA used in calcium determination.
2. Also find out the volume of EDTA used in hardness (Ca^{++} + Mg^{++}) determination with same, volume of the sample as taken in the calcium determination.

Calculation

$$\text{Magnesium (mg/l)} = \frac{y - x \times 400.8}{\text{ml of sample} \times 1.645}$$

where,

y = EDTA used in hardness determination

x= EDTA used in calcium determination for the same volume of the sample.

(b) Mg^{++}, (mg/l) = Total hardness (as mg/l $CaCO_3$)-Calcium hardness (as mg/l $CaCO_3$) x .244

Where, calcium hardness (as mg/l. $CaCO_3$) = Ca, mg/l × 2.497.

Sodium (mg/l)For 1, 2 and 3

A characteristic light is produced due to excitation of electron when the sample with sodium was sprayed into a flame. The intensity of this characteristic radiation is proportional to the conc. of sodium and can be read at 589 nm by using suitable filter devices. The standard curve is linear one at the lower concentration of sodium, however, at higher concentrations it has got a tendency to level off.

Reagent and Apparatus

(A) Flame photometer,

(B) Stock sodium solution (1000mg/l Na): Dissolved 2.542gm NaCl dried at 140°C in distilled water to make 1 liter of solution.

(C) Intermediate sodium solution (100mg/l Na): Dilute the stock sodium solution 10 times.

(D) Standard sodium solution (10mg/l Na): Dilute the intermediate solution further 10 times.

(E) Nitric acid conc.

(F) Hydrochloric acid conc.

(G) Hydrogen peroxide, 30 per cent.

(H) NH_4OH conc.

Procedure

Filtered the sample through a filter paper to remove any suspended matter which other wise may clog the capillary of the instruments. Find out the concentration of sodium using flame photometer follow the instructions provided by the manufacture for use, because of the differences in the makes and models of different instruments. All the instruments require preparation of a calibration curve in the ranges of 0 to 1, 0 to 10, 0 to 100 mg/l of sodium by using the various standard solution of sodium use any of these curve for determination of Na for accurate results, it is better to use smaller ranges and if sample is having more Na, it can diluted to come in the range of determination.

Calculation

Sodium (mg/l) = (mg/l Na in diluted aliquot) × Dilution factor.

Potassium (mg/l)For 1, 2 and 3

Potassium was determined accurately by flame photometer. The characteristic radiation for potassium is 768 nm, the intensity of which read on a scale by using a filter for this wavelength.

Apparatus and Reagents

(A) Flame Photometer.

(B) *Stock potassium solution (1000mg/l K)*: Dissolved 1.907 gm of KCl, dried at 110°C, in distilled water to prepare 1 liter of solution.

(C) *Intermediate potassium solution (100mg/l K)*: Dilute the stock solution 10 times.

(D) *Standard potassium solution (10mg/l K)*: Dilute the intermediate solution further 10 times.

Procedure

The same procedure as described for determination of sodium was followed except that to use filter for potassium at 768 nm wavelengths instead of sodium filter.

Calculation

$$K\ (mg/l) = (mg/l\ K\ in\ diluted\ aliquot) \times (Diluted\ factor).$$

Water Holding Capacity (Per cent)[For 4, 5 and 6]

Water holding capacity is a property as to how much water is absorbed or retained by given amount of soil. It significant that the tendency of soil of absorbing water depends upon that how much water is absorbed or retained by given amount of soil.

Procedure

Taken a keen box and fill soil in it and after than it keen box take in the water for 24 hour for saturation. Now weights of keen water with box and after it taken in oven at 105° C for 24 hour and weight it.

Calculation

$$\text{Water Holding Capacity (\%)} = \frac{a-b}{a} \times 100$$

where,

a: Weight in soil with water,

b: Weight of dried soil

Bulk Density (mg/cm³)[For 4, 5 and 6]

The bulk density of soil is defined as the dry weight of a unit volume of it and it is expressed as mg/cm^3 for medium to fine textured soil and from 1.2 to 1.65 mg/cm^3 for coarse textured soil but it is slightly higher in case of alkaline saline soils.

The bulk density is inversely related to pore space of soil.

Procedure

Dried the soil sample in an oven at 105 °C until a constant weight was attained. Transfer a little dried soil to a measuring cylinder and not the volume. Recorded the weight of this volume of soil of a balance.

Calculation

$$\text{Bulk Density (mg/cm}^3\text{)} = \frac{\text{Weight of soil (mg)}}{\text{Volume of water (cm}^3\text{)}}$$

Moisture Content (Per cent)For 4, 5 and 6

These samples were collected rapidly in the moisture boxes and after weighing allowed to dry in electric oven 105°C for 48 hours. After proper drying sample were weighed again. With the help of recorded value (Excluding weight of moisture box), content of moisture was determined by using the formula:

$$\text{Moisture Content (\%)} = \frac{\text{Fresh moisture weight of soil \& oven dry weight of soil}}{\text{Oven dry weight of soil}} \times 100$$

Hydrogen Ion Concentration (pH)For 4, 5 and 6

pH of the soil is the measure of the hydrogen zone activity. It depends extremely on relative amounts of the absorbed hydrogen and metallic zones.

pH of the soil suspension highly depends on the soil water ratio and increase with dilution. It is measured on a long scale and equal to negative log 10 of hydrogen zone concentration.

$$\text{pH} = -\log\,[H^+]$$

$$\text{pH} = -\log 10\,[1/H^+]$$

A neutral soil has a pH of 7.0 while a pH less than 7.0 renders it acidic and pH more than 7.0 alkaline.

Apparatus and Reagents

(a) *The pH Meter*: pH was estimated at the sampling site by help of pH strip and than confirmed by pH meter.

(b) *Potassium Hydrogen Phthalate Buffer*: Dissolve 10.2 gm of potassium hydrogen phthalate in water to prepare 1000 ml buffer.

(c) *Phosphate Buffer*: Dissolve 3.40 gm of KH_2PO_4 and 4.45 gm of $Na_2HPO_4.2H_2O$ in water to prepare 1000 ml of buffer.

(d) *Borax Buffer*: Dissolve 3.81 gm of $Na_2\,B_4O_7.10\,H_2O$ in water to prepare 1000 ml of buffer.

Procedure

pH is to be determined in 1:5 soil suspension. Taken 20 gm of soil and 100ml distilled water to it. Stirred for about on hour at regular intervals and then filter then pH is determined using pH strip and electrometrically using glass electrode pH meter. Result is expressed directly in pH units associated with the specific dilution suspension.

Electrical Conductivity (mho/cm)For 4, 5 and 6

Conductivity is a measure of the current carrying capacity, thus gives a clear idea of soluble salts present in the soil. Conductivity values depends on the dilution of the soil suspension and most of the laboratories it is conventionally determined in 1:5 soil suspension. Conductivity measurement should be made within few hours of the preparation of soil solution; otherwise microbial activity can vary the results significantly. The results of the conductivity are also sometimes converted into per cent soluble salts by multiplying the conductivity values in mS at 20°C with a factor 375. However, the validity of such conversion is doubtful (Trivedi and Goel, 1984). Like water, the conductivity of the soil suspension is also measured with the help of a conductivity meter.

Apparatus

Conductivity meter, thermometer, glassware to prepare soil suspension conductors depend on the area of the metallic electrodes and distance between them. The factor used to convert the observed conductance into conductivity is called as the cell constant is the ratio of the specific conductance and observed conductance and can be calculated by finding out the conductance of different conc. of KCl solution.

Procedure

1. Prepared 1:5 soil suspensions by taking 20g of soil in 100 ml of aerated distilled water. Shake mechanically for one hour. Measure the conductivity of the soil suspension with conductivity meter by direct dipping the cell into the suspension.
2. Taken temperature of the soil suspension, and convert the results at 25°C according to conversion table.

Calculation

Expressed the results of conductivity at 25 °C specifying the dilution of the soil suspension.

Conductivity (mho/cm) = observed conductance × cell constant × temperature (at 25°C)

Rests of the soil parameters were analyzed as per the methodology adopted for water samples after make a suspension of soil and sediment into distilled water. Sodium, Potassium, Calcium, Magnesium, Chlorides, and heavy metals were detected using soil suspension of soil.

Sodium and Potassium (μg/g)For 4, 5 and 6

Exchangeable sodium and potassium were also determined in ammonium acetate leachate, following flame photometry method as described for water.

Reagents

A. Ethyl alcohol, 40 per cent
   (As for Ca and Mg determination)

B. Absolute alcohol

C. Ammonium acetate solution (1N)

Diluted 57 ml of glacial acetic acid to 800 ml with distilled water and then neutralize to pH 7.0 with concentrated NH_4Cl. Make up final volume to one liter.

Procedure

1. Prepared the soil extract by leaching with 1 M ammonium acetate solution following the procedure as described in determination of exchangeable Ca and Mg.
2. Find out the concentrations of sodium and potassium following flame photometric method as described for the water analysis.
3. In case of total sodium and potassium (exchangeable + soluble) or when soluble fraction is very less, do not apply the alcohol washing for the extraction. Add directly the ammonium acetate solution to the soil.
4. If calcium, magnesium, sodium and potassium all are to be determine in a sample of soil, make the extract only once, and determine all of them in same extract.

Calculation:

$$\text{Sodium (\%)} = \frac{\text{mg Na/L of soil extract} \times V}{10000 \times S}$$

$$\text{Sodium (mg/100g)} = \frac{\text{mg Na/L of soil extract} \times V}{10 \times S}$$

$$\text{Potassium (\%)} = \frac{\text{mg K/L of soil extract} \times V}{10000 \times S}$$

$$\text{Potassium (mg/100g)} = \frac{\text{mg K/L of soil extract} \times V}{10 \times S}$$

where,

V: Total volume of soil extract prepared (500 ml in this case)

S: Weight of soil taken (50 gm in this case)

2.10 Data Handling and Statistics

AAS output data and physico-chemical parameter's data were stored and handled with the help of EXEL (Microsoft Office) and formed in to tables and graphs after some calculation using statistics of MS-EXEL and supporting software like Systat, SPSS, Sigmaplot, Mathtype, etc. Some main statistical tools of statistics were applied in the formation of suitable presentation of raw data of the present study. Some tools are given as below (Prasad, 2003):

The data for all physico-chemical and heavy metals parameters were collected and analyzed for calculating mean, standard deviations of each parameter and correlation as per (Prasad, 2003) with the help of computer on Microsoft excel.

Arithmetic Mean

$$x = \frac{\Sigma x}{n}$$

where,

x: Individual reading of parameter.

Σ X: Total of all readings for parameter.

n: Number of samples.

Standard Deviation (SD or σ)

$$SD(\sigma) = \sqrt{\frac{\Sigma(x - \bar{x})^2}{(n-1)}}$$

where,

x: Individual reading of parameter

$\bar{x}$: Mean Σ X

n: Number of samples.

Coefficient of Correlation (r)

$$r = \sqrt{\frac{\Sigma(x - \bar{x})(y - \bar{y})}{\Sigma(x - x)^2 (y - y)^2}}$$

where,

x: Individual reading of 1st parameter

$\bar{x}$: Mean of Σ X

y: Individual reading of 2nd parameter

$\bar{y}$: Mean of Σ Y

Transfer Factor (Coefficient)

Transfer coefficient is the indication of metal deposition or transportation rate of any metal from one medium to another. Metals have a tendency to accumulate or migrate from one component to other component. Accumulation period or rate may influence the metal availability and balance among both components. In the present study, abiotic components such as textile industrial effluent, pond water, groundwater, agricultural soils and bed sediments of pond and drain are taken for the evaluation of transfer factor.

"Transfer (coefficient) factor is used for determination the rate of metal transferred from one environmental background to another or from one trophic level to another".

Transfer factor was calculated according to Lokeshwari and Chandrappa (2006).

$$\text{Transfer Factor} = \frac{\text{Ab(ppm)}}{\text{Bt(ppm)}}$$

where,

A*b* is the detected metal content in A component originating from the component B, and B*t* is the total metal content in the component B.

Enrichment Factor

The primary objectives of the sediment sampling study were to estimate how sediment chemistry changes throughout a watershed and to relate these changes to natural and manmade sources of pollution. The iron normalized metal enrichment in the bed sediments was calculated following the method given by Jain (2004). Environmental factor is a general factor (coefficient) comparatively with iron for every metal ranging between from 1 to 1000 on a simple scale.

"Enrichment factor is used for the determination of the metal concentration in sediment in comparison to iron, when the value of iron is 1000."

$$\text{Metal Enrichment Factor} = \frac{\text{M}}{\text{Fe}} \times 1000$$

where,

M is the mean metal concentration in the sediment (μg/g), and Fe is the mean concentration of iron in the sediments (μg/g).

Geo-accumulation Index

To assess the anthropogenic influence in the study area, the evaluation of sediments contamination with heavy metals was made using the index of accumulation, I_{geo} which permits assessment of the extent of sediment contamination with respect to the global standards.

"I_{geo} is to know the metal accumulation status in sediment in comparison to natural background level (shell values) of a particular metal."

I_{geo} was calculated by the method of Muller (1979) using the following mathematical relation:

$$I_{geo} = \log_2 \frac{C_n}{1.5 \times B_n}$$

where,

C_n is the mean measured total concentration of the examined element 'n' in the studied bottom sediment, B_n is the average (crustal) geochemical background value for concentration of the element 'n' in fossil argillaceous sediment ("average shale") and 1.5 is the factor compensating the background data (correlation factor) due to lithogenic effect (Taylor, 1964). The I_{geo} can be classified in to seven grades (0-6). The highest grade, 6 reflect 100 fold enrichment and 0 reflects the background concentration. The following description classification is given for the index of geo-accumulation by Forstner *et al.* (1990) as follows:

<0	Partially unpolluted
0-1	Unpolluted to moderately polluted
1-2	Moderately polluted
2-3	Moderately to strongly polluted
3-4	Strongly polluted
4-5	Strong to very strongly polluted
5-6	Very strongly polluted

Metal Accumulation Factor

Toxic pollutants especially pesticides and heavy metals have a tendency to accumulate in various interrelated environmental components and also at different trophic levels of food chain of biological system. Metal accumulation index indicates the degree of metal pollution in the experimental component by comparing the control site. Metal accumulation factor 1 indicates the 100 per cent similarity in the metal concentrations in experimental and control sites, while accumulation factor 2 indicates the two-fold contamination of metal on experimental site in comparison with control. Metal accumulation factor was calculated as described by Fakayode and Onianwa (2002):

$$\text{Metal Accumulation Factor} = \frac{M_{Exp}}{M_C}$$

where,

M_{Exp}: Concentrations of metal on experimental site,

M_C: Concentrations of metal on control site.

Graphical Presentation

Different types of graphs were applied for the presentation of data based on investigation. Some popularly used graphs were bar diagram, line diagram, line-bar graph etc.

Chapter 3

Chemical Characteristics of Textile Effluents

3.1 Textile Industrial Effluent

The untreated effluents of various textile industries can alter the physico-chemical properties of freshwater resources and increase the pollution load in environment of aquatic ecosystem.

The textile effluents may affect different parts of environment in various ways. So, the impact assessment of textile effluents on physico-chemical characteristics of surface and groundwater, agriculture soil, plants and other biota must be required. A lot of researchers are engaged in bio-monitoring and evaluate the controlling strategies to checkout the negative impacts on different components of environment.

Textiles industries mostly used vat and azo dyes as a constituent for dyeing and discharging the effluent directly into a common composite drain. This drainage system joined the effluent to the pond thus influencing the surface water and land quality of area. After delay of effluent in pond for a prolong time period, leaching of pollutants especially heavy metals would started and contaminate the groundwater of this particular area.

3.2 Historical Resume

Different workers from India and abroad have been discussed various aspects of pollution by industrial efflents in term of characteristics of water bodies, affected agricultural soils and bottom sediments from time-to-time. This review includes some of important contribution made in the field of industrial pollution, with references to characterization of textile industrial effluents and bottom sediment by different authors of country and abroad.

Aggarwal and Mehrotra (1952) surveyed the soil and soil water affected by effluent in some selected regions of Uttar Pradesh. Southern and Alsapaugh (1957) made some suggestion for the abatement of pollution in the textile industry *e.g.* process chemical substitution, closer process control and good housekeeping.

Porter (1970) highlighted the changing nature of textile processing and wastewater treatment technology according to time. Walker (1970) described the different metabolism of azo compounds used in textile dyeing unit. Nemerow (1971) described the origin of industrial water pollution, characteristics and treatment of industrial wastewater. Eckenfelder and Bornad (1971) justified the cost of waste treatment and relationship with its application for industrial waste.

Porter (1972) used the activated carbon for the treatment of textile waste. Considine (1974) described the various activities and functions of different chemicals effluents and their related internal chemical processes. Verma and Mathur (1974) studied the toxicity of industrial wastes to some organisms such as *Macrobranchium deyanum*. Widyanto (1975) conducted a survey on textiles factories and revealed that using the wastewater from textile factory a 1:1 dilution of the concentrate may stimulate the growth of water hyacinth after 10 days and suggested to utilize as a bio-agent to purify wastewater of a textile factory before reusing. Dulka and Risby (1976) stated ultra trace metals in some environmental and biological systems in the vicinity of textile industrial sector. Horing (1976) evaluated the characterization and treatment of textile dyeing wastewaters with the help of a designed model. Poots *et al.* (1976) indicated the removal of acid dye from effluent using natural absorbents.

Judlins and Hornsby (1978) carried out an observation on color removal from textile dye waste using magnesium carbonate. Hitz *et al.* (1978) stated the absorption of dyes on activated sludge. Govindan and Sunderlingam (1979) recognized the treatment of textile effluent in admixture with sewage (1:5) by the waste stabilization pond with the detention time of 8-12 days, using acclimated algae culture *Chlorella pyreoidosa, Microcystis, Aeruginosa* and *Euglena*. McKay (1979) described the color removal method from textile waste effluents.

Samy and Gananarethinam (1980) found that industrial effluents were toxic to the aquatic plants like *Eicchornia* even after dilution. Synthetic compounds were found toxic to *Spirodela polyrhiza* and *Lemna aquinoctialis*. Watler and Wint (1981) described the treatment processes of industrial effluent step by step. Wint (1981) indicated the disposal and handling problem of toxic wastes from industrial wastewater.

Chung (1983) highlighted the significance of azo-reduction in the mutagenesis and carcinogenesis of azo dyes. Ajmal and Khan (1985) analysed the effects of textile waste on physico-chemical properties of soil and seed germination. The BOD, COD and total solids, Cl, SO_4, Na, Ca and Mg concentrations in the effluents were found to be very high. Michaels and Lewis (1985) postulated the sorption and toxicity of azo and triphenyl methane dues to aquatic microbial populations.

Dayama (1987) reported that in their experimental studies on *Spinacea oleracea, Triticum aestivum, Cymopsis tetragonloba* and *Cicer arietinum* that effluent emerging from textiles industries affects the growth of plants. Zilliox (1989) determined the industrial impacts on the quality of groundwater in a large basin as a case of the Rhine aquifer in France and noticed it harmful for public consumption. Nawar and Doma (1989) studied on the removal of dyes from effluents using low-cost agricultural byproducts.

Navarro *et al.* (1993) reported the waste dumping and accumulation of industrial and domestic waste in an industrial area, and impacts of untreated textile wastewater respectively. Goltenboth (1994) evaluated the impact of wastewater of textile factory in Salatiga on the fresh water aquatic ecosystem of Ledok river system at Central Java. Srinivasan and Murali (1996) carried out some studies on effect of industrial pollutants on properties of soils.

Sharma *et al.* (1999) and Sastry and Rathee (1999) reported the impact of industrial effluents on fresh water system of three villages in Rohtak district. Al-Degs, *et al.* (2000) described the effect of carbon surface chemistry on the removal of reactive dyes from textile effluent and observed a method for decolorization of wastewater of textile industry having reactive dyes with azo bond and associated chromophores.

Sheth and Patel (2001) studied on characterization, treatment and comparative cost analysis of textile processing wastewater of Vatva industrial complex. Kumar *et al.* (2001) studied on the impact of textile industry on groundwater quality of Sanagar, Jaipur and found the high amount of chemicals (e.g. sulphuric acid and acetic acid) and dyes (e.g. chrome and metal dyes) suspended solids and toxic metals in the waste which may enter in the groundwater and deteriorate its quality. Thorat and Pathade (2001) described the characterization, processes, pollution and the treatment of wastewater generated from each step of textile industry and suggested processes for the neutralization of alkaline waste and their chemical and biological treatments.

Srinivas *et al.* (2002) pointed out the groundwater pollution due to the industrial effluents in Kothur industrial area, Mahaboonagar, (A.P.). Annadurai *et al.* (2002) used the cellulose-based wastes for adsorption of dyes from aqueous solutions, from the dye manufacturing industry, textile industry, and pulp and paper industry.

Mohanasundaram (2003) suggested the affordable and effective textile processing effluent treatment plants. Ameta *et al.* (2003) observed the prominent growth and increased sugar and protein percentage and chlorophyll content in onion grown in photo-catalytically treated effluent and described the photo-catalytic treatment of wastewater can be considered as an effective method, which will help in reusing the effluent from dye industry for irrigation purposes. Khan *et al.* (2003) harvested some *Solanum melongena* plant in a laboratory experiment using five different levels of textile wastewater and observed the contamination of heavy metals in the soil of pots treated with highest ratio of distilled water and wastewater.

Lee *et al.* (2004) used nitrogen for colour and COD removal by fungi in textile effluent. Bousselmi *et al.* (2004) studied textile wastewater treatment by solar catalysis from a plant in Tunisia. Krull and Dopkens (2004) recycled the dye house effluents by biological and chemical treatment in Germany. Bae *et al.* (2004) used Fenton process for textile dyeing wastewater treatment in Korea.

Yu *et al.* (2005) studied color, dye and DOC removal, and acid generation during Fenton oxidation of dyes and investigated discoloration using the synthetic dye wastewater containing various dyes. Sevimli *et al.* (2005) carried out the work on investigating the effect of some operational parameters on the decolorization of textile effluent and dye solution by ozonation. Dhar *et al.* (2005) studied the effect of acid activated sawdust on the removal of different dissolved tannery dyes (Acid dye) from aqueous solution and designed acid activated sawdust in absorbing D-Brown EGP and Lurazol Brown PM dyes from aqueous solution. Saw dust is an excellent low cost adsorbent of colored organic anions and may have significant potential as a color removal from tannery wastewater.

Aboulhassan *et al.* (2005) suggested the treatment process of textile wastewater using a natural flocculants. Bhat and Kulkarni (2005) studied on chemical oxygen demand reduction of dye industry effluent having COD in excess of 100000 mg/l and are extremely difficult to treat and pose severe pollution problems if discharged directly to treatment plants and used rice husk, saw dust and charcoal as adsorbents for the reduction of COD load of dye waste. It was found that activated charcoal is suitable adsorbent under specified condition. Also, cheaply available adsorbent such as saw dust and rice husk also show some good results.

Sheth and Desai (2005) developed the Bench Scale Completely Mixed Activated Sludge Reactor Model in the laboratory and observed the changes in various treatment parameters and terminate the experiments when reactor has reached to a steady state. Using the concept of the Monod expression by a graphical method all the bio-kinetic parameters for textile wastewater was determined.

Malik *et al.* (2006) reported the impact of textile effluents on surface water quality of Panipat industrial area. Gharaibeh *et al.* (2007) found the impact of field application of treated wastewater on hydraulic properties of vertisols.

It is necessary to reduce the levels of toxic metals from industrial effluents before discharging into surface waters. The general methods of treating wastewater-containing metals are coagulation, ion exchange, reverse osmosis and adsorption. Adsorption is a highly effective physico-chemical treatment for removing heavy metals for effluents. Due to high cost of activated carbon, efforts are being directed towards finding low cost adsorbents. Recently much interest has been shown for removal of heavy metals using a large number of low cost materials. Coal fly ash, the solid waste of power plants is one of the cheapest and non-conventional adsorbents (De, 2005).

Textile industry requires different chemical dyes in dying during wet processing and release them with the wastewater discharged through industrial outlet to effluent drain. A common effluent drain contains all types of dye residuals in a composite form, which is very toxic in nature due to the presence of different synthetic dyes, which contain various toxic heavy metals. In the textile dying processes for every color, a different metal compound is required and used for dying and ultimately washed out with the effluents of dye houses in Panipat industrial area. The common effluent drain ends up in to a pond near village Binjhole so the future of these heavy metals is not so managed and predicted. Most of the villagers think that the dye houses are the only responsible for the water quality degradation of regional groundwater system.

According to the voice of time, it is require assessing the groundwater quality of village Binjhole, industrial area and some local surroundings of dye houses. The effluents discharged from dye houses are quite important for temporal monitoring. Because the effluents store pond is highly contaminated and is a depot of heavy metals due to the presence of chemical dye portions continuously since a long period of time. Heavy metals remain in water ecosystem, but these may travel through food web at various trophic levels and environmental components as well as. Thus, metals moves pond sediments and reach in shallow water table by leaching processes and the metal contamination may disperse in a wide area of the region. Contamination of soils of close areas occurs when the effluents or polluted water comes under the contact of soil. Thus, the evaluation of textile effluents, pond water, groundwater, sediments and soils is essential to know the pollution load of heavy metals on hydrosphere and lithosphere and to determine the drinking water quality of Panipat industrial area for the assessment of its potability, suitability and evaluate some possible impacts on human health.

In the present study, water, soil and sediment samples were collected and analyzed for determination of physico-chemical parameters and heavy metals. The parallel control samples of groundwater and soil were also collected and analyzed from control site, pointed at Jatal road and 10 Km far from industrial area of Panipat city. The results are depicted in the form of table and graphs here below.

The findings of seasonal observations are given here on of experimental and control sites during year 2005-2007.

3.3 Observations

Experimental Sites

Physico-chemical Parameters of Textile Effluents

Temperature (°C)

The temperature of textile industrial effluents was recorded maximum 32.83 ± 2.65 in monsoon 2006 and minimum 25.74 ± 1.04 in winter 2005-06 ranged between 24.8 to 35.7 °C. The highest value of temperature in textile effluents was found 35.7 °C during July 2005 and lowest 24.8 °C in January 2006. The temperature variation during the study is given Table 3.1. Monthly variation in temperature is given in Figure 3.1

pH

The maximum pH value of textile industrial effluents were recorded 10.72 ± 0.75 in monsoon 2006 and minimum 9.65 ± 0.07 in summer 2007 ranged between 9.5 and 11.5. The highest value of pH in textile effluents was found 11.5 during July 2006 and lowest 9.5 in April 2005. The variation in the pH during the study is given Table 3.1. Monthly variation in pH is given in Figure 3.1

Table 3.1: Seasonal Variation of Physico-chemical Characteristics of Textile Effluents

Para-meters	Unit	Summer 2005		Monsoon 2005		Winter 2005-06		Summer 2006		Monsoon 2006		Winter 2006-07		Summer 2007	
		Mean	*SD*	*Mean*	*SD*	*Mean*	*SD*	*Mean*	*SD*	*Mean*	*SD*	*Mean*	*SD*	*Mean*	*SD*
Temp.	°C	31.967	3.027	32.733	2.857	25.740	1.038	30.875	3.425	32.833	2.650	26.125	1.044	26.000	0.707
pH		9.817	0.388	10.537	0.667	10.014	0.232	9.913	0.397	10.720	0.752	10.163	0.293	9.650	0.071
TS	mg/l	2465.000	115.763	2579.667	248.902	2210.400	39.202	2479.250	109.113	2643.667	220.731	2262.750	31.245	2370.500	55.861
TSS	mg/l	281.667	33.606	534.000	122.340	141.800	86.187	259.500	66.841	570.667	94.875	165.750	83.699	184.000	43.841
TDS	mg/l	2183.333	82.513	2045.667	184.028	2068.600	93.420	2219.750	55.187	2073.000	193.915	2097.000	52.896	2186.500	12.021
EC	mho/cm	3.411	0.129	3.196	0.288	3.232	0.146	3.450	0.096	3.270	0.312	3.243	0.135	3.415	0.049
DO	mg/l	0.867	0.777	2.867	1.457	3.356	0.649	1.200	0.909	3.233	1.250	3.525	0.479	3.150	0.495
Free CO_2	mg/l	2.800	0.300	0.900	1.559	0.380	0.531	2.500	0.716	0.733	1.270	0.425	0.492	0.750	1.061
BOD	mg/l	170.000	6.245	163.000	22.113	138.600	16.517	162.500	23.116	158.333	14.572	157.000	21.432	160.000	7.071
COD	mg/l	422.333	31.565	445.000	40.927	367.000	46.179	415.000	48.710	437.333	20.404	386.500	47.339	351.500	14.849
Total alkalinity	mg/l	836.333	52.975	865.667	70.359	633.600	69.551	813.500	77.350	889.000	47.885	642.500	73.532	715.000	14.142
Ca	mg/l	21.333	3.055	25.333	6.506	10.580	3.200	20.250	6.946	27.367	6.783	11.450	3.844	13.000	1.414
Mg	mg/l	16.067	3.465	15.200	3.119	6.580	3.353	13.275	6.734	16.067	2.676	8.950	2.431	4.700	2.121
K	mg/l	7.667	0.764	8.567	0.404	7.716	0.278	7.750	0.714	8.833	0.577	7.950	0.238	7.400	0.141
Cl	mg/l	648.333	95.144	808.667	35.921	652.200	54.843	657.250	95.503	825.667	29.939	691.750	55.030	588.500	9.192
Na	mg/l	153.000	4.359	174.000	10.817	149.600	4.219	152.750	8.302	177.000	7.211	153.750	3.500	145.000	4.243

Panipat dye houses
Pond
Pond
Image © 2006 DigitalGlobe
© 2006 Europa Technologies
Google
Pointer 29°21'57.31" N 76°56'23.23" E elev 764 ft
Streaming 100%
Eye alt 3410 ft

B

Total Solids (mg/l)

The maximum total solids in textile industrial effluents were recorded 2643.67 ± 220.73 in monsoon 2006 and minimum 2210.40 ± 39.20 in winter 2005-06 ranged between 2183 to 2780 mg/l. The highest value of total solids in textile effluents was found 2780 mg/l during August 2006 and lowest 2183 mg/l in January 2006. The variation in the total solids during the study is given Table 3.1. Monthly variation in total solids is given in Figure 3.2

Total Dissolved Solids (mg/l)

Total dissolved solids in textile industrial effluents were recorded maximum 2219.75 ± 55.18 in summer 2006 and minimum 2045.67 ± 184.03 in monsoon 2005 ranged between 1842 to 2298 mg/l. The highest value of total dissolved solids in textile effluents was found 2298 mg/l during June 2006 and lowest 1842 mg/l in September 2005. The variation in the total dissolved solids during the study is given Table 3.1. Monthly variation in total dissolved solids is given in Figure 3.2.

Total Suspended Solids (mg/l)

Total suspended solids in textile industrial effluents were recorded maximum 570.67 ± 94.87 in monsoon 2006 and minimum 141.80 ± 86.18 in winter 2005-06 ranged between 49 to 680 mg/l. The highest value of total suspended solids in textile effluents was found 680 mg/l during August 2006 and lowest 49 mg/l in January 2006. The variation in the total suspended solids during the study is given Table 3.1. Monthly variation in total suspended solids is given in Figure 3.2.

Electrical Conductivity (mho/cm)

Maximum electric conductivity in textile industrial effluents was recorded 3.45 ± .096 in summer 2006 and minimum 3.196 ± 0.29 in monsoon 2005 ranged between 2.88 to 3.58 mho/cm. The highest value of electric conductivity in textile effluents was found 3.58 mho/cm during June 2006 and lowest 2.88 mho/cm in September 2005. The variation in the electric conductivity during the study is given Table 3.1. Monthly variation in electric conductivity is given in Figure 3.3.

Dissolved Oxygen (mg/l)

Dissolved oxygen in textile industrial effluents was recorded maximum 3.52 ± 0.48 in winter 2006-07 and minimum 0.87 ± 0.77 in summer 2005 ranged between 0.00 to 4.5 mg/l. The highest value of dissolved oxygen in textile effluents was found 4.5 during January 2006 and lowest 0.00 mg/l in May 2005. The variation in the dissolved oxygen content during the study is given Table 3.1. Monthly variation in dissolved oxygen is given in Figure 3.3.

Free CO_2 (mg/l)

Free carbon dioxide in textile industrial effluents was recorded maximum 2.8 ± 0.30 in summer 2005 and minimum 0.38 ± 0.53 in winter 2005-06 ranged between 0.00 to 3.1 mg/l. The highest value of free carbon dioxide in textile effluents was found 3.1 mg/l during May 2005 and lowest 0.00 in January 2006. The variation in the free carbon dioxide content during the study is given Table 3.1. Monthly variation in carbon dioxide is given in Figure 3.3.

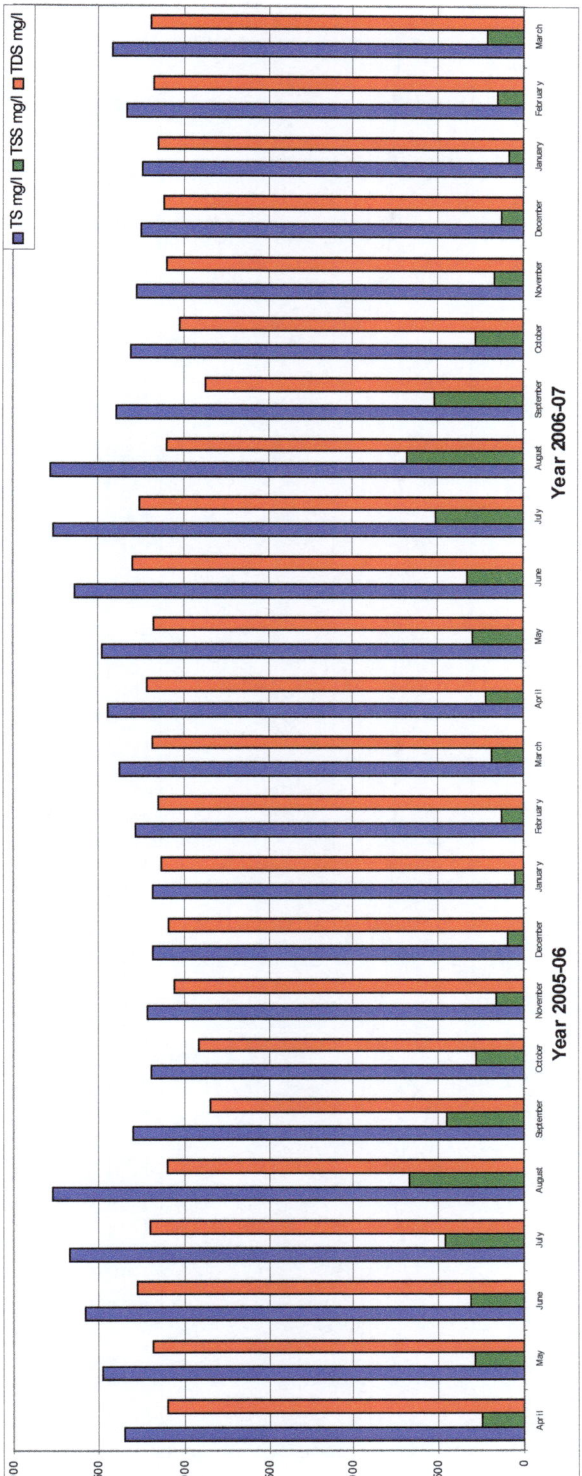

Figure 3.2: Monthly Variation of TS, TDS and TSS in Textile Industrial Effluent

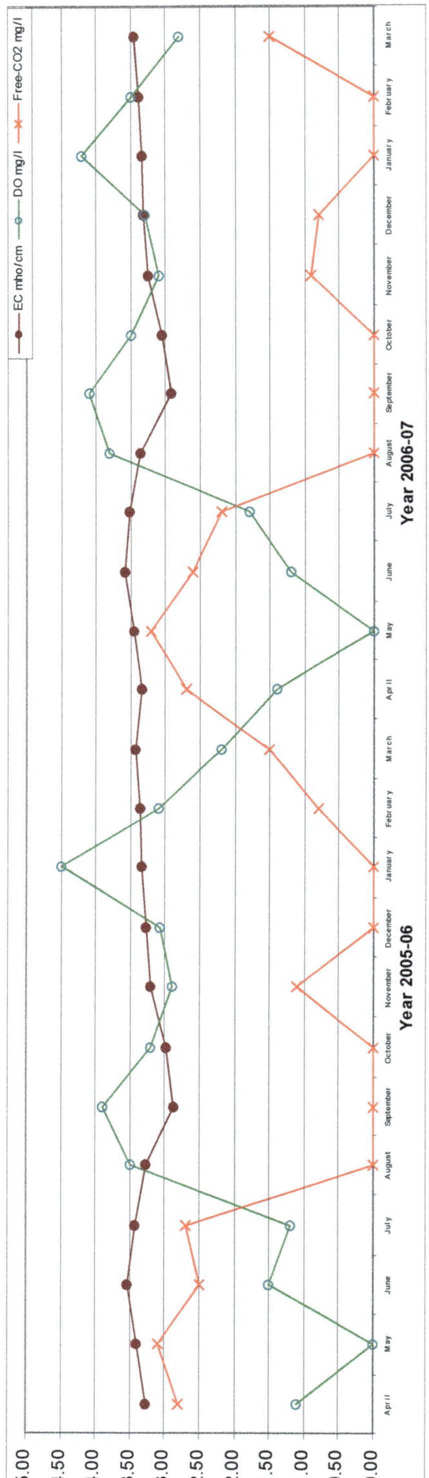

Figure 3.3: Monthly Variation of EC, DO and Free CO_2 in Textile Industrial Effluent

Biochemical Oxygen Demand (mg/l)

The biochemical oxygen demand in textile industrial effluents was recorded maximum 170.0 ± 6.25 in summer 2005 and minimum 138.6 ± 16.52 in winter 2005-06 ranged between 122.0 to 188.0 mg/l. The highest value of biochemical oxygen demand in textile effluents was found 188.0 mg/l during July 2005 and lowest 122.00 mg/l in January 2006. The variation in biochemical oxygen demand during the study is given Table 3.1. Monthly variation in biochemical oxygen demand is given in Figure 3.4.

Chemical Oxygen Demand (mg/l)

The chemical oxygen demand in textile industrial effluents was recorded maximum 445.0 ± 40.93 in monsoon 2005 and minimum 351.5 ± 14.85 in summer 2007 ranged between 310 to 490 mg/l. The highest value of chemical oxygen demand in textile effluents was found 490 mg/l during July 2005 and lowest in 310 mg/l January 2006. The variation in chemical oxygen demand during the study is given Table 3.1. Monthly variation in chemical oxygen demand is given in Figure 3.4.

Total Alkalinity (mg/l)

Total alkalinity in textile industrial effluents was recorded maximum 889.0 ± 47.88 in monsoon 2006 and minimum 633.60 ± 69.55 in winter 2005-06 ranged between 535 to 946 mg/l. The highest value of total alkalinity in textile effluents was found 946 mg/l during July 2005 and lowest 535 in December 2005. The variation in total alkalinity during the study is given Table 3.1. Monthly variation in total alkalinity is given in Figure 3.5.

Calcium (mg/l)

The concentrations of calcium in textile industrial effluents were recorded maximum 27.37 ± 6.78 in monsoon 2006 and minimum 10.58 ± 3.2 in winter 2005-06 ranged between 5.6 to 34.5 mg/l. The highest value of calcium in textile effluents was found 34.5 mg/l during July 2006 and lowest 5.6 mg/l in January 2006. The variation in calcium during the study is given Table 3.1. Monthly variation in calcium is given in Figure 3.6.

Magnesium (mg/l)

The concentrations of magnesium in textile industrial effluents were recorded maximum 16.07 ± 3.46 in summer 2005 and minimum 4.70 ± 2.12 in summer 2007 ranged between 2.4 to 18.9 mg/l. The highest value of magnesium in textile effluents was found 18.9 mg/l during July 2006 and lowest 2.4 mg/l in February 2006. The variation in magnesium during the study is given Table 3.1. Monthly variation in magnesium is given in Figure 3.6.

Potassium (mg/l)

The concentration of potassium in textile industrial effluents was recorded maximum 8.83 ± 0.58 in monsoon 2006 and minimum 7.40 ± 0.14 in summer 2007 ranged between 7.0 to 9.5 mg/l. The highest value of potassium in textile effluents was found 9.5 mg/l during July 2006 and lowest 7.0 mg/l in April 2005. The variation

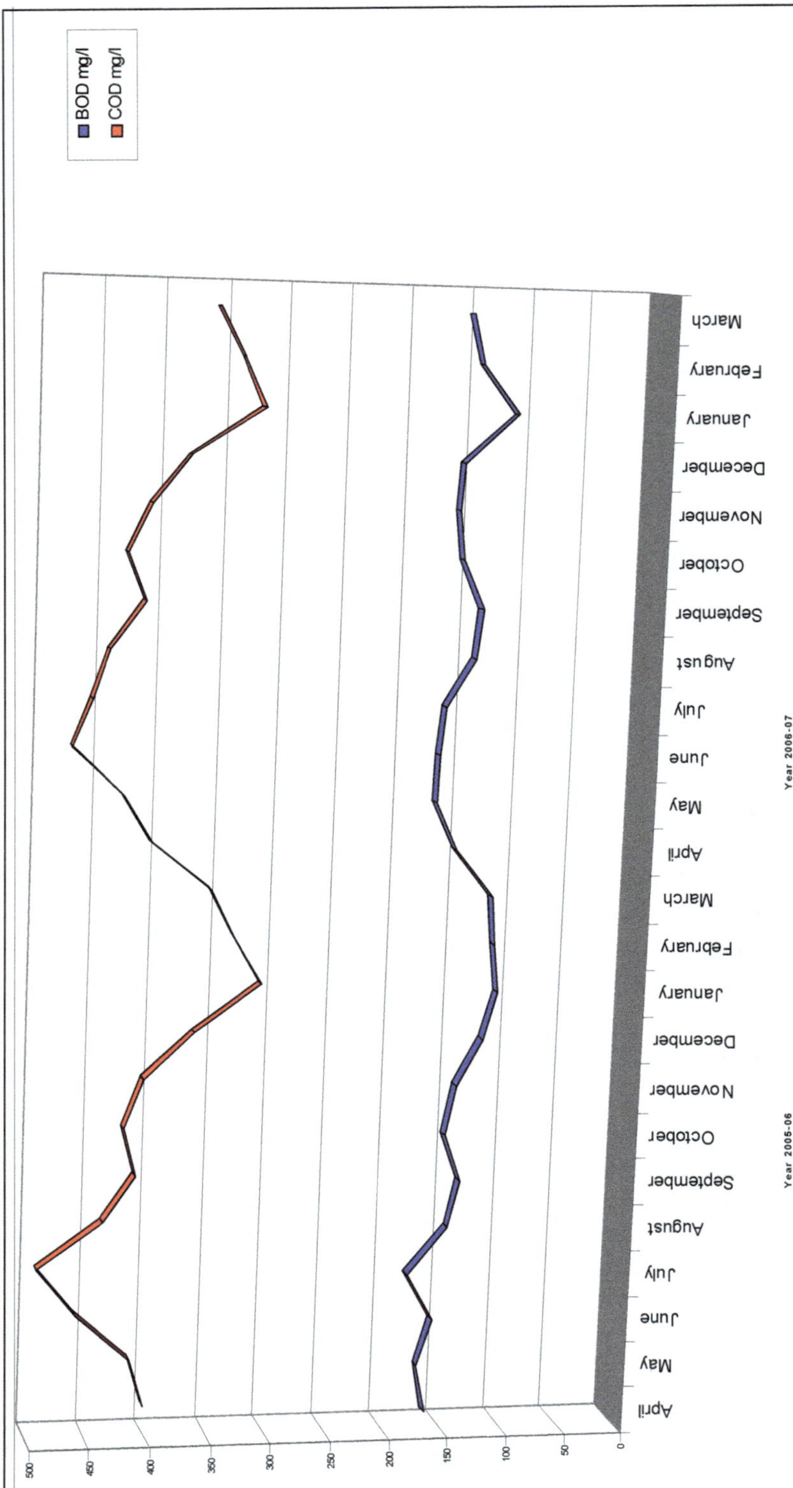

Figure 3.4: Monthly Variation of BOD and COD in Textile Industrial Effluent

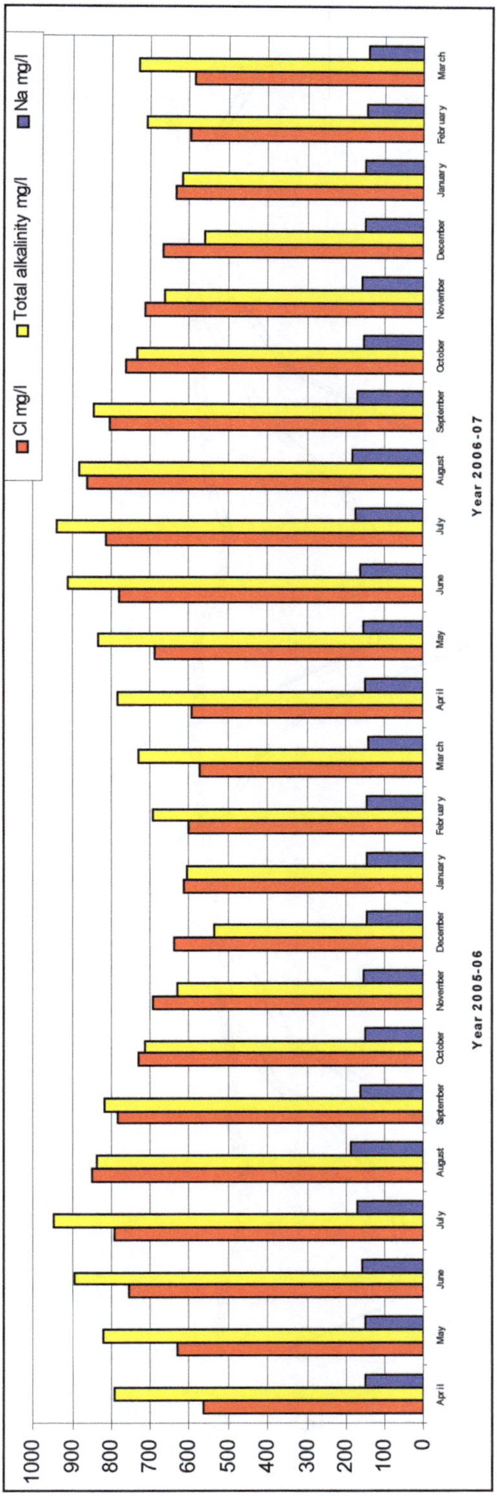

Figure 3.5: Monthly Variation of Chloride, Total Alkalinity and Sodium in Textile Industrial Effluent

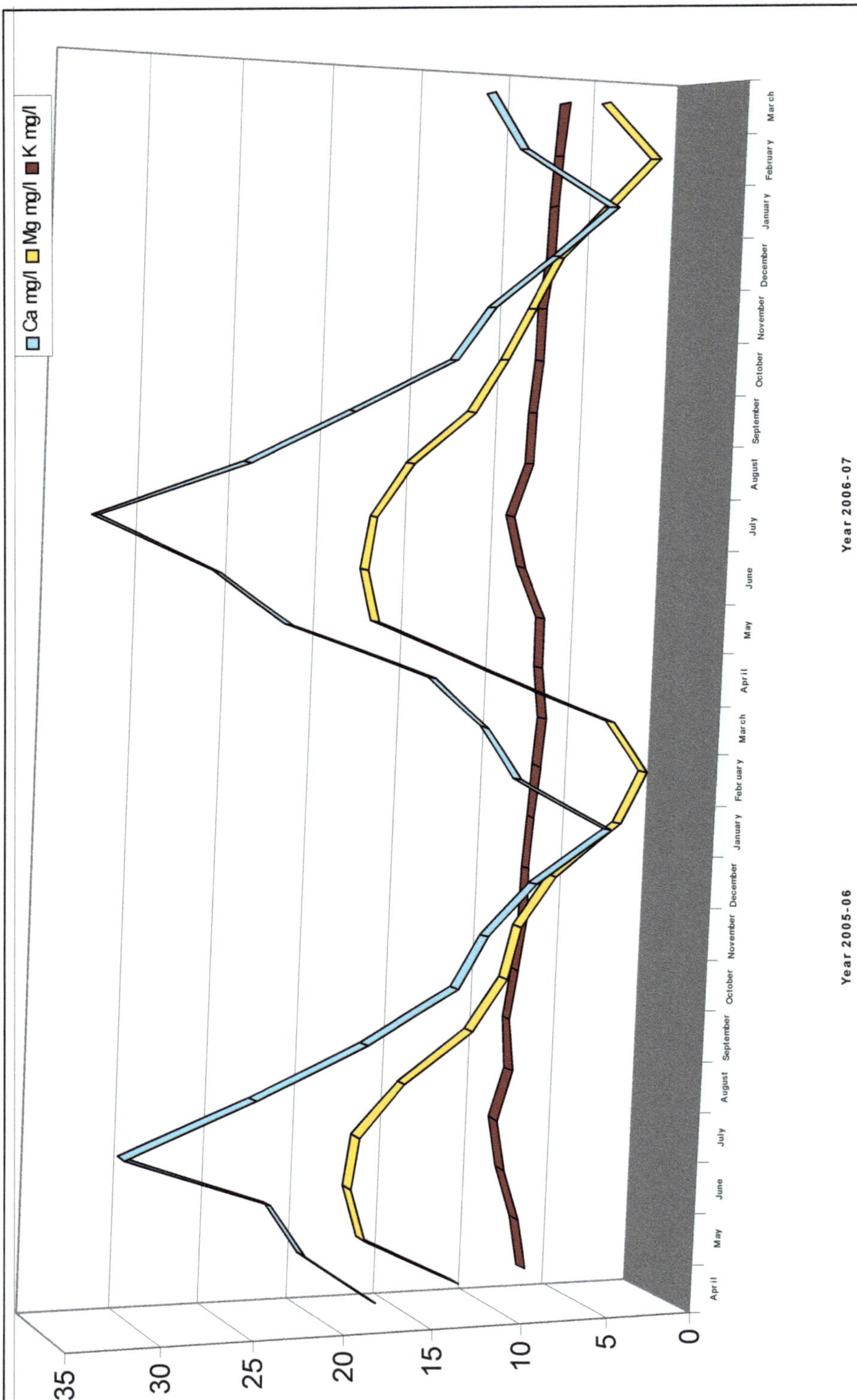

Figure 3.6: Monthly Variation of Calcium, Magnesium and Potassium in Textile Industrial Effluent

in potassium during the study is given Table 3.1. Monthly variation in potassium is given in Figure 3.6.

Sodium (mg/l)

Sodium in textile industrial effluents was recorded maximum 177.0 ± 7.21 in monsoon 2006 and minimum 145.0 ± 4.24 in summer 2007 ranged between 142.0 to 186.0 mg/l. The highest value of Sodium in textile effluents was found 186.0 mg/l during August 2005 and lowest 142.0 mg/l in March 2006 . The variation in Sodium during the study is given Table 3.1. Monthly variation in Sodium is given in Figure 3.5.

Chloride (mg/l)

Chloride in textile industrial effluent was recorded maximum 825.67 ± 29.94 in monsoon 2006 and minimum 588.50 ± 9.19 in summer 2007 ranged between 565.0 to 860 mg/l. The highest value of Chloride in textile effluents was found 860.0 during August 2006 and lowest 565.0 mg/l in April 2005. The variation in Chloride during the study is given Table 3.1. Monthly variation in Chloride is given in Figure 3.5.

Physico-chemical Parameters of Drain Sediment

Temperature (°C)

Temperature of drain sediment was recorded maximum 24.93 ± 3.49 in summer 2005 and minimum 15.46 ± 3.13 in winter 2005-06 ranged between 10.5 to 28.1 °C. The highest value of temperature in drain sediment was found 28.1 °C during July 2005 and lowest 10.5 °C in January 2006. The temperature variation during the study is given Table 3.2. Monthly variation in temperature is given in Figure 3.7.

Water Holding Capacity (Per cent)

Water holding capacity of drain sediment was recorded maximum 31.71 ± 0.51 in summer 2005 and minimum 24.50 ± 5.11 in monsoon 2005 ranged between 20.23 to 32.25 per cent . The highest value of water holding capacity in drain sediment was found 32.25 per cent during June 2006 and lowest 20.23 per cent during September 2005. The water holding capacity variation during the study is given Table 3.2. Monthly variation in water holding capacity is given in Figure 3.7.

Bulk Density (g/cm^3)

The bulk density of drain sediment was recorded maximum 1.45 ± 0.07 in summer 2005 and minimum 1.33 ± 0.03 in winter 2005-06 ranged between 1.29 to 1.51 g/cm^3. The highest value of bulk density in drain sediment was found 1.51 g/cm^3 during June 2005 and lowest 1.29 g/cm^3 in January 2006. The bulk density variation during the study is given Table 3.2. Monthly variation in bulk density is given in Figure 3.8.

Soil Moisture (per cent)

The soil moisture of drain sediment was recorded maximum 20.85 ± 2.25 in monsoon 2006 and minimum 13.33 ± 1.15 in summer 2005 ranged between 12.5 to 22.3 per cent. The highest value of soil moisture in drain sediment was found 22.3 per

Table 3.2: Seasonal Variation of Physico-chemical Characteristics of Drain Sediments

Para-meters	*Unit*	*Summer 2005*		*Monsoon 2005*		*Winter 2005-06*		*Summer 2006*		*Monsoon 2006*		*Winter 2006-07*		*Summer 2007*	
		Mean	*SD*	*Mean*	*SD*	*Mean*	*SD*	*Mean*	*SD*	*Mean*	*SD*	*Mean*	*SD*	*Mean*	*SD*
Temp.	°C	24.933	3.485	23.433	2.259	15.460	3.131	22.838	3.772	23.767	2.250	15.700	2.988	17.600	1.980
WHC	%	31.710	0.505	24.500	5.109	25.670	2.836	31.305	0.865	24.777	5.133	24.888	2.496	29.675	0.247
Bulk density	mg/cm³	1.453	0.067	1.433	0.015	1.332	0.034	1.428	0.063	1.450	0.030	1.353	0.046	1.360	0.028
Soil moisture	%	13.333	1.154	20.723	2.409	15.298	1.163	13.713	0.831	20.850	2.252	15.963	1.436	14.500	0.707
pH		7.840	0.115	8.090	0.165	8.310	0.074	7.970	0.268	8.067	0.153	8.318	0.047	8.325	0.035
EC	mho/cm	32.500	0.889	31.233	1.250	27.860	0.918	32.050	1.529	31.267	1.286	28.100	0.829	29.150	1.344
Cl	µg/g	836.667	145.060	785.333	81.831	613.000	36.173	772.000	168.740	797.667	77.520	637.250	40.434	609.000	33.941
Na	µg/g	490.667	114.605	157.000	98.853	326.400	161.347	515.750	94.979	168.333	89.002	254.000	115.360	587.500	41.719
K	µg/g	31.000	1.000	27.333	1.528	30.800	1.304	31.750	1.258	28.000	1.000	31.000	1.155	33.500	0.707

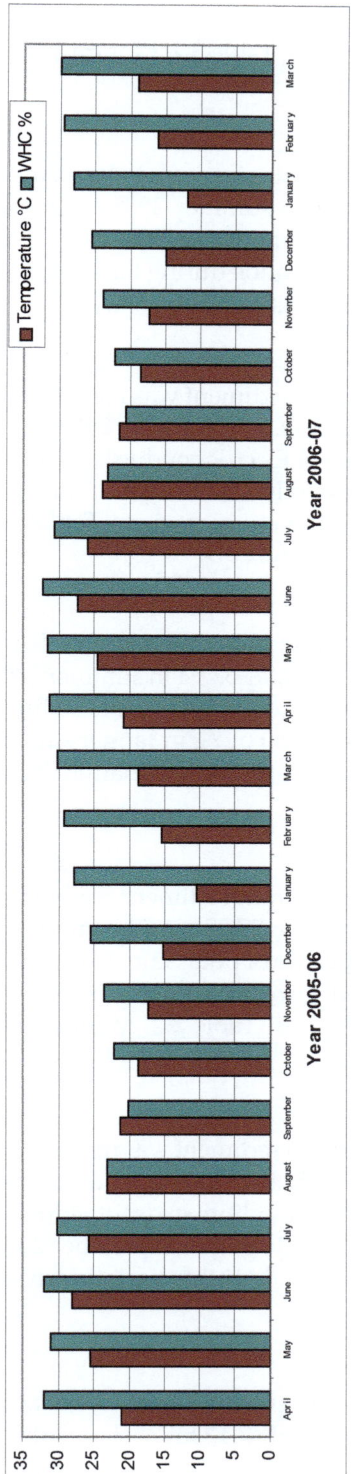

Figure 3.7: Monthly Variation of Temperature and WHC in Effluent Drain Sediment

cent during August 2005 and lowest 12.5 per cent in April 2005. The soil moisture variation during the study is given Table 3.2. Monthly variation in soil moisture is given in Figure 3.8.

pH

pH of drain sediment on pH scale was recorded maximum 8.32 ± 0.04 in monsoon 2006 and minimum 7.84 ± 0.12 in summer 2005 ranged between 7.7 to 8.5. The highest value of pH in drain sediment was found 8.5 during February 2006 and lowest 7.7 in June 2006. The pH variation during the study is given Table 3.2. Monthly variation in pH is given in Figure 3.8.

Electrical Conductivity (mho/cm)

Electric conductivity of drain sediment was recorded maximum 32.50 ± 0.89 in summer 2005 and minimum 27.86 ± 0.92 in winter 2005-06 ranged between 26.8 and 33.7 mho/cm. The highest value of Electric conductivity in drain sediment was found 33.7 mho/cm during June 2006 and lowest 26.8 mho/cm in January 2006. The Electric conductivity variation during the study is given Table 3.2. Monthly variation in Electric conductivity is given in Figure 3.9.

Chloride (µg/g)

Chloride concentration in drain sediment was recorded maximum 836.67 ± 145.1 in summer 2005 and minimum 609.0 ± 33.94 in summer 2007 ranged between 580.0 to 998.0 µg/g. The highest value of chloride in drain sediment was found 998.0 µg/g during June 2005 and lowest 580.0 µg/g in January 2006. The chloride variation during the study is given Table 3.2. Monthly variation in chloride is given in Figure 3.9.

Potassium (µg/g)

Potassium concentration in drain sediment was recorded maximum 33.50 ± 0.71 in summer 2007 and minimum 27.33 ± 1.53 in monsoon 2005 ranged between 26.0 to 34.0 µg/g. The highest value of potassium in drain sediment was found 34.0 µg/g during March 2007 and lowest 26.0 µg/g in August 2005. The potassium variation during the study is given Table 3.2. Monthly variation in potassium is given in Figure 3.9.

Sodium (µg/g)

Sodium concentration in drain sediment was recorded maximum 587.50 ± 41.72 in summer 2007 and minimum 157.0 ± 98.85 in monsoon 2005 ranged between 95.0 to 617.0 µg/g. The highest value of sodium in drain sediment was found 617.0 µg/g during March 2007 and lowest 95.0 µg/g in September 2005. The sodium variation during the study is given Table 3.2. Monthly variation in sodium is given in Figure 3.9.

Correlation Coefficient

Correlation between physico-chemical parameters and heavy metals of textile industrial effluent are given in Table 7.17 for each metal. For cadmium correlation was found maximum 0.5954 and 0.6627 with temperature and free carbon dioxide

Table 3.3: Seasonal Variations of Heavy Metals in Textile Effluents

Heavy Metals	Unit	Summer 2005		Monsoon 2005		Winter 2005-06		Summer 2006		Monsoon 2006		Winter 2006-07		Summer 2007	
		Mean	SD	Mean	SD	Mean	SD	Mean	SD	Mean	SD	Mean	SD	Mean	SD
Cadmium (Cd)	ppm	0.052	0.042	0.025	0.037	0.012	0.008	0.028	0.008	0.009	0.001	0.005	0.004	0.019	0.001
Copper (Cu)	ppm	0.440	0.085	0.343	0.058	0.165	0.066	0.430	0.127	0.367	0.091	0.180	0.030	0.340	0.079
Iron (Fe)	ppm	0.527	0.093	0.253	0.130	0.266	0.079	0.543	0.170	0.287	0.131	0.240	0.100	0.350	0.085
Manganese (Mn)	ppm	0.153	0.050	0.127	0.047	0.224	0.072	0.176	0.060	0.110	0.020	0.215	0.034	0.310	0.014
Nickel (Ni)	ppm	0.037	0.016	0.016	0.015	0.031	0.013	0.067	0.037	0.007	0.003	0.029	0.013	0.045	0.007
Lead (Pb)	ppm	0.500	0.207	0.250	0.085	0.422	0.133	0.475	0.111	0.330	0.028	0.408	0.158	0.485	0.092
Zinc (Zn)	ppm	0.173	0.059	0.110	0.012	0.238	0.132	0.222	0.091	0.165	0.064	0.338	0.146	0.275	0.035

Table 3.4: Seasonal Variations of Heavy Metals in Drain Sediments

Heavy Metals	*Unit*	*Summer 2005*		*Monsoon 2005*		*Winter 2005-06*		*Summer 2006*		*Monsoon 2006*		*Winter 2006-07*		*Summer 2007*	
		Mean	*SD*	*Mean*	*SD*	*Mean*	*SD*	*Mean*	*SD*	*Mean*	*SD*	*Mean*	*SD*	*Mean*	*SD*
Cadmium (Cd)	ppm	5.783	0.231	5.103	0.220	4.250	0.544	5.468	0.553	5.157	0.475	5.110	0.266	5.020	0.184
Copper (Cu)	ppm	49.247	4.963	45.897	6.950	46.454	6.678	54.825	7.333	47.527	10.398	49.435	7.303	45.865	5.819
Iron (Fe)	ppm	8467.667	2448.183	5656.000	1216.719	8090.000	938.127	6469.500	1179.615	6362.000	501.728	7921.250	1751.800	5813.500	185.969
Manganese (Mn)	ppm	1047.333	76.002	791.333	51.733	978.600	52.624	1105.000	107.083	699.333	124.243	1028.250	67.869	1021.000	36.770
Nickel (Ni)	ppm	75.000	3.000	58.000	3.000	60.600	5.941	74.250	4.113	58.667	6.028	58.250	8.139	68.000	4.243
Lead (Pb)	ppm	70.000	4.359	60.667	4.509	47.200	3.834	64.000	11.916	60.667	2.517	52.500	2.646	49.500	10.607
Zinc (Zn)	ppm	260.000	83.540	208.000	87.000	151.400	47.184	251.000	96.620	174.667	67.899	143.500	32.296	196.000	19.799

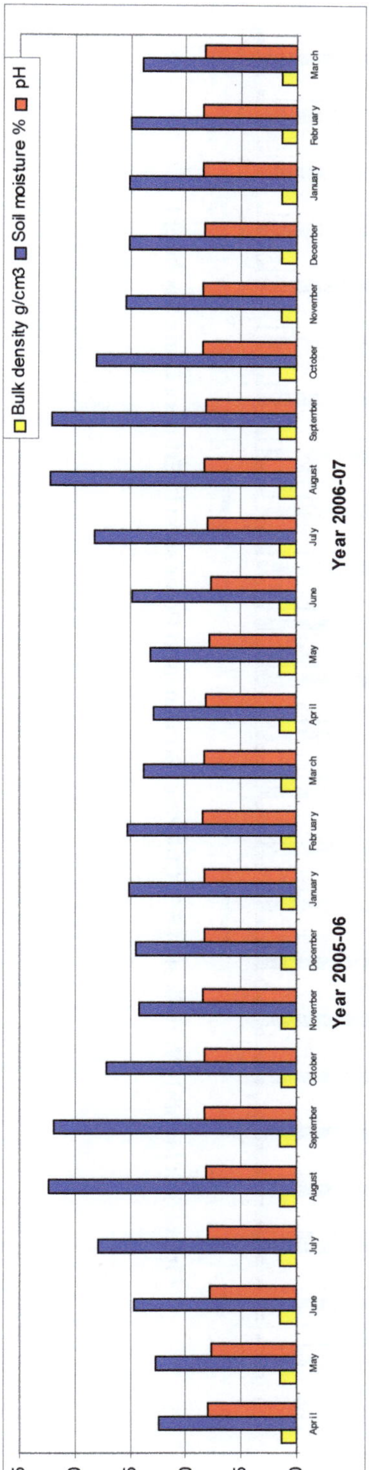

Figure 3.8: Monthly Variation of Bulk Density, Soil Moisture and pH in Effluent Drain Sediment

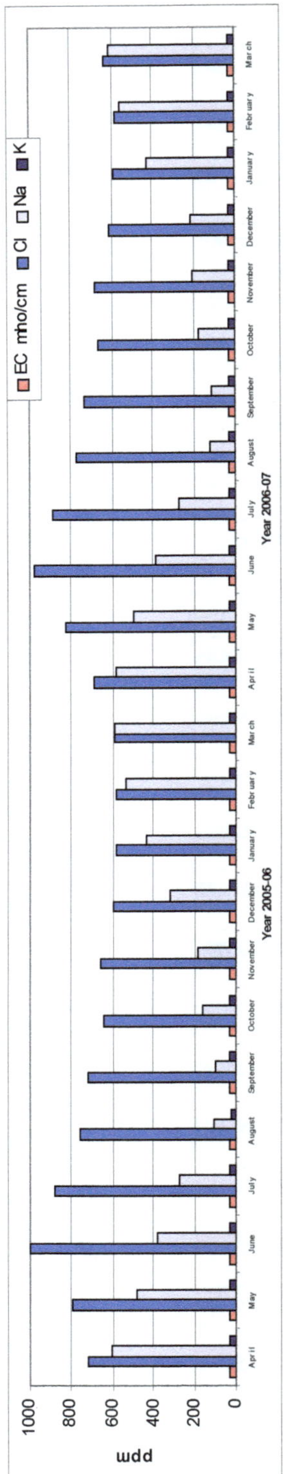

Figure 3.9: Monthly Variation of EC, Chloride, Sodium and Potassium in Effluent Drain Sediment

respectively. Correlation for copper was found maximum 0.6126 and 0.6112 with total dissolved solids and free carbon dioxide respectively. Correlation for iron was found maximum 0.6848 and 0.7573 with total dissolved solids and free carbon dioxide respectively. Correlation for manganese was found maximum 0.3101 and 0.2820 with total dissolved solids and Electric conductivity respectively. Correlation for nickel was found maximum 0.3455 and 0.4410 with total dissolved solids and free carbon dioxide respectively. Correlation for lead was found maximum 0.7373 and 0.7303 with total dissolved solids and Electric conductivity respectively. Correlation for zinc was found maximum 0.0579 and 0.0685 with total dissolved solids and DO respectively.

Correlation between heavy metals of textile industrial effluent and heavy metals of pond water are given in Table 7.18 for each metal. For cadmium correlation was found maximum 0.6257 and 0.5244 with copper and manganese respectively. Correlation for copper was found maximum 0.7762 and 0.7975 with nickel and lead respectively. Correlation for iron was found maximum 0.8110 and 0.8506 with nickel and lead respectively. Correlation for manganese was found maximum 0.4070 and 0.8434 with iron and zinc respectively. Correlation for nickel was found maximum 0.5772 and 0.5855 with cadmium and lead respectively. Correlation for lead was found maximum 0.6835 and 0.5946 with cadmium and nickel respectively. Correlation for zinc was found maximum 0.0683 and 0.4195 with cadmium and zinc respectively.

Correlation between physico-chemical parameters and heavy metals of effluent drain sediment are given in Table 7.23a for each metal. For cadmium correlation was found maximum 0.7226 and 0.7258 with bulk density and electric conductivity respectively. Correlation for copper was found maximum 0.2637 and 0.3623 with temperature and bulk density respectively. Correlation for iron was found maximum 0.1399 and 0.1114 with chloride and potassium respectively. Correlation for manganese was found maximum 0.6220 and 0.7253 with sodium and potassium respectively. Correlation for nickel was found maximum 0.8752 and 0.7610 with water holding capacity and sodium respectively. Correlation for lead was found maximum 0.8233 and 0.8331 with temperature and electric conductivity respectively. Correlation for zinc was found maximum 0.7598 and 0.7570 with water holding capacity and chloride.

Correlations between heavy metals of drain sediment are given in Table 7.23b for each metal and correlations between heavy metals of pond sediment are given in Table 7.23d for each metal. However for both correlation values were not found highly positive and significant.

3.4 Discussion

With the increase in industrialization our health problems are increasing day by day. Today the pollution especially of water resources is increasing steadily due to rapid population growth, industrial proliferation, urbanization increasing living standards and wide sphere of human activities. The greatest danger to the quality of environment is posed by human activities such as poorly organized and unregulated disposal of industrial, domestic waste and crude oil spillage are the major causes of the environmental deterioration. The industrial effluent over the years has caused serious damages of the surrounding soil, groundwater and agriculture crops (Chao *et al.*, 2005).

Most of the dye houses of the Panipat region discharge their effluent directly into a common effluent drain, which carry the total of composite effluents towards the village pond at Binjhole adjoining to industrial area, ultimately has altered the water quality and biotic and nutrient dynamic of the pond ecosystem in a tremendous way and further may affect the groundwater quality. Due to post irrigation practices, the soil turns to barren or unfertile and affected agriculture productivity.

Generally, textile industry consumes substantial volume of water and chemicals in wet processing of textile materials. Wet processing of cloth or fibres dying requires huge amount of different chemical dyes and wash away them with the wastewater discharged through industrial outlet. Thus, the common effluent drain consist all types of dye residuals in a composite form. This composite effluent is very toxic in nature due to the presence of different synthetic dyes, which contain various toxic heavy metals.

Further future of these heavy metals is not so managed because the common effluent drain ends up in to a pond near village Binjhole, Industrial area and some local surroundings of dye houses. The effluents discharged from dye houses are quite important for temporal monitoring. Because the effluents store pond is highly contaminated and is a depot of heavy metals due to the presence of chemical dye portions continuously since a long period of time. Heavy metals remain in water ecosystem, but these may travel through food web at various trophic levels and environmental components as well as. Metals moves pond sediments and reach in shallow water table by leaching processes and the metal contamination may disperse in a wide area of the region. Contamination of soils of close areas occurs when the effluents or polluted water comes under the contact of soil. The regular contact of polluted water with soil makes the sodicity and alters the physico-chemical characteristics, texture and profile of soil stratum.

Thus, the quality assessment of textile effluents, pond water, groundwater, sediments and soils required to know the pollution load of heavy metals on hydrosphere and lithosphere and to determine the drinking water quality of Panipat industrial zone in term of suitability and possible impacts on human health. In the present study, water, soil and sediment samples were collected and analyzed for determination of physico-chemical characteristics and distribution of heavy metals from April 2005 to March 2007 regularly. The results are depicted in the form of tables and graphs. During the study, physico-chemical characteristics and heavy metals in textile effluents, surface (pond) water, groundwater, soils and sediments were found with some fluctuation according to seasons, place and other factors.

3.5 Physico-chemical Parameters of Textile Effluents

In the present investigation of water resources, physico-chemical characteristics of water directly reflect its quality for irrigation, bathing, drinking and industrial uses. Physico-chemical parameter like total dissolved solids, pH, chloride, total alkalinity, biochemical oxygen demand, chemical oxygen demand, pH are the some important parameters for assessing the quality of water for irrigation and drinking purposes.

Temperature mostly fluctuates according to climatic conditions of the region and directly depends upon seasonal air temperature. However, groundwater temperature remains comparatively lower (25.5-31.6 °C) due to the geothermal energy. The temperature of textile common effluents (25.7-32.8 °C) and pond water (25.5-32.4 °C) has fluctuated significantly due to the lentic behaviour of aquatic ecosystem. Temperature of composite effluent drops slightly while reaching in the pond (25.5-32.4 °C). Pathe *et al.* (2005) reported the temperature fluctuation trend from cluster of small cotton textile units from different sector, which follow permissible limit of B.I.S and WHO and remain always below 40°C during study. During the study period temperature of textile industrial effluent never cross the standards limits. Temperature slightly influences some parameters of water and affects the concentration of dissolved oxygen contents and can also influence the activity of bacteria and toxic chemicals in water. Textile industrial effluent may cause problems with high temperature (Groff, 1993). Temperature facilitates the oxidation process biochemical oxygen demand and chemical oxygen demand achieved maximum reduction at high temperature as compared to low temperature (Aslam *et al.*, 2004). Temperature of water depends upon the meteorological characteristics of existing environment; especially air temperature. Temperature fluctuated with air temperature with positive relation. However due to some processes in textile units, the temperature of textile effluents was slightly higher than pond water and groundwater.

In the present study, the pH of the textile effluents was recorded mostly alkaline. The pH of the composite effluents was highly alkaline, which slightly decreases in pond water system. Extreme high pH or rapid pH change can exert stresses conditions or kill aquatic life. Changes in pH can also affect aquatic system indirectly by altering other aspects of water chemistry. Low pH levels accelerate the release of metals from rocks or sediments in the water. Groff (1993) reported that textile wastewater may cause problem used as untreated effluent consist higher pH and temperature. pH has significant importance in all chemicals and biochemical reactions in an aquatic system (Basu *et al.*, 1973).

pH of textile industrial effluent was found mostly basic (9.6-10.7) due to the high salt load and high alkalinity. Pond water's pH was also alkaline (8.2-8.6) due to the directly mixing with effluent drains with pond. The pH of both, drain sediment (7.8-8.3) and pond sediments (7.9-8.4) were also noticed alkaline in nature. Similar trend is in agreement with that reported by Sial *et al.* (2006), Pathe *et al.* (2005) in textile industrial effluent, Singh *et al.* (1994) in effluent drain and Mor *et al.* (2003).

Total solids were found very high in textile industrial effluent (2210.4-2643.7 mg/l) due to the high degree of pollution. Different dyes and fibres are present in the textile wastewater, which generally released dissolved solids into common effluent drain. The dissolved solids in wastewater are mainly inorganic salts and relatively unaffected by biological treatment processes and can accumulate in water recirculation system. Failure to remove them may lead to an increase in total solids content of surface water sources. The dissolved solids in wastewater may be harmful to vegetation and also preclude use in various irrigation processes. Total dissolved solids in the textile effluents were observed very high than the prescribed limits of C.P.C.B. and B.I.S values. In the pond, the values of total solids and total dissolved solids were noticed comparatively high, however, these were not found higher than textile industrial effluent. Dissolved solids vary qualitatively and quantitatively in different water depending upon the season, location and other factors. In the present study, the total dissolved solids values in effluents (2045.7-2219.7 mg/l) and pond water (1861.0-2037.5 mg/l) were not so high according to CPCB standards. Similar observations were also reported by Yusuf and Sonibare (2004), Garg and Kaushik (2006) in textile industrial effluent.

Total dissolved solids are the measure of the quantity of material dissolved in water that can pass through a filter (usually with a pore size of 0.45 micrometers). The material can include carbonate, bicarbonate, chloride, sulfate, phosphate, nitrate, calcium, magnesium, sodium, organic ions, and other ions. Similar to total suspended solids, high concentrations of total dissolved solids may also reduce water transparency, combine with toxic trace compounds and heavy metals, and lead to an increase in water temperature. Total dissolved solids are used to estimate the quality of drinking water, because it represents the amount of ions in the water. High dissolved solids concentrations can produce laxative effects and can give an unpleasant mineral taste to water. High dissolved solids concentrations in water are also unsuitable for water recycling processes and many industrial applications. The values of total dissolved solids concentrations recorded in the present study are comparable to those obtain by Pathe *et al.* (2005), Yusuf and Sonibare (2004), Garg and Kaushik (2006) in textile industrial effluent.

Total suspended solids include both organic and inorganic materials, which adversely affect the aquatic life, especially fishes and bottom fauna of surface water ecosystems. These produce carbon dioxide, hydrogen sulphide, methane gases, which are very noxious in nature. They increase the turbidity of water, reduce light penetration and impair photosynthetic activity of aquatic plants in surface water system. Total suspended solids in the effluents (141.8-570.7 mg/l) and pond water (96.8-487.7 mg/l) were found slightly high than recommended by C.P.C.B. and B.I.S. Chavan *et al.* (2005) reported the adverse impacts of total suspended solids on the surroundings environment of the industrial effluent near Sukhna river at Midc area, Aurangabad.

Because the suspended particles absorb heat from sunlight and changed dissolved oxygen levels to fall even further because warmer waters can hold less dissolved oxygen. High total suspended solids in a water body can often mean higher concentrations of bacteria, nutrients, pesticides, and metals in the water. These pollutants may attach to sediment particles on the land and be carried into water bodies with storm water. In the water, the pollutants may be released from the sediment or travel farther. The flow rate of the effluent drain is a primary factor in total suspended solids concentrations. Fast running water can carry more particles and larger-sized sediment. Heavy rains can pick up sand, silt, clay, organic particles and heavy rubbish particles from the land and carry it to surface water. A change in flow rate can also affect total suspended solids; if the speed or direction of the water current increases, particulate matter from bottom sediments may be re-suspended.

Total solids were found very high in textile industrial effluent due to the dirt release during the washing of textile materials. The ratio of total dissolved solids and total suspended solids with total solids was found quite similar during the study. Such types of trends in determining the total suspended solids load have been reported by Garg and Kaushik (2006), Yusuf and Sonibare (2004) in textile industrial effluent. Liquid wastewater from textile industries are not fit for irrigation regarding electric conductivity and total suspended solids (Bohn *et al.*, 1976, Abbas, 1991).

Conductance is a measurement of an electrical current with increasing amount and mobility of ions. These ions, which come from the breakdown of compounds, conduct electricity because they are negatively or positively charged when dissolved in water. Therefore, conductance is an indirect measure of the presence of dissolved solids such as chloride, nitrate, sulfate, phosphate, sodium, magnesium, calcium, and iron, and can be used as an indicator of water pollution.

It is not necessary that the water having high electric conductivity is not fit for irrigation. Only one parameter can't ensure the suitability of the water quality. However, the salt composition is high in the water having high Electric conductivity, but it may be beneficial for those soils having a lack or deficiency of some salts. Electric conductivity has a positive correlation with total dissolved solids in textile industrial effluent and groundwater also. Generally, there are not regulatory levels for electric conductivity. Instead, the concentration of total dissolved solids is often regulated. However, electric conductivity is an easily-obtained parameter that is a good indicator of the amount of dissolved solids in water, and thus can be used to detect contaminants in water. Total dissolved solids have a direct positive relationship with electric conductivity. If, total dissolved solids of water are high, electric conductivity will be more definitely (Mor *et al.*, 2003).

Electric conductivity was recorded very highly positive correlated to total dissolved solids due to the direct relation of total dissolved solids with electric conductivity in textile industrial effluent (3.1-3.4 mho/cm). Sial *et al.* (2006), Garg and Kaushik (2006) were made similar observations in textile industrial effluent and Singh *et al.* (1994) in drain. which support the result of present study.

The dissolved oxygen is often attributed to the fact the oxygen is dissolved more during the period of active photosynthesis. Hutchinson (1957) reported that the water has low temperature has greater capacity for holding gases. The value of dissolved oxygen in the all sampling sites was found out to be low dissolved oxygen. The DO was found nil or low in textile effluent (0.87-3.5 mg/l), however, in some months dissolved oxygen was found nil in textile industrial effluent due to the increasing rate of biodegradation of different compounds in water column. As the matter of fact that the effluents carried highly pollution load, which could not allow any flora and fauna to survive in unfavourable condition and there will be no photosynthesis process in polluted habitat. The dissolved oxygen showed an inverse relationship with temperature as reported by Das and Srivastava (1956) and Dobriyal (1985) in aquatic system.

The value of dissolved oxygen recorded in the present study are comparable to results obtained by Yusuf and Sonibare (2004), Garg and Kaushik (2006) in textile industrial effluent and Singh *et al.* (1994) in effluent drain.

Free carbon dioxide in most textile industrial effluents was fluctuated in the range of 0.38 to 2.8 mg/l. However, the dissolved oxygen contents was also not found high in textile industrial effluent (0.87-3.5 mg/l). Free carbon dioxide mainly depends upon some physico-chemical parameter like dissolved oxygen, pH, temperature, etc. (ww.holar.is). Carbon dioxide has highly positive correlations (0.6627, 0.6112, 0.7573, -0.0984, 0.4410, 0.3252, -0.0395) in textile industrial effluents with almost all heavy metals (Cd, Cu, Fe, Mn, Ni, Pb and Zn) respectively.

Biochemical oxygen demand and chemical oxygen demand are the measurement of the oxygen consuming capabilities of organic matter. Textile effluent during their process of deposition exerts biochemical oxygen demand/chemical oxygen demand, which have a catastrophic effect on the ecosystem by depleting the oxygen supply. Decay process causes the production of noxious gases as hydrogen sulphide and methane. Water with high biochemical oxygen demand/chemical oxygen demand indicates presence of decomposing organic matter and subsequent high bacterial counts that degrade its quality and potential uses (Verma and Kharat, 2000). Biochemical oxygen demands of the effluents drain (138.6-170.0 mg/l) were found high as given in standards of C.P.C.B., B.I.S. and WHO. This trend is an agreement with that reported by Aslam *et al.* (2004), Yusuf and Sonibare (2004), Sial *et al.* (2006), Pathe *et al.* (2005) in textile industrial effluent, and Singh *et al.* (1994) in drain.

Chemical oxygen demand is directly proportional to biochemical oxygen demand. Dilution of effluents for biochemical oxygen demand analysis always depends upon the value of chemical oxygen demand in a wastewater. In fact, chemical oxygen demand explains the load of pollutants and degree of pollution in effluent drain. Effluents of dye houses having a tremendous value of COD in the range of 351.5-445.0 mg/l due to the high rate of decomposition of organic material in polluted and stored water. COD shows a positive relationship with BOD and negative with DO (Khanna, 1993) and mostly remains in the ratio of approximately 1:3 ± 0.5 in the case of wastewater. Sevimli *et al.* (2005) observed the increasing trends of biochemical oxygen demand and chemical oxygen demand together while performing an experiment on decolorization of textile effluent and dye solution by ozonation. Senthinathan and Azeez (1999) and Groff (1993) reported that textile wastewater may cause problem used as untreated effluent consist higher COD, pH, temperature, Phosphorous and absorbable organic halogen. COD of textile wastewater may be reduced by using some appropriate technology Bhat *et al.* (2005). Aslam *et al.* (2004), Yusuf and Sonibare (2004), Sial *et al.* (2006), Pathe *et al.* (2005) were also reported the similar values in textile industrial effluent. Waste containing high biochemical oxygen demand and chemical oxygen demand are responsible for a heavy depletion of oxygen levels in the particular sector of water body or soil (Sial *et al.*, 2006).

Alkalinity is caused by strong bases and the salts of strong alkalis and weak acids. Carbonates are added to a water system if the water passes through soil and rock that contain carbonate minerals, such as calcite ($CaCO_3$). Where limestone and sedimentary rocks and carbonate-rich soils are predominant, waters will often have high alkalinity. Where igneous rocks (such as granite) and carbonate-poor soils are predominant waters will have low alkalinity. Alkalinity is a measure of the buffering capacity of water, or the capacity of bases to neutralize acids. Waters with low alkalinity are very susceptible to changes in pH. Waters with high alkalinity are able to resist major shifts in pH. As increasing amounts of acid are added to a water body, the pH of the water decreases, and the buffering capacity of the water is consumed. Because alkalinity and pH are so closely related, changes in pH can also affect alkalinity, especially in a poorly buffered system. Above pH 8.3, alkalinity is mostly in the form of carbonate (CO_3^{2-}); below 8.3, alkalinity is present mostly as bicarbonate (HCO_3^-).

The values of alkalinity in the effluents (633.6-889.0 mg/l) were found very high. Bhat *et al.* (2005) observed the increasing trend in alkalinity of various textile effluents according to discharge points. Alkalinity not only helps regulate the pH of a water body, but also the metal content. Alkalinity has found with positive correlations with Cd (0.0237), Cu (0.809), Fe (0.8896), Mn (0.7664), Ni (0.7918), Pb (0.5841) and Zn (0.7069) Bicarbonate and carbonate ions in water can remove toxic metals (such as lead, arsenic, and cadmium) by precipitating the metals out of solution. Pathe *et al.* (2005), Garg and Kaushik (2006) have been made some observations in textile industrial effluent; which support the results of present study.

Hardness generally represents the concentration of calcium and magnesium ions, because these are the most common polyvalent cations. Other ions, such as iron and manganese, may also contribute to the hardness of water, but are generally present in much lower concentrations. Calcium, magnesium, and other polyvalent cations such as iron and manganese may be added to a natural water system as it passes through soil and rock containing large amounts of these elements in mineral deposits. Hardness mainly depends on the amount of calcium and magnesium salts or both present in the groundwater, water hardness is not health hazard but its values should remain below permissible level to restore the taste to water (WHO, 1984).

Some textile industrial processes may also produce significant amounts of calcium and manganese that are later discharged as industrial effluents. Calcium was found 10.58-27.4 mg/l in textile industrial effluent. Similar conditions were also reported by Garg and Kaushik (2006), Yusuf and Sonibare (2004) and Helena *et al.* (1999) in textile industrial effluent.

When hardness equals alkalinity, the only cations present in significant concentrations in the water are calcium and magnesium. When hardness is greater than alkalinity, the waters may contain considerable amounts of other cations. Magnesium was found low in textile industrial effluent (4.7-16.1 mg/l). The Mg concentrations levels were comparable with those found by Yusuf and Sonibare (2004) and Helena *et al.* (1999) in textile effluents.

Potassium is one of the important element among ten chemical element required in rather large quantities fir growth of all crops. Plants growth may be limited by a deficient available supply of potassium but its high some quantity may alter the soil composition. High K in water used in irrigation may alter the soil quality (Miller and Turk, 2002). Potassium was found low in textile industrial effluent (7.4-8.8 mg/l). Similar observation were made by Pathe *et al.* (2005) and Garg and Kaushik (2006) in textile industrial effluent.

Sodium salts in the water are much more objectionable than salts of calcium and magnesium because of the tendency of sodium to cause deflocculation of the colloidal fraction of the soil and developed an undesirable structure water contain sufficient heavy metals to prove toxic to continue use by plants. As sodium chloride used in textile industry so the concentration of sodium and chloride will be high in the wastewater (Aslam *et al.,* 2004). Sodium was found very high in groundwater of industrial area, because sodium should not be higher than 20 mg/l in drinking water (WHO, 2006). Sodium values were found very high in textile industrial effluent (145.0-177.0 mg/l). Helena *et al.* (1999) and Garg and Kaushik (2006) were observed the similar observations in textile industrial effluent.

Chloride is one of the most conservative constituents in wastewater, being affected very little either by vegetative growth or other biological processes, or by precipitation or anion exchange reaction. Chloride concentrations in the effluents (588.5-825.7 mg/l) were also expected to be high by the permissible limit of C.P.C.B. A work conducting on the effect of natural flocculants on textile wastewater, chloride concentration is increased in wastewater used in textile industries and delayed later (Aboulhassan *et al.,* 2005).

Chloride does not have a health-based standard for drinking water. Chloride has been suggested as a standard value for drinking purposes (EPA, 1989). The Maximum Contaminant Level (MCL) of 250000 μg/L is based on poor taste chloride imparts to water. Chloride is not affected by adsorption or degradation processes and is therefore a conservative chemical in soil and groundwater. Chloride concentrations begin to increase at some depth due to dissolution reactions within the aquifer. Chloride is therefore a good indicator of human impacts in the environment (MPCA, 2001). In the wastewater the concentration of chloride will be high due to the use of sodium chloride in textile industry. Chloride concentration in irrigation water should be less then 75 ppm, otherwise it may be harmful for soil structure. Textile effluents are high in toxicity not only for human beings of surroundings but also a serious threat to groundwater and surface water resources (Aslam *et al.,* 2004). The chloride concentrations levels were comparable with those found by Helena *et al.* (1999), Garg and Kaushik (2006) in textile industrial effluent.

3.6 Physico-chemical Parameters of Effluent Drain Sediment

Physico-chemical parameters of soil may demonstrate the pollution status in the industrial region, so, physical and chemical characteristics are important for the assessment of contamination and pollution load. Sediment characteristics are the good indicator for the pollution load in effluent drain.

The temperature of soil greatly affects the physical, biological and chemical processes occurring in soil system. Biological decomposition can come to a near stand still, thereby limiting the rate at which nutrients such as nitrogen, phosphorus, sulphur and calcium are made available. Absorption and transport of water and nutrient ions by higher plants are adversely affected by low temperature (Brady, 1995). The temperature of field soils is not subject to radical human regulation. However, two kinds of management practices have significant effects on soil temperature, those that keep some type of cover or mulch on the soil and those that reduce excess soil moisture. There is not any significant variation in the temperature of drain sediment (15.5–24.9°C) and other soil and sediment samples. The chemical processes and activities of microorganism which convert plant nutrients into available forms are also materially influenced by temperature. Removal of excess water from soil may facilitate changes in soil temperature. Humus content, good tillage, dark color of soil are the some other factors which influence the soil temperature (Miller and Turk, 2002).

Water holding capacity is a property as to water quantity is absorbed or retained by given amount of soil (Ramani and Srivastava, 1989). Water held in the soil pores with varying degrees of tenacity depending on the amount of water present and the size of pores in soils. Depending on the soil, as much as one forth to two thirds of the moisture may remain in the soil after the plants have wilted or died due to the lack of soil water.

The phenomenon of capillarity is a common one, the classic example being the movement of water up a wick when the lower end is immersed in water. Capillary forces are at work in all moist soils. However, the rate of movement and the rise in height are less than one would expect on the basis of soil pore size. One reason is that soil pores are filled with air, which may be entrapped, showing down or preventing the movement of water by capillary. Usually the height of the rise resulting from capillarity is greater with fine textured soils if sufficient time is allowed and the pores are not too small. This is readily explained on the basis of the capillary size and the continuity of the pores. With sandy soils the adjustment is rapid but so many of the pores are non-capillary that the height of rise can not be great (Brady, 1995). Water holding capacity was found similar in drain sediment (24.5-31.7 per cent) and other soil and sediment samples with much difference between soil and sediment. Yuandong *et al.* (2006) described the similar trend of water holding capacity on the basis of soil texture and chemical composition in the agricultural soil system.

Bulk density relates to the combined volumes of the solids and pore spaces, soils with a high proportion of pore space to solids have lower bulk densities than those that are more compact and have less pore space. Consequently, any factor that influences soil pore space will affect bulk density. Mainly soil texture and compactness strongly influence the bulk density of soils (Brady, 1995). The pore space of a soil is that portion of the soil volume occupied by air and water. The amount of pore space is determined by the arrangement of the solid particles. If the particles lie close together, as in stands or compact sub-soils, the total porosity is low. If they are arranged in porous aggregates, as is often the case in medium textured soils high in organic matter, the pore space per unit volume will be high. Soil aeration may influence the availability of nutrients to plants (Miller and Turk, 2002). The bulk density was found similar between drain sediment (1.33-1.45 g/cm^3) and pond sediment (1.36-

1.49 g/cm^3). Bulk density was similarly described by Yuandong *et al.* (2006) and Krebs *et al.* (1999) in soil environment.

The structure of the soil influences its moisture content. A well granulated soil has more total pore space than one with poor granulation or one that has been compacted. The reduced pore space may be reflected in a lower water holding capacity. The compacted soil also has to higher proportion of small and medium sized pores, which tend to hold the water with greater tenacity than do the larger pores. Moisture content of drain sediment (13.3-20.8 per cent) and other soil and sediment samples were found similar during the study. Rainwater leaching have made more efficient if antecedent moisture content of the soil profile was increased by saline water irrigation just before the onset of monsoons. On a wet soil, a rainfall event is expected to displace the salts to a greater depth. Conservation and proper utilization of rainwater for leaching salts is also utmost important.

The measurement of soil pH shows the acidity and basic nature of the soil. The soil is acidic if the pH is lower, while the higher pH means that soil is basic. If the decomposition of organic matter is there in soil, it will turn the pH low, other wise some foreign substances may alter the soil nature and properties towards alkaline (Nongkynrih *et al.*, 1996). Initial soil pH is an important property of the soil because of its influence on the infiltration rate of the soil (Gupta *et al.*, 1989). There are several causes of minor fluctuation in soil pH. For instance, the movements of salts into and out of soil moisture moves up and down the profile will influence pH. Similarly, the pH of mineral soils declines during the crop-growing season as a result of the acids produced by microorganisms and by the roots of higher plants. When soil temperatures are low, an increase in pH often is noted because biotic activities during these times are considerably slower.

Soil pH significantly influences other soil chemical properties as well as biological organisms. High pH of textile industrial effluent can hampers plant crop growth (Sial *et al.*, 2006). The soil pH significantly affects the availability of most of the chemical element of importance to plants and microbes. Likewise, the tendency for toxic levels of element such as iron, manganese, aluminum to be established at low pH and deficiencies of iron and manganese at high pH have been mentioned. Continuous application of effluents with high pH will render the soils unfit for further cultivation (Sial *et al.*, 2006). Soil pH was found alkaline in drain sediment (7.8-8.3) and pond sediment.

Electric conductivity is the measure of current carrying capacity, which gives a clear idea of soluble salts present in soil. Conductivity generally depends upon the dilution of suspension. However, the salt composition is high in the water having high electric conductivity, but it may be beneficial for those soils having a lack or deficiency of some salts. The electric conductivity of drain sediment was found in the range of 27.8-32.5 mho/cm with some minor fluctuations.

Chloride was found very high in effluent drain sediment and slightly low as with similar fluctuation in pond bed sediment, but in agricultural soil and soil near textile industries chloride was found comparatively low because of the high settlement of chloride in drain sediment from textile industrial effluent; while rest was reach in agricultural soil by various contacts. The chloride in drain sediment was found 609.0-836.6 μg/g with some minor fluctuations. Manchanda (1990) has also reported that pulse crops such as pea, mungbean, gram, lentil, and broadbean suffered more due to toxic accumulation of chloride, when irrigated with Cl ion then SO_4^{-2} dominant (70 per cent) water of more salinity.

In most cases, the exchangeable potassium is the chief source of available potassium in soil (Jha and Pandey, 1984). Potassium plays many essential roles in plants. It is an activator of dozens of enzymes responsible for such plants processes as energy metabolism, starch synthesis, nitrate reduction, and sugar degradation. Potassium is extremely mobile with in the plant and helps regulate the opening and closing of stomata in the leaves and the uptake of water by root cells. At any one time most of the potassium is in the primary mineral and non-exchangeable forms. To maintain the level of exchangeable and soil solution potassium high enough to encourage good crop production, chemical fertilizer are used. Potassium may deficient in alkaline soil and an excess of potassium may delay maturity of plant. In some crops it may tend to accumulate in the leaf and stem rather then in the grain (Miller and Turk, 2002). The potassium in drain sediment was found 27.3-33.5 μg/g with some minor fluctuations. Potassium has slightly positive correlation with many heavy metals *viz.* Cd (0.6852) Cu (-0.1689), Fe (0.7116), Mn (0.0611), Ni (0.2623), Pb (0.2369) and Zn (-0.6269) in agricultural soil. Korfali and Davies (2004) and Soltan (1999) have been exhibited a clear idea about the fluctuations in potassium concentrations in soil system.

Sodium in alkali soils results in a deflocculation of the colloids and hence in a breaking down of the soil structural units. This puddle condition renders the soil more or less impervious and retards entrance of irrigation and rain water impedes drainage. In heavy textured soil, the penetration of roots may be restricted by the deflocculated zone. Aeration is also much reduced, setting up aeration conditions and resulting in the formation of reduced compounds which are toxic to plants (Miller and Turk, 2002). The highly alkaline soil was due to the presence of sodium carbonate. Large quantity of adsorbed sodium repressed the availability of several nutrients especially, iron, manganese and phosphorous (Miller and Turk, 2002). The occurrence of sodium in drain sediment was found very high 157.0-587-5 μg/g respectively with some minor fluctuations. Desalination of soil is usually accomplished by impounding of water at the surface of the alkaline soil. The relationship in between reduction of salts and exchangeable sodium from a given depth of soil and the amount of leaching water for the sandy loam soil (Minhas and Khosla, 1986). The amount of soluble salts in soil can be reduce to safe limits for agriculture crops by leaching (Rao *et al.*, 1987).

In most experiments, usually the effects of one or more water quality parameters on the changes in physico-chemical characteristics of soils and crop productivity have been studied and evaluating the suitability of water for irrigation, however, integration of the interactive and interdependent processes of various water quality indices with soils have required to obtain comprehensive knowledge of the whole system. It requires the development of computerized models for providing a quantitative description of the dynamic behavior of soils upon interaction with irrigation water quality and its relation with crop productivity. Modeling efforts should go a long way in making generalizations based on the experience gained under specific conditions and extrapolation of experimental findings to different locations and seasons in study area.

Chapter 4

Impact Assessment on Surface Water

4.1 Surface Water

As the water, the most abundant and vital natural resource is extremely essential for survival of all living organisms, but today clean water has become a precious commodity and its quality is threatened by numerous sources of pollution. Water pollution creates physical, chemical and biological changes in the water, and further affects the water quality for a long time of period. As the heavy metals enter in the food chains of aquatic ecosystem, ultimately affect the health of higher animals of that particular ecosystem. Different heavy metals like Mercury, Lead, Chromium, and Nickel, Cadmium etc. release from various sources like agriculture, urban and industrial uses, and these factors degrade the surface water quality.

There is a main source of water in India by the rivers. In India fourteen major river systems share about 83 per cent of the drainage basis *i.e.* Indian rivers carry about 16,45,000 million cubic meter of water annually (Rao, 1975). In India all 14 major rivers including Coocum, Ganga, Gomti, Kaveri, Damodar and Mini-Mahi have become polluted due to various anthropogenic factors especially for the development activities. The river Damodar is perhaps the most heavily polluted river. Many parts of river do not seem to have any dissolved oxygen and no wonder that they fail to support the growth of desirable aquatic fauna and flora.

The Ganga is a holy river of India. Many Indians mainly Hindus worship it like a Goddess. Ganga is also known as Bhagarithi Ganga. Ganga is one of the four mighty rivers of world with its tributaries, and it is a symbol of purity. There are more than 50 big cities situated along with 12,500 km long bank of river Ganga. According to Deming (1975) if the entire ice over of the world would keep flow for 830 years. But now a day due to the various natural and human activities including burning of fossil fuels, natural activities like volcanic eruption, deforestation etc. the earth atmosphere is warming regularly due to the increase in the CO_2 level in the atmosphere. The Himalayas are the cradle of a large number of streams and mighty rivers. These hill streams and rivers ensure all the year round availability of water for irrigating crops as well as for meeting domestic and other needs. Agriculture being the main source of livelihood, water resources generated in the Himalayas are therefore, their greatest gift to this country particularly so far the vast areas in the Indo-gangetic plains in the north. Everything originated in the water and everything is sustained by water. All life on earth depends on water. Man uses water for many purposes drinking, irrigation, fisheries, industrial purposes, transportation and recreations.

The Himalayan rivers receive 20-30 per cent of their water from melting of snow and ice during summer and monsoon seasons. Thus a high altitude Himalayan river basin provides a natural unit within which one can examine natural weathering and agrochemical processes. However, even today the study and management of the Himalayan River is unfortunately constrained by a lock of reliable hydrological data. It is either confined to a hear source region or is based on only few samples. Adequate understanding of the Himalayan proglacial streams is extremely important for the development of a realistic program for utilizing the potential of water that exits in the form of snow and ice in this area. Several studies have been carried out in the lower part of the Ganga river system (Abbas and Subramaniam, 1984, Sarin *et al.*, 1989 and Handa, 1972). But very little information is available on agrochemical and biological aspects of the Ganga headwater.

The sustainable use of river water in agriculture resources is an essential component in the future development of Asian third world countries, especially those regions bordering or connecting the Himalayas, where population growth is rapid and predicted to continue. This population pressure results in an intensification of agriculture through the adoption of multiple annual crop solution and supported by large application of inorganic and organic fertilizers and the expansion of cultivated land on to steeper will stapes and previously forested areas (Schreier *et al.*, 1994) continued and excessive application of mineral and organic fertilizers to increase crop field may however, lead to detrimental effects in the form of soil and water acidification as well as downstream eutrophication. The impact on water chemistry of anthropogenic pollution from industrial, agricultural, sources, quarrying and tourists activity has concerned environmentalists and scientists for the past two decades. In consequences, the chemistry of headwater streams, lakes and lowland rivers has been monitored widely in Europe and North America in an effort to define and quantifying these anthropogenic impacts. An understanding of now anthropogenic pollutants affect water chemistry and there by the ability to predict the location and severity of potential impact can be achieved only through knowledge of hydro-chemical processes in pristine waters of similar physico-chemical characteristics (Bruns *et at.* 1991). If a sufficiently detailed knowledge has pristine, or baseline, water chemistry is available, the effects of local pollution, stemming form land use and atmospheric deposition of acidic oxides for example can be quantified.

Heavy metals are introduced into aquatic system from various anthropogenic sources. Heavy metals constitute a special group of contaminants of aquatic systems and deserve special attention. Since natural degradation processes do not remove metals, they may become enriched in sediments over time. Metal contamination of sediments is an issue of growing concerns worldwide (NRC, 1989). Both natural processes and anthropogenic activities are responsible for introducing metals in to the aquatic system (Nriagu, 1989). Many contaminants discharged into surface waters rapidly become associated with the particulate matter and incorporated in sediments. Metals in aquatic systems become part of the water-sediment system and their distribution is controlled by a dynamic set of physical-chemical interactions and equilibria, largely governed by pH and type of legends and chelating agents, oxidation state of the mineral components and the redox conditions of the system. Metal contaminated sediments may release heavy metals back to the overlying water column and, thus, pose risk to aquatic life and ecosystems. Due to their particle reactivity, heavy metals tend to accumulate in sediment as a result may persist in the environment long after their primary sources have been removed (Forstner and Wittman, 1981).

Textile effluent pollution may also alter the characteristics of sediment and soil of industrial area directly or indirectly. Composite effluents from textile industries and dye houses of Panipat industrial area normally discharged openly into a common effluent drain without any adequate treatment and commences in a large pond near Binjhole village near to industrial area, which may change the quality of bottom sediments of drain and pond sediment quality and also the physico-chemical characteristics.

The sediments existing as the bottom of the water column play a major role in the pollution scheme of an aquatic system by heavy metals (Forstner, 1985). They reflect the current quality of the water system and can be used in detecting the presence of contaminants that do not remain soluble after discharge into surface water. As a result of complex physical, chemical and biological processes a major fraction of heavy metals (contributed naturally as well as through various anthropogenic activities) is found to be associated with bottom sediments (Baruah *et al.*, 1996). Bed sediments in surface water systems thus act both as a sink and source of metals. Metal accumulation in sediments provides a record of the spatial and temporal history of pollution (Martin and Whitfield, 1983). Hence, sediment monitoring can provide important information on various pollution events.

Beside rivers, large ponds and lakes are the important surface water resources of ecological and socio-economical values. But due to the increasing rate of industrialization almost ponds are going polluted day by day. In the present work, textile industrial effluent drained out into a common effluent drain in Panipat industrial area, which flows alongwith industrial region and ultimately joins a large pond (approx. 12 hac) near Binjhole village.

4.2 Historical Resume

Different workers from India and abroad have been discussed various aspects of water pollution in term of characteristics of water bodies time to time. This review includes some of important contribution made in the field of industrial pollution, with references to physical, chemical and biological parameters of surface water by different authors of country and abroad.

Hyene (1970) observed the biology of polluted water of a water body near industrial area in New York. Katz (1975) highlighted the effect of heavy metals on some environmental components and biotic organisms of aquatic environment in New York.

Dulka and Risby (1976) stated ultra trace metals in some environmental and biological systems in the vicinity of textile industrial sector. Samy and Gananarethinam (1980) found that industrial effluents were toxic to the aquatic plants like *Eicchornia* even after dilution. Synthetic compounds were found toxic to *Spirodela polyrhiza* and *Lemna aquinoctialis*. Wint (1981) indicated the disposal and handling problem of toxic wastes from industrial wastewater.

Martin and Cougherty (1982) stated the basic monitoring techniques for detection of heavy metals pollution through the biological agents. Ireland (1983) described the heavy metals uptake and their distribution in tissues of organisms. Numberg (1984) studied on the volumetric approach in trace metal chemistry of natural waters and atmospheric precipitation. Nriagu and Pacyna (1988) carried out a study on the quantitative assessment of worldwide contamination of air, water and soil with trace metals.

Goltenboth (1994) evaluated the impact of wastewater of textile factory in Salatiga on the fresh water aquatic ecosystem of Ledok river system at Central Java. Singh *et al.* (1994) highlighted the degradation of water and soil quality of Parwanoo area with respect to heavy metals. Khurshid and Shabeer (1997) observed the degradation of water quality due to heavy metal pollution.

Sharma *et al.* (1999) and Sastry and Rathee (1999) reported the impact of industrial effluents on fresh water system of three villages in Rohtak district. Pujari and Sinha (1999) studied on the water and soil quality of some villages of Attabira area irrigated by bargarg main canal originated from Hirakund reservoir of Orissa.

Chao *et al.* (2005) reported the heavy metal contaminated site in Taiwan and described the status and distribution of heavy metals in environment. Reduce the levels of toxic metals from industrial effluents is necessary before discharging into surface waters. The general methods of treating wastewater-containing metals are coagulation, ion exchange, reverse osmosis and adsorption. Adsorption is a highly effective physico-chemical treatment for removing heavy metals for effluents. Due to high cost of activated carbon, efforts are being directed towards finding low cost adsorbents. Recently much interest has been shown for removal of heavy metals using a large number of low cost materials. Coal fly ash, the solid waste of power plants is one of the cheapest and non-conventional adsorbents (De, 2005).

Saed *et al.* (2005) prepared a novel adsorbent from sugarcane bagasse for removing dye from water and adsorption process was carried out in a batch process with different concentrations of the dye concentrations of the dye solution, different adsorbent doses, adsorbent with different particle size at different pH values and contact times. Kumar *et al.* (2005) described the status of water resources of India and highlighted the hydrological cycle, monsoon and precipitation, surface and groundwater resources, water requirements in domestic use, irrigation, hydroelectric power, industrial requirement and groundwater management in India.

Malik *et al.* (2006) reported the impact of textile effluents on surface water quality of Panipat industrial area. Lokeshwari and Chandrappa (2006) evaluated the impact of heavy metals contamination of a Lake on soil and cultivated vegetation near Bangalore and found the presence of some heavy metals in rice and vegetables, beyond the limit of Indian Standards and calculated metal transfer factors from soil to vegetation are found significant for Zn, Cu, Pb and Cd. Hedge *et al.* (2006) revealed the provenance of heavy minerals with special reference to ilmenite of the Honnevar beach in central west coast of India and observed some trace heavy elements like Co, Cr, V and Ni in the sandy soil of the beach.

Trevors and Saier (2007) described the regulation of pollution in various environmental components *viz.* water, air and soil.

4.3 Observations

Physico-chemical Parameters of Pond Water

Temperature (°C)

Temperature of pond water was recorded maximum 32.4 ± 2.5 in monsoon 2006 and minimum 25.3 ± 1.2 in winter 2005-06 ranged between 24.0 to 35.2 °C. The highest value of temperature in pond water was found 35.2 °C during July 2005 and lowest 24.0 °C in January 2006. The temperature variation during the study is given Table 4.1. Monthly variation in temperature is given in Figure 4.1.

pH

The maximum pH value of pond water was recorded 8.63 ± 0.15 in summer 2005 and minimum 8.17 ± 0.12 in monsoon 2005 ranged between 8.1 and 8.8 on pH scale. The highest value of pH in pond water was found 8.8 during May 2005 and lowest 8.1 in September 2006. The variation in the pH during the study is given Table 4.1. Monthly variation in pH is given in Figure 4.1.

Total Solids (mg/l)

The maximum total solids in pond water were recorded 2365.67 ± 104.18 in monsoon 2006 and minimum 2033.20 ± 30.38 in winter 2005-06 ranged between 2012.0 to 2476.0 mg/l. The highest value of total solids in pond water was found 2476 mg/l July 2006 during and lowest 2012 mg/l in January 2006. The variation in the total solids during the study is given Table 4.1. Monthly variation in total solids is given in Figure 4.2.

Total Dissolved Solids (mg/l)

Total dissolved solids in pond water were recorded maximum 2037.50 ± 44.46 in summer 2006 and minimum 1861.00 ± 102.62 in monsoon 2005 ranged between 1776.0 to 2093.0 mg/l. The highest value of total dissolved solids in pond water was found 2093.0 mg/l during June 2006 and lowest 1776.0 mg/l in September 2005. The

Table 4.1: Seasonal Variation of Physico-chemical Characteristics of Pond Water

Para-meters	*Unit*	*Summer 2005*		*Monsoon 2005*		*Winter 2005-06*		*Summer 2006*		*Monsoon 2006*		*Winter 2006-07*		*Summer 2007*	
		Mean	*SD*	*Mean*	*SD*	*Mean*	*SD*	*Mean*	*SD*	*Mean*	*SD*	*Mean*	*SD*	*Mean*	*SD*
Temp.	°C	31.733	3.035	32.133	3.002	25.300	1.204	30.050	3.730	32.400	2.506	25.625	0.854	25.500	0.707
pH		8.633	0.153	8.167	0.115	8.360	0.114	8.550	0.129	8.233	0.153	8.450	0.129	8.500	0.000
TS	mg/l	2251.333	73.009	2308.667	109.792	2033.200	30.376	2255.000	103.418	2365.667	104.175	2102.250	36.854	2155.500	6.364
TSS	mg/l	230.333	32.868	447.667	67.002	96.800	76.810	217.500	60.688	487.667	23.007	158.750	96.555	141.500	37.477
TDS	mg/l	2021.000	40.951	1861.000	102.621	1936.400	73.040	2037.500	44.463	1878.000	91.831	1943.500	81.127	2014.000	31.113
EC	mho/cm	3.158	0.064	2.908	0.160	3.026	0.114	3.184	0.069	2.934	0.143	3.037	0.127	3.147	0.049
DO	mg/l	1.233	0.252	3.433	0.814	3.360	0.513	1.600	0.707	3.067	0.751	3.250	0.526	3.000	0.283
Free CO_2	mg/l	2.767	0.252	0.833	1.443	0.500	0.458	2.250	0.794	0.700	1.212	0.650	0.443	0.800	0.141
BOD	mg/l	136.000	4.583	129.000	23.302	106.800	16.423	125.250	19.687	134.333	23.438	117.250	17.095	99.000	4.243
COD	mg/l	359.667	19.655	335.000	52.431	282.600	28.387	333.000	45.233	346.667	51.404	298.250	24.918	274.500	4.950
Total alkalinity	mg/l	696.667	56.607	725.000	46.872	563.400	45.114	696.500	52.399	732.333	50.213	571.250	51.370	647.500	36.062
Ca	mg/l	180.000	30.000	216.667	65.064	104.000	11.402	172.500	51.235	236.667	51.316	115.000	28.868	110.000	14.142
Mg	mg/l	141.667	41.016	140.000	34.699	61.000	31.297	114.500	62.804	157.000	20.298	78.500	25.891	42.500	16.263
K	mg/l	7.300	0.854	8.200	0.265	7.540	0.207	7.500	0.668	8.433	0.208	7.675	0.206	7.350	0.071
Cl	mg/l	609.333	105.235	762.667	33.262	633.200	58.538	610.250	91.482	770.667	21.127	664.250	57.570	576.500	17.678
Na	mg/l	137.667	2.517	152.333	11.676	130.600	4.879	134.500	11.387	158.667	8.622	137.750	6.076	128.500	2.121

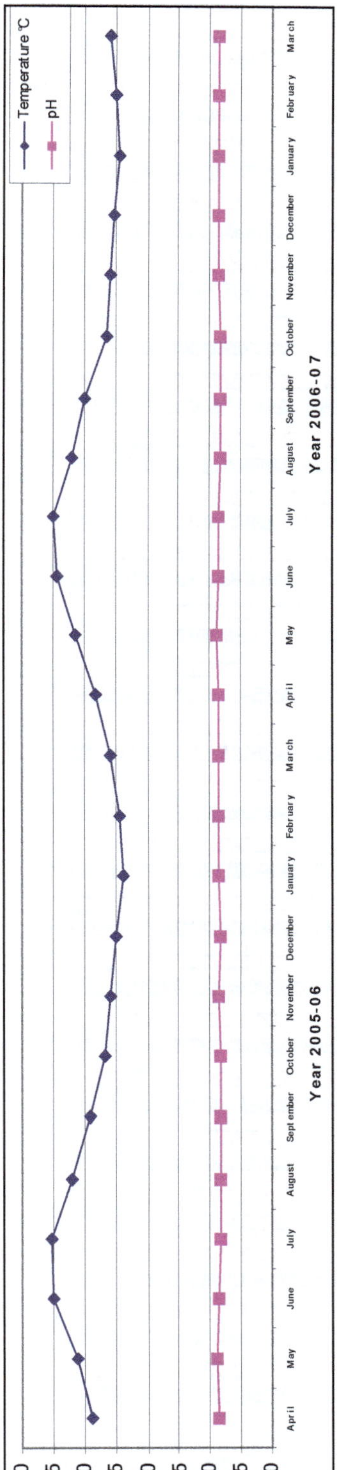

Figure 4.1: Monthly Variation of Temperature and pH in Pond Water

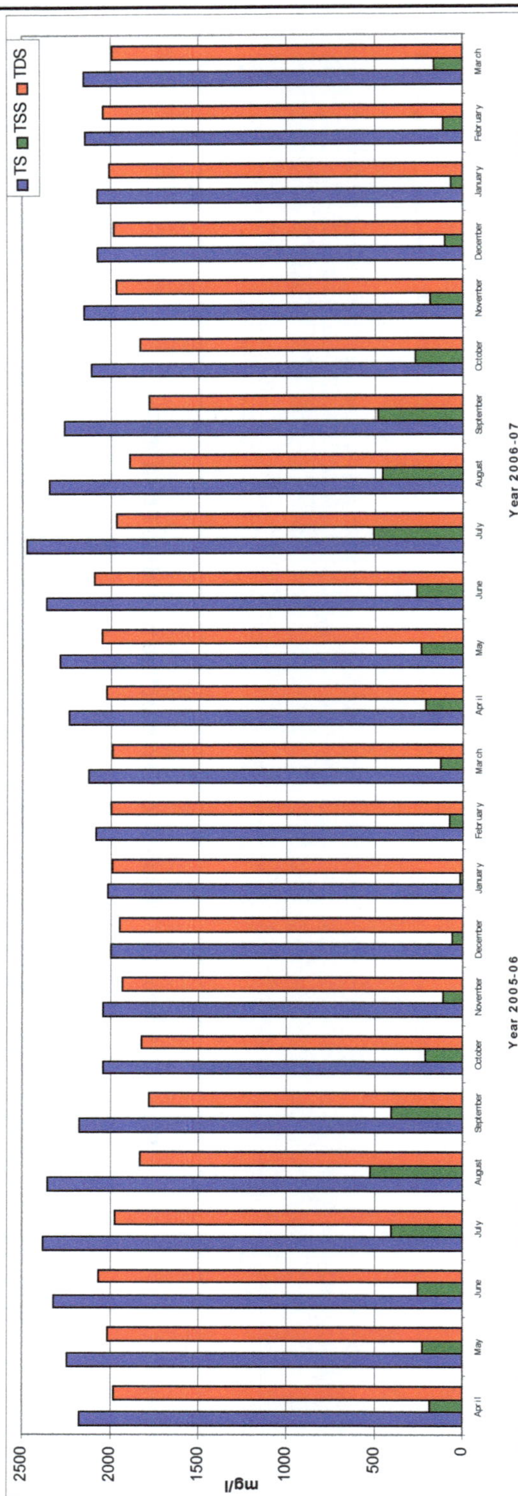

Figure 4.2: Monthly Variation of TS, TDS and TSS in Pond Water

variation in the total dissolved solids during the study is given Table 4.1. Monthly variation in total dissolved solids is given in Figure 4.2.

Total Suspended Solids (mg/l)

Total suspended solids in pond water were recorded maximum 487.67 ± 23.0 in monsoon 2006 and minimum 96.80 ± 76.81 in winter 2005-06 ranged between 20.0 to 525.0 mg/l. The highest value of total suspended solids in pond water was found 525.0 mg/l during August 2005 and lowest 20.0 mg/l in January 2006. The variation in the total suspended solids during the study is given Table 4.1. Monthly variation in total suspended solids is given in Figure 4.2.

Electrical Conductivity (mho/cm)

Maximum electric conductivity in pond water was recorded 3.18 ± 0.07 in summer 2006 and minimum 2.91 ± 0.16 in monsoon 2005 ranged between 2.78 and 3.27 mho/cm. The highest value of electric conductivity in pond water was found 3.27 mho/cm during June, 2006 and lowest 2.78 mho/cm in September 2005. The variation in the electric conductivity during the study is given Table 4.1. Monthly variation in electric conductivity is given in Figure 4.3.

Dissolved Oxygen (mg/l)

Dissolved oxygen in pond water was recorded maximum 3.43 ± 0.81 in monsoon 2005 and minimum 1.23 ± 0.25 in summer 2005 ranged between 0.9 to 4.2 mg/l. The highest value of dissolved oxygen in pond water was found 4.2 mg/l during January 2006 and lowest 0.9 mg/l in June 2006. The variation in the dissolved oxygen content during the study is given Table 4.1. Monthly variation in dissolved oxygen is given in Figure 4.3.

Free CO_2 (mg/l)

Free carbon dioxide in pond water was recorded maximum 2.77 ± 0.25 in summer 2005 and minimum 0.50 ± 0.46 in winter 2005-06 ranged between 0.00 to 3.1 mg/l. The highest value of free carbon dioxide in pond water was found 3.1 mg/l during June, 2006 and lowest 0.00 mg/l in August 2005. The variation in the free carbon dioxide content during the study is given Table 4.1. Monthly variation in carbon dioxide is given in Figure 4.3.

Biochemical Oxygen Demand (mg/l)

The biochemical oxygen demand in pond water was recorded maximum 136.0 ± 4.58 in summer 2005 and minimum 99.0 ± 4.24 in summer 2007 ranged between 90.0 to 161.0 mg/l. The highest value of biochemical oxygen demand in pond water was found 161.0 mg/l during July 2006 and lowest 90.0 mg/l in January 2006. The variation in biochemical oxygen demand during the study is given Table 4.1. Monthly variation in biochemical oxygen demand is given in Figure 4.4.

Chemical Oxygen Demand (mg/l)

The chemical oxygen demand in pond water was recorded maximum 359.67 ± 19.65 in summer 2005 and minimum 274.50 ± 4.95 in summer 2007 ranged between 252.0 to 405.0 mg/l. The highest value of chemical oxygen demand in pond water

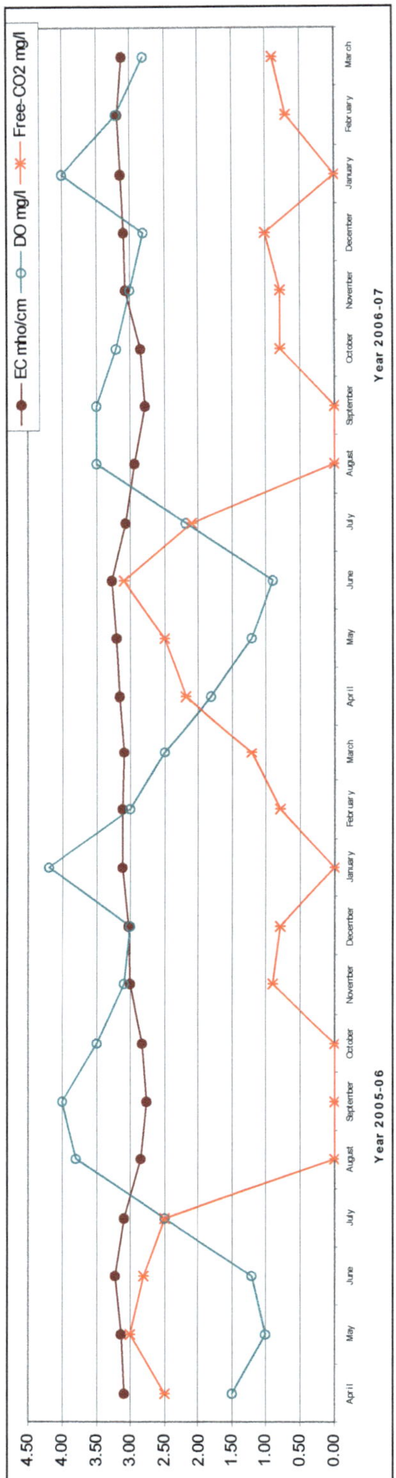

Figure 4.3: Monthly Variation of EC, DO and Free CO_2 in Pond Water

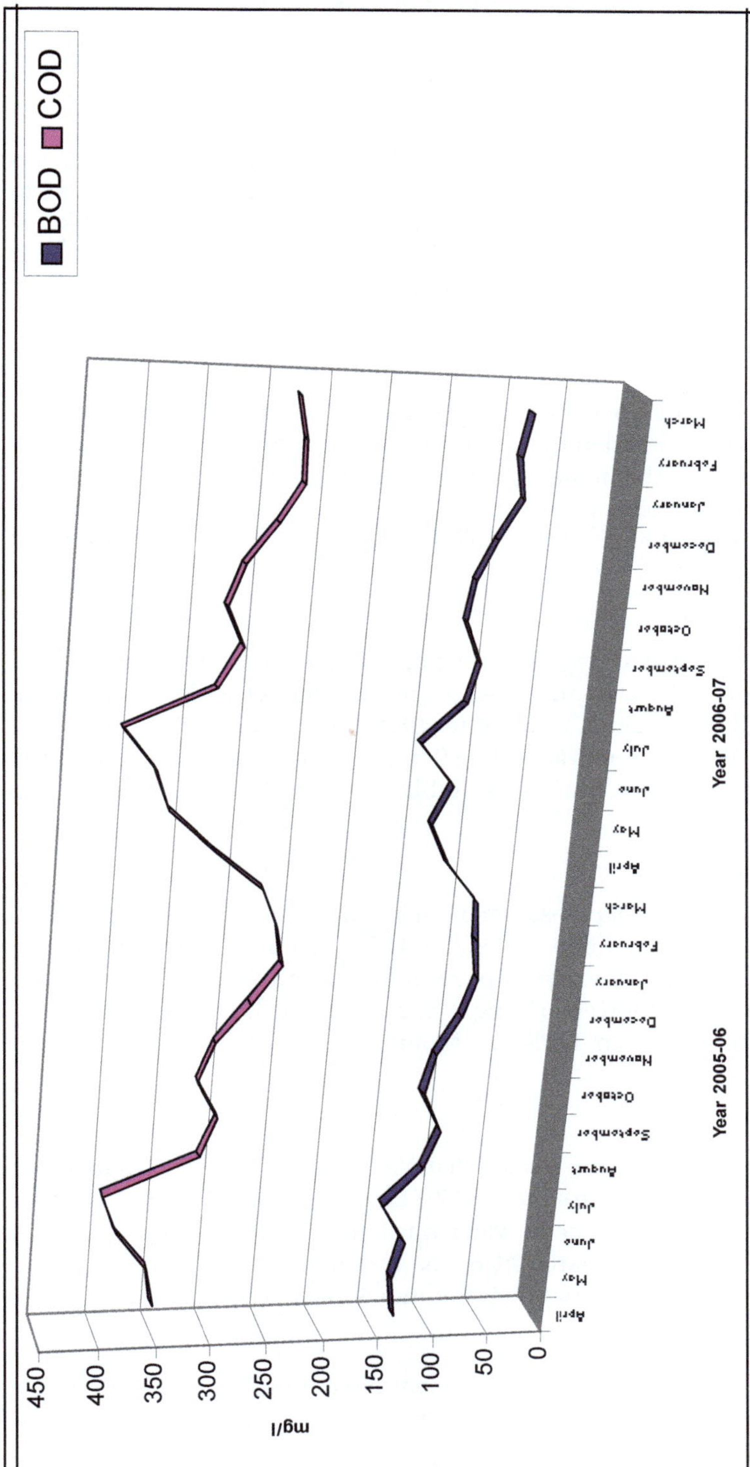

Figure 4.4: Monthly Variation of BOD and COD in Pond Water

was found 405.0 mg/l during July 2006 and lowest 252.0 mg/l in January 2006. The variation in chemical oxygen demand during the study is given Table 4.1. Monthly variation in chemical oxygen demand is given in Figure 4.4.

Total Alkalinity (mg/l)

Total alkalinity in pond water was recorded maximum 732.33 ± 50.21 in monsoon 2006 and minimum 563.40 ± 45.11 in winter 2005-06 ranged between 505.0 to 785.0 mg/l. The highest value of total alkalinity in pond water was found 785.0 mg/l during July 2006 and lowest 505.0 mg/l in December 2005. The variation in total alkalinity during the study is given Table 4.1. Monthly variation in total alkalinity is given in Figure 4.6.

Calcium (mg/l)

The concentration of calcium in pond water was recorded maximum 236.7 ± 51.3 in monsoon 2006 and minimum 104.0 ± 11.4 in winter 2005-06 ranged between 80.0 to 280.0 mg/l. The highest value of calcium in pond water was found 280.0 mg/l during July 2005 and lowest 80.0 mg/l in December 2006. The variation in calcium during the study is given Table 4.1. Monthly variation in calcium is given in Figure 4.5.

Magnesium (mg/l)

The concentration of magnesium in pond water was recorded maximum 141.7 ± 41.0 in summer 2005 and minimum 42.5 ± 16.3 in summer 2007 ranged between 25.0 to 178.0 mg/l. The highest value of magnesium in pond water was found 178.0 mg/l during July 2005 and lowest 25.0 mg/l in February 2006. The variation in magnesium during the study is given Table 4.1. Monthly variation in magnesium is given in Figure 4.5.

Potassium (mg/l)

The concentration of potassium in pond water was recorded maximum 8.43 ± 0.21 in monsoon 2006 and minimum 7.3 ± 0.85 in summer 2005 ranged between 6.5 to 8.6 mg/l. The highest value of potassium in pond water was found 8.6 mg/l during July 2006 and lowest 6.5 mg/l in April 2005. The variation in potassium during the study is given Table 4.1. Monthly variation in potassium is given in Figure 4.5.

Sodium (mg/l)

Sodium in pond water was recorded maximum 158.67 ± 8.62 in monsoon 2006 and minimum 128.5 ± 8.62 in summer 2007 ranged between 120.0 to 168.0 mg/l. The highest value of Sodium in pond water was found 168.0 mg/l during August 2006 and lowest 120.0 mg/l in March 2006. The variation in Sodium during the study is given Table 4.1. Monthly variation in Sodium is given in Figure 4.6.

Chloride (mg/l)

Chloride in pond water was recorded maximum 770.67 ± 21.13 in monsoon 2006 and minimum 576.50 ± 17.68 in summer 2007 ranged between 512.0 to 793.0 mg/l. The highest value of Chloride in pond water was found 793.0 mg/l during

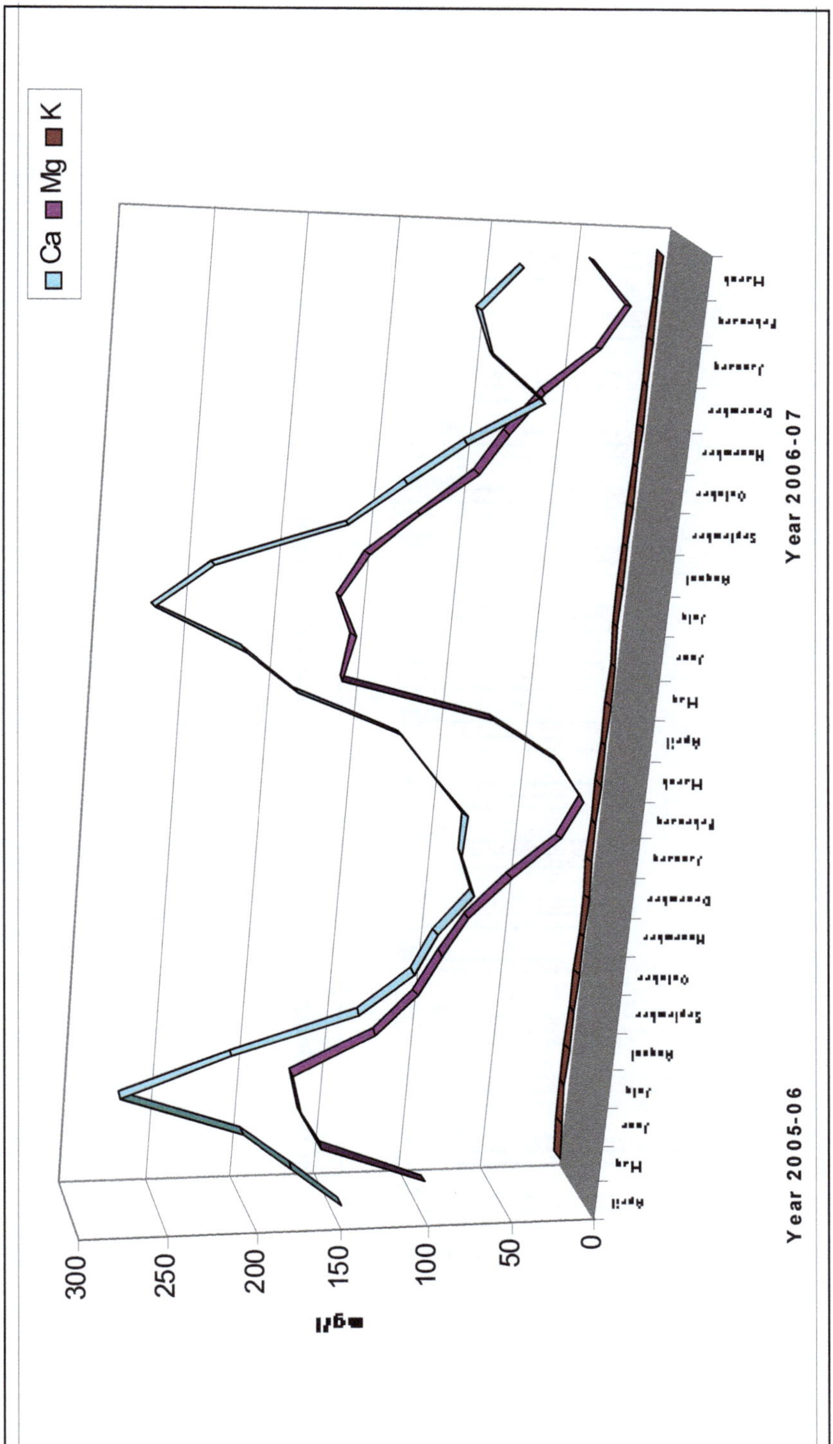

Figure 4.5: Monthly Variation of Calcium, Magnesium and Potassium in Pond Water

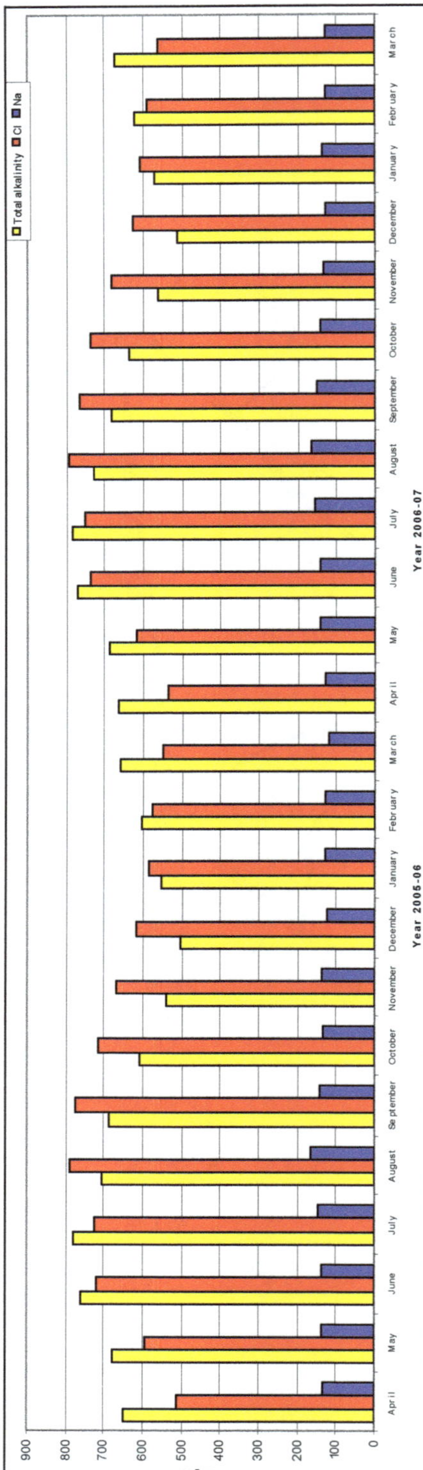

Figure 4.6: Monthly Variation of Chloride, Total Alkalinity and Sodium in Pond Water

August 2006 and lowest 512.0 mg/l in April 2005. The variation in Chloride during the study is given Table 4.1. Monthly variation in Chloride is given in Figure 4.6.

Textile effluent pollution may also alter the characteristics of sediment and soil of industrial area directly or indirectly. Composite effluents from textile industries and dye houses of Panipat industrial area normally discharged openly into a common effluent drain without any adequate treatment and commences in a large pond near Binjhole village near to industrial area, which may change the quality of bottom sediments of pond sediment quality and also the physico-chemical characteristics. The sediments existing as the bottom of the water column play a major role in the pollution scheme of an aquatic system by heavy metals (Forstner, 1985). They reflect the current quality of the water system and can be used in detecting the presence of contaminants that do not remain soluble after discharge into surface water. As a result of complex physical, chemical and biological processes a major fraction of heavy metals (contributed naturally as well as through various anthropogenic activities) is found to be associated with bottom sediments (Baruah *et al.*, 1996). Bed sediments in surface water systems thus act both as a sink and source of metals. Metal accumulation in sediments provides a record of the spatial and temporal history of pollution (Martin and Whitfield, 1983). Hence, sediment monitoring can provide important information on various pollution events.

Physico-chemical Parameters of Pond Sediment

Temperature (°C)

Temperature of pond sediment was recorded maximum 24.83 ± 3.48 in summer 2005 and minimum 15.35 ± 3.19 in winter 2005-06 ranged between 10.25 to 28.0 °C. The highest value of temperature in pond sediment was found 28.0 °C during June 2005 and lowest 10.25 °C in January 2006. The temperature variation during the study is given Table 4.2. Monthly variation in temperature is given in Figure 4.7.

Water Holding Capacity (per cent)

Water holding capacity of pond sediment was recorded maximum 36.36 ± 0.77 in summer 2005 and minimum 29.52 ± 4.99 in monsoon 2005 ranged between 25.33 to 37.75 per cent . The highest value of water holding capacity in pond sediment was found 37.75 per cent during June 2006 and lowest 25.33 per cent in September 2005. The water holding capacity variation during the study is given Table 4.2. Monthly variation in water holding capacity is given in Figure 4.7.

Bulk Density (g/cm^3)

The bulk density of pond sediment was recorded maximum 1.49 ± 0.05 in monsoon 2006 and minimum 1.36 ± 0.03 in winter 2005-06 ranged between 1.32 to 1.58 g/cm^3. The highest value of bulk density in pond sediment was found 1.58 g/cm^3 during June 2006 and lowest 1.32 g/cm^3 in January 2006. The bulk density variation during the study is given Table 4.2. Monthly variation in bulk density is given in Figure 4.8.

Table 4.2: Seasonal Variation of Physico-chemical Characteristics of Pond Sediments

Parameters	*Unit*	*Summer 2005*		*Monsoon 2005*		*Winter 2005-06*		*Summer 2006*		*Monsoon 2006*		*Winter 2006-07*		*Summer 2007*	
		Mean	*SD*	*Mean*	*SD*	*Mean*	*SD*	*Mean*	*SD*	*Mean*	*SD*	*Mean*	*SD*	*Mean*	*SD*
Temp.	°C	24.833	3.485	23.317	2.251	15.350	3.185	23.450	3.903	23.733	2.359	15.775	2.777	17.150	2.333
WHC	%	36.360	0.777	29.517	4.998	30.544	2.767	36.275	1.253	30.367	4.849	29.958	2.349	34.425	1.011
Bulk density	mg/cm³	1.483	0.071	1.460	0.046	1.366	0.034	1.463	0.085	1.493	0.051	1.373	0.040	1.385	0.007
Soil moisture	%	13.543	1.135	20.900	2.433	15.456	1.097	13.965	0.731	20.943	2.376	16.150	1.666	14.740	0.099
pH		7.923	0.093	8.167	0.153	8.370	0.077	8.063	0.180	8.100	0.100	8.350	0.058	8.350	0.071
EC	mho/cm	31.933	0.808	30.833	1.350	27.300	0.914	31.450	1.605	31.233	1.250	27.500	0.963	28.650	1.626
Cl	µg/g	827.000	147.553	775.333	81.513	603.000	34.792	761.500	171.148	798.333	71.459	608.000	46.890	588.500	10.607
Na	µg/g	472.000	115.052	143.667	93.002	310.200	160.243	495.250	93.053	151.000	73.750	267.500	126.234	556.000	32.527
K	µg/g	32.000	1.000	28.333	1.528	31.400	1.342	32.750	0.500	28.667	1.528	31.500	1.291	33.500	0.707

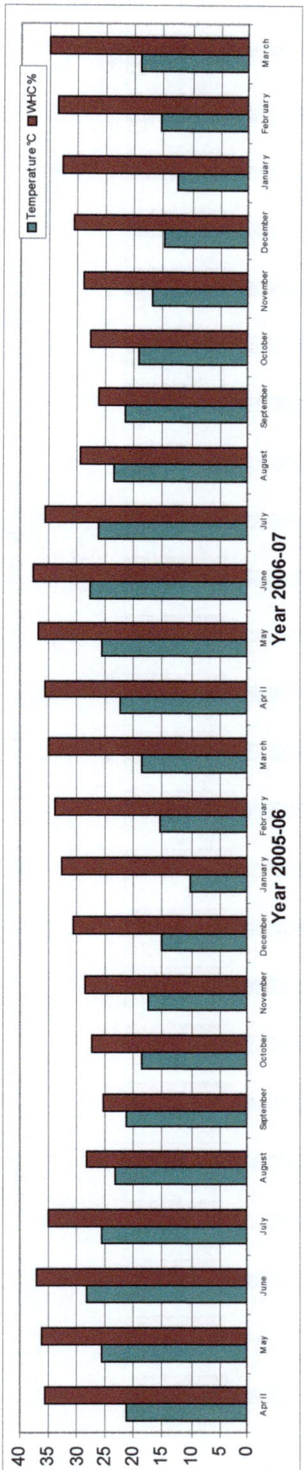

Figure 4.7: Monthly Variation of Temperature and WHC in Pond Sediment

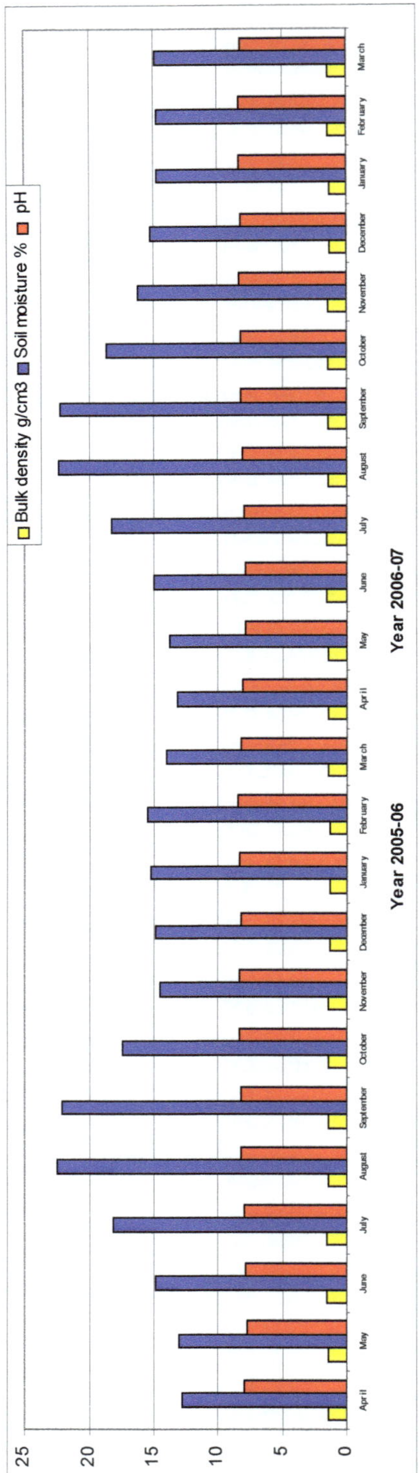

Figure 4.8: Monthly Variation of Bulk Density, Soil Moisture and pH in Pond Sediment

Soil Moisture (per cent)

The soil moisture of pond sediment was recorded maximum 20.94 ± 2.37 in monsoon 2006 and minimum 13.54 ± 1.14 in summer 2005 ranged between 12.8 to 22.5 per cent . The highest value of soil moisture in pond sediment was found 22.5 per cent during August 2005 and lowest 12.8 per cent in April 2005. The soil moisture variation during the study is given Table 4.2. Monthly variation in soil moisture is given in Figure 4.8.

pH

pH of pond sediment on pH scale was recorded maximum 8.37 ± 0.08 in winter 2005-06 and minimum 7.92 ± 0.09 summer 2005 ranged between 7.82 to 8.45 on pH scale. The highest value of pH in pond sediment was found 8.45 during February 2006and lowest 7.82 in May 2005. The pH variation during the study is given Table 4.2. Monthly variation in pH is given in Figure 4.8.

Electrical Conductivity (mho/cm)

Electric conductivity of pond sediment was recorded maximum 31.93 ± 0.81 in summer 2005 and minimum 27.30 ± 0.91 in winter 2005-06 ranged between 26.3 and 33.0 mho/cm. The highest value of Electric conductivity in pond sediment was found 33.0 mho/cm during June 2006 and lowest 26.3 mho/cm in January 2006. The Electric conductivity variation during the study is given Table 4.2. Monthly variation in Electric conductivity is given in Figure 4.9.

Chloride (µg/g)

Chloride concentration in pond sediment was recorded maximum 827.0 ± 147.55 in summer 2005 and minimum 588.50 ± 10.61 summer 2007 ranged between 561.0 to 991.0 µg/g. The highest value of chloride in pond sediment was found 991.0 µg/g during June 2005 and lowest 561.0 µg/g in January 2007. The chloride variation during the study is given Table 4.2. Monthly variation in chloride is given in Figure 4.9.

Potassium (µg/g)

Potassium concentration in pond sediment was recorded maximum 33.50 ± 0.71 in summer 2007 and minimum 28.33 ± 1.53 monsoon 2005 ranged between 27.0 to 34.0 µg/g. The highest value of potassium in pond sediment was found 34.0 µg/g during March 2007 and lowest 27.0 µg/g in August 2005. The potassium variation during the study is given Table 4.2. Monthly variation in potassium is given in Figure 4.9.

Sodium (µg/g)

Sodium concentration in pond sediment was recorded maximum 556.00 ± 32.53 in summer 2007 and minimum 143.67 ± 93.0 in monsoon 2005 ranged between 87.0 and 589.0 µg/g. The highest value of sodium in pond sediment was found 589.0 µg/g during April 2005 and lowest 87.0 µg/g in September 2005. The sodium variation during the study is given Table 4.2. Monthly variation in sodium is given in Figure 4.9.

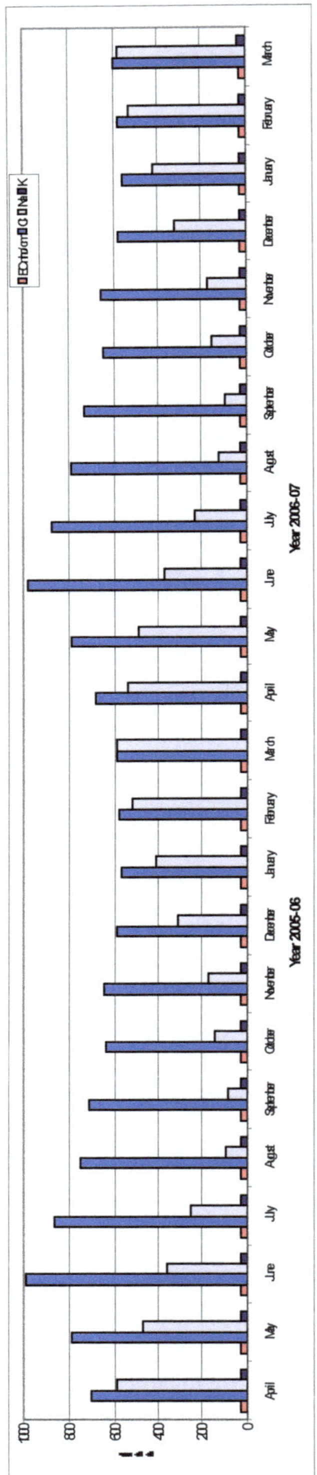

Figure 4.9: Monthly Variation of EC, Chloride, Sodium and Potassium in Pond Sediment

Correlation Coefficient

Correlation between heavy metals of pond water and heavy metals of groundwater are given in Table 7.18 for each metal. For cadmium correlation was found maximum 0.7155 and 0.7618 with manganese and lead respectively. Correlation for copper was found maximum 0.9116 and 0.8906 with copper and manganese respectively. Correlation for iron was found maximum 0.9580 and 0.9422 with iron and zinc respectively. Correlation for manganese was found maximum 0.8076 and 0.7998 with nickel and lead respectively. Correlation for nickel was found maximum 0.9223 and 0.9085 with nickel and zinc respectively. Correlation for lead was found maximum 0.9387 and 0.8863 with iron and lead respectively. Correlation for zinc was found maximum 0.6681 and 0.5405 with cadmium and zinc respectively.

Correlation between heavy metals of pond sediment and heavy metals of effluent drain sediment are given in Table 7.22 for each metal. For cadmium correlation was found maximum 0.9493 and 0.7144 with cadmium and lead respectively. Correlation for copper was found maximum 0.3647 and 0.9273 with cadmium and copper respectively. Correlation for iron was found maximum 0.9466 and 0.1694 with iron and manganese respectively. Correlation for manganese was found maximum 0.9223 and 0.5129 with manganese and nickel respectively. Correlation for nickel was found maximum 0.8060 and 0.9359 with nickel and zinc respectively. Correlation for lead was found maximum 0.6434 and 0.9388 with cadmium and lead respectively. Correlation for zinc was found maximum 0.7559 and 0.9563 with nickel and zinc respectively.

Correlation between physico-chemical parameters and heavy metals of pond sediment are given in Table 7.23c for each metal. For cadmium correlation was found maximum 0.6199 and 0.6032 with electric conductivity and chloride respectively. Correlation for copper was found maximum 0.3937 and 0.3652 with temperature and electric conductivity respectively. Correlation for manganese was found maximum 0.6170 and 0.7442 with sodium and potassium respectively. Correlation for nickel was found maximum 0.8307 and 0.6717 with water holding capacity and chloride respectively. Correlation for lead was found maximum 0.7460 and 0.7158 with temperature and chloride respectively. Correlation for zinc was found maximum 0.8126 and 0.7137 with water holding capacity and chloride respectively.

4.4 Discussion

Dye houses of the Panipat region discharge their effluents directly into a common effluent drain, which carry the total of composite effluents towards the village pond at Binjhole adjoining to industrial area, ultimately has altered the water quality and biotic and nutrient dynamic of the pond ecosystem in a tremendous way and further may affect the groundwater quality. Due to post irrigation practices, the soil turns to barren or unfertile and affected agriculture productivity. Further future of these heavy metals is not so managed because the common effluent drain ends up in to a pond near village Binjhole, Industrial area and some local surroundings of dye houses. The effluents discharged from dye houses are quite important for temporal monitoring. Because the effluents store pond is highly contaminated and is a depot of heavy

Table 4.3: Seasonal Variations of Heavy Metals in Pond Water

Heavy Metals	*Unit*	*Summer 2005*		*Monsoon 2005*		*Winter 2005-06*		*Summer 2006*		*Monsoon 2006*		*Winter 2006-07*		*Summer 2007*	
		Mean	*SD*	*Mean*	*SD*	*Mean*	*SD*	*Mean*	*SD*	*Mean*	*SD*	*Mean*	*SD*	*Mean*	*SD*
Cadmium (Cd)	ppm	0.009	0.001	0.001	0.002	0.004	0.002	0.012	0.006	0.002	0.003	0.004	0.002	0.007	0.001
Copper (Cu)	ppm	0.427	0.045	0.397	0.136	0.252	0.084	0.423	0.053	0.403	0.150	0.245	0.070	0.370	0.014
Iron (Fe)	ppm	9.700	3.035	4.233	1.450	4.820	2.961	9.450	1.529	4.567	1.501	3.850	2.111	8.750	0.636
Manga-nese (Mn)	ppm	0.530	0.061	0.463	0.070	0.172	0.062	0.490	0.122	0.470	0.056	0.170	0.059	0.295	0.035
Nickel (Ni)	ppm	0.127	0.025	0.073	0.033	0.054	0.045	0.130	0.014	0.082	0.022	0.047	0.038	0.115	0.007
Lead (Pb)	ppm	1.800	0.557	0.610	0.412	0.504	0.364	1.725	0.287	0.727	0.448	0.503	0.346	1.165	0.332
Zinc (Zn)	ppm	0.156	0.031	0.110	0.020	0.158	0.066	0.165	0.017	0.113	0.012	0.183	0.025	0.195	0.007

Table 4.4: Seasonal Variations of Heavy Metals in Pond Sediments

Heavy Metals	*Unit*	*Summer 2005*		*Monsoon 2005*		*Winter 2005-06*		*Summer 2006*		*Monsoon 2006*		*Winter 2006-07*		*Summer 2007*	
		Mean	*SD*	*Mean*	*SD*	*Mean*	*SD*	*Mean*	*SD*	*Mean*	*SD*	*Mean*	*SD*	*Mean*	*SD*
Cadmium (Cd)	ppm	5.717	0.451	4.780	0.455	4.054	0.595	5.410	0.677	4.993	0.359	5.073	0.153	4.795	0.120
Copper (Cu)	ppm	51.520	5.098	43.880	6.877	43.834	7.332	52.463	6.505	43.477	12.357	44.450	13.422	38.645	9.228
Iron (Fe)	ppm	5640.333	2179.050	4312.667	463.092	6013.400	856.579	4356.500	998.172	4157.667	403.026	5844.000	1878.244	3803.500	113.844
Manga-nese (Mn)	ppm	942.000	48.135	724.000	56.471	836.400	38.708	928.000	56.291	605.000	161.707	883.500	27.791	927.500	4.950
Nickel (Ni)	ppm	63.667	8.083	52.333	9.074	52.800	3.768	63.750	11.117	54.333	7.024	49.000	5.944	56.500	2.121
Lead (Pb)	ppm	69.000	6.245	55.333	9.292	44.000	5.385	62.000	13.491	57.333	6.028	49.250	3.403	44.500	10.607
Zinc (Zn)	ppm	240.000	92.601	142.333	59.341	123.200	42.032	207.250	86.496	153.667	62.963	116.250	32.755	166.500	12.021

metals due to the presence of chemical dye portions continuously since a long period of time. Heavy metals remain in water ecosystem, but these may travel through food web at various trophic levels and environmental components as well as. Metals moves pond sediments and reach in shallow water table by leaching processes and the metal contamination may disperse in a wide area of the region. Contamination of soils of close areas occurs when the effluents or polluted water comes under the contact of soil. The regular contact of polluted water with soil makes the sodicity and alters the physico-chemical characteristics, texture and profile of soil stratum.

The quality assessment of pond water required to know the alteration in physico-chemical characteristics and pollution load of heavy metals on hydrosphere and lithosphere and to determined the drinking water quality of Panipat industrial zone in term of suitability and possible impacts on human health. The results are depicted in the form of tables and graphs. During the study, physico-chemical characteristics and heavy metals in surface (pond) water were found with some fluctuation according to seasons, place and other factors.

4.5 Physico-chemical Parameters of Pond Water

In the present investigation of water resources, physico-chemical characteristics of water directly reflect its quality for irrigation, bathing, drinking and industrial uses. Physico-chemical parameter like total dissolved solids, pH, chloride, total alkalinity, biochemical oxygen demand, chemical oxygen demand, pH are the some important parameters for assessing the quality of water for irrigation and drinking purposes.

Temperature mostly fluctuates according to climatic conditions of the region and directly depends upon seasonal air temperature. The temperature of pond water (25.5-32.4°C) has fluctuated significantly due to the lentic behavior of aquatic ecosystem. Temperature of composite effluent drops slightly while reaching in the pond (25.5-32.4°C). Pathe *et al.* (2005) reported the temperature fluctuation trend from cluster of small cotton textile units from different sector, which follow permissible limit of B.I.S and WHO and remain always below 40°C during study. During the study period temperature of textile industrial effluent never cross the standards limits. Temperature slightly influences some parameters of water and affects the concentration of dissolved oxygen contents and can also influence the activity of bacteria and toxic chemicals in water. Textile industrial effluent may cause problems with high temperature (Groff, 1993). Temperature facilitates the oxidation process biochemical oxygen demand and chemical oxygen demand achieved maximum reduction at high temperature as compared to low temperature (Aslam *et al.*, 2004). Temperature of water depends upon the meteorological characteristics of existing environment; especially air temperature. Temperature fluctuated with air temperature with positive relation. However, due to some processes in textile units, the temperature of textile effluents was slightly higher than pond water and groundwater. In drain sediments, temperature has positive relation with lead (0.8233), Zn (0.673), Cd (0.6827); while temperature has not any significant correlation with heavy metal in pond sediment'.

Extreme high pH or rapid pH change can exert stresses conditions or kill aquatic life. Changes in pH can also affect aquatic system indirectly by altering other aspects of water chemistry. Low pH levels accelerate the release of metals from rocks or sediments in the water. Groff (1993) reported that textile wastewater may cause problem used as untreated effluent consist higher pH and temperature. pH has significant importance in all chemicals and biochemical reactions in an aquatic system (Basu *et al.*, 1973). Pond water's pH was also alkaline (8.2-8.6) due to the directly mixing with effluent drains with pond. The pH of both, drain sediment (7.8-8.3) and pond sediments (7.9-8.4) were also noticed alkaline in nature.

The total solids in pond surface water were fluctuated in the range of 2033.2 to 2365.7 mg/l. Similar alterations in surface water quality of the pond were observed by Malik *et al.* (2006). The dissolved solids in wastewater are mainly inorganic salts and relatively unaffected by biological treatment processes and can accumulate in water recirculation system. Failure to remove them may lead to an increase in total solids content of surface water sources. The dissolved solids in wastewater may be harmful to vegetation and also preclude use in various irrigation processes. In the pond, the values of total solids and total dissolved solids were noticed comparatively high, however, these were not found higher than textile industrial effluent. Dissolved solids vary qualitatively and quantitatively in different water depending upon the season, location and other factors. In the present study, the total dissolved solids values in pond water (1861.0-2037.5 mg/l) were not so high according to CPCB standards.

Total dissolved solids are the measure of the quantity of material dissolved in water that can pass through a filter (usually with a pore size of 0.45 micrometers). The material can include carbonate, bicarbonate, chloride, sulfate, phosphate, nitrate, calcium, magnesium, sodium, organic ions, and other ions. Similar to total suspended solids, high concentrations of total dissolved solids may also reduce water transparency, combine with toxic trace compounds and heavy metals, and lead to an increase in water temperature. Total dissolved solids are used to estimate the quality of drinking water, because it represents the amount of ions in the water. High dissolved solids concentrations can produce laxative effects and can give an unpleasant mineral taste to water. High dissolved solids concentrations in water are also unsuitable for water recycling processes and many industrial applications.

Total suspended solids include both organic and inorganic materials, which adversely affect the aquatic life, especially fishes and bottom fauna of surface water ecosystems. These produce carbon dioxide, hydrogen sulphide, methane gases, which are very noxious in nature. They increase the turbidity of water, reduce light penetration and impair photosynthetic activity of aquatic plants in surface water system. Total suspended solids in the pond water (96.8-487.7 mg/l) were found slightly high than recommended by C.P.C.B. and B.I.S.

Because the suspended particles absorb heat from sunlight and changed dissolved oxygen levels to fall even further because warmer waters can hold less dissolved oxygen. High total suspended solids in a water body can often mean higher concentrations of bacteria, nutrients, pesticides, and metals in the water. These pollutants may attach to sediment particles on the land and be carried into water bodies with storm water. In the water, the pollutants may be released from the sediment or travel farther. Fast running water can carry more particles and larger-sized sediment. Heavy rains can pick up sand, silt, clay, organic particles and heavy rubbish particles from the land and carry it to surface water. A change in flow rate can also affect total suspended solids; if the speed or direction of the water current increases, particulate matter from bottom sediments may be re-suspended.

In pond water, total solids were slightly low due to the settlement of heavy particles on bottom sediment. The ratio of total dissolved solids and total suspended solids with total solids was found quite similar during the study.

Conductance is a measurement of an electrical current with increasing amount and mobility of ions. These ions, which come from the breakdown of compounds, conduct electricity because they are negatively or positively charged when dissolved in water. Therefore, conductance is an indirect measure of the presence of dissolved solids such as chloride, nitrate, sulfate, phosphate, sodium, magnesium, calcium, and iron, and can be used as an indicator of water pollution.

It is not necessary that the water having high electric conductivity is not fit for irrigation. Only one parameter can't ensure the suitability of the water quality. However, the salt composition is high in the water having high Electric conductivity, but it may be beneficial for those soils having a lack or deficiency of some salts. Generally, there are not regulatory levels for electric conductivity. Instead, the concentration of total dissolved solids is often regulated. However, electric conductivity is an easily-obtained parameter that is a good indicator of the amount of dissolved solids in water, and thus can be used to detect contaminants in water. Total dissolved solids have a direct positive relationship with electric conductivity. If, total dissolved solids of water are high, electric conductivity will be more definitely (Mor *et al.*, 2003). Electric conductivity was recorded very highly positive correlated to total dissolved solids due to the direct relation of total dissolved solids with electric conductivity in pond water (2.9-3.2 mho/cm). The dissolved oxygen in water is often attributed to the fact the oxygen is dissolved more during the period of active photosynthesis. Hutchinson (1957) reported that the water has low temperature has greater capacity for holding gases. The value of dissolved oxygen in the all sampling sites was found out to be low dissolved oxygen. The DO was found low in pond water (1.2-3.4 mg/l), however, in some months dissolved oxygen was found very low in pond water due to the increasing rate of biodegradation of different compounds in water column. As the matter of fact that the effluents carried highly pollution load, which could not allow any flora and fauna to survive in unfavourable condition and there will be no photosynthesis process in polluted habitat. The value of dissolved oxygen was recorded even less values (2.56±0.21 mg/l) in the summer month. This trend was stated by Khanna *et al.* (1992) and Joshi (1996) in surface water system. The dissolved oxygen showed an inverse relationship with temperature as reported by Das and Srivastava (1956) and Dobriyal (1985) in aquatic system.

Biochemical oxygen demand and chemical oxygen demand are the measurement of the oxygen consuming capabilities of organic matter. Textile effluent during their process of deposition exerts biochemical oxygen demand/chemical oxygen demand, which have a catastrophic effect on the ecosystem by depleting the oxygen supply. Decay process causes the production of noxious gases as hydrogen sulphide and methane. Water with high biochemical oxygen demand/chemical oxygen demand indicates presence of decomposing organic matter and subsequent high bacterial counts that degrade its quality and potential uses (Verma and Kharat, 2000). Biochemical oxygen demands of the pond water (99.0-136.0 mg/l) were found high as given in standards of C.P.C.B., B.I.S. and WHO.

Chemical oxygen demand is directly proportional to biochemical oxygen demand. In fact, chemical oxygen demand explains the load of pollutants and degree of pollution in aquatic system. COD was observed 274.5-359.7 mg/l in pond water due to the high rate of decomposition of organic material in polluted and stored water. COD shows a positive relationship with BOD and negative with DO (Khanna, 1993) and mostly remains in the ratio of approximately 1:3 ± 0.5 in the case of wastewater. Water containing high biochemical oxygen demand and chemical oxygen demand are responsible for a heavy depletion of oxygen levels in the particular sector of water body or soil (Sial *et al.*, 2006).

Water alkalinity is caused by strong bases and the salts of strong alkalis and weak acids. Carbonates are added to a water system if the water passes through soil and rock that contain carbonate minerals, such as calcite ($CaCO_3$). Where limestone and sedimentary rocks and carbonate-rich soils are predominant, waters will often have high alkalinity. Where igneous rocks (such as granite) and carbonate-poor soils are predominant waters will have low alkalinity. Alkalinity is a measure of the buffering capacity of water, or the capacity of bases to neutralize acids. Waters with low alkalinity are very susceptible to changes in pH. Waters with high alkalinity are able to resist major shifts in pH. As increasing amounts of acid are added to a water body, the pH of the water decreases, and the buffering capacity of the water is consumed. Because alkalinity and pH are so closely related, changes in pH can also affect alkalinity, especially in a poorly buffered system. Above pH 8.3, alkalinity is mostly in the form of carbonate (CO_3^{2-}); below 8.3, alkalinity is present mostly as bicarbonate (HCO_3^-).

The values of alkalinity in the pond water (563.4-732.3 mg/l) were found very high. Alkalinity not only helps regulate the pH of a water body, but also the metal content. Alkalinity has found with positive correlations with Cd (0.0237), Cu (0.809), Fe (0.8896), Mn (0.7664), Ni (0.7918), Pb (0.5841) and Zn (0.7069) Bicarbonate and carbonate ions in water can remove toxic metals (such as lead, arsenic, and cadmium) by precipitating the metals out of solution.

Hardness represents the concentration of calcium and magnesium ions, because these are the most common polyvalent cations. Other ions, such as iron and manganese, may also contribute to the hardness of water, but are generally present in much lower concentrations. Calcium, magnesium, and other polyvalent cations such as iron and manganese may be added to a natural water system as it passes through soil and rock containing large amounts of these elements in mineral deposits. Hardness mainly depends on the amount of calcium and magnesium salts or both present in the groundwater, water hardness is not health hazard but its values should remain below permissible level to restore the taste to water (WHO, 1984)

Some textile industrial processes may also produce significant amounts of calcium and manganese that are later discharged as industrial wastewater. Calcium was found 10.4-23.6 mg/l in pond water. When hardness equals alkalinity, the only cations present in significant concentrations in the water are calcium and magnesium. When hardness is greater than alkalinity, the waters may contain considerable amounts of other cations. Magnesium was found low in pond water (4.2-14.1 mg/l). Potassium is one of the important element among ten chemical element required in rather large quantities fir growth of all crops. Plants growth may be limited by a deficient available supply of potassium but its high some quantity may alter the soil composition. High K in water used in irrigation may alter the soil quality (Miller and Turk, 2002). Potassium was found low in pond water (7.3-8.4 mg/l).

Sodium salts in the water are much more objectionable than salts of calcium and magnesium because of the tendency of sodium to cause deflocculation of the colloidal fraction of the soil and developed an undesirable structure water contain sufficient heavy metals to prove toxic to continue use by plants. As sodium chloride used in textile industry so the concentration of sodium and chloride will be high in the wastewater (Aslam *et al.*, 2004). Sodium was found very high in groundwater of industrial area, because sodium should not be higher than 20 mg/l in drinking water (WHO, 2006). Sodium values were found very high in pond water (128.5-158.7 mg/l).

Chloride is one of the most conservative constituents in wastewater, being affected very little either by vegetative growth or other biological processes, or by precipitation or anion exchange reaction. Chloride concentrations in pond water (576.5-770.7 mg/l) were also expected to be high by the permissible limit of C.P.C.B. Chloride does not have a health-based standard for drinking water. Chloride has been suggested as a standard value for drinking purposes (EPA, 1989). The Maximum Contaminant Level (MCL) of 250000 µg/L is based on poor taste chloride imparts to water. Chloride is not affected by adsorption or degradation processes and is therefore a conservative chemical in soil and groundwater. Chloride concentrations begin to increase at some depth due to dissolution reactions within the aquifer. Chloride is therefore a good indicator of human impacts in the environment (MPCA, 2001). Chloride concentration in irrigation water should be less then 75 ppm, otherwise it may be harmful for soil structure. Textile effluents are high in toxicity not only for human beings of surroundings but also a serious threat to groundwater and surface water resources (Aslam *et al.*, 2004).

4.6 Physico-chemical Parameters of Pond Sediment

Physico-chemical characteristics of soil may demonstrate the pollution status in the industrial region, so, physical and chemical characteristics are important for the assessment of contamination and pollution load. Sediment characteristics are the good indicator for the pollution load in pond bottom. The temperature of soil greatly affects the physical, biological and chemical processes occurring in soil system. Biological decomposition can come to a near stand still, thereby limiting the rate at which nutrients such as nitrogen, phosphorus, sulphur and calcium are made available. Absorption and transport of water and nutrient ions by higher plants are adversely affected by low temperature (Brady, 1995). The temperature of field soils is not subject to radical human regulation. However, two kinds of management practices have significant effects on soil temperature, those that keep some type of cover or mulch on the soil and those that reduce excess soil moisture. There is not any significant variation in the temperature of drain sediment (15.5-24.9°C) and pond sediment (15.3-24.8 °C) and also between agricultural soil (17.4-27.8°C) and control soil (17.5-27.8°C). The chemical processes and activities of microorganism which convert plant nutrients into available forms are also materially influenced by temperature. Removal of excess water from soil may facilitate changes in soil temperature. Humus content, good tillage, dark color of soil are the some other factors which influence the soil temperature (Miller and Turk, 2002).

Water holding capacity is a property as to water quantity is absorbed or retained by given amount of soil (Ramani and Srivastava, 1989). Water held in the soil pores with varying degrees of tenacity depending on the amount of water present and the size of pores in soils. Depending on the soil, as much as one forth to two thirds of the moisture may remain in the soil after the plants have wilted or died due to the lack of soil water. The phenomenon of capillarity is a common one, the classic example being the movement of water up a wick when the lower end is immersed in water. Capillary forces are at work in all moist soils. However, the rate of movement and the rise in height are less than one would expect on the basis of soil pore size. One reason is that soil pores are filled with air, which may be entrapped, showing down or preventing the movement of water by capillary. Usually the height of the rise resulting from capillarity is greater with fine textured soils if sufficient time is allowed and the pores are not too small. This is readily explained on the basis of the capillary size and the continuity of the pores. With sandy soils the adjustment is rapid but so many of the pores are non-capillary that the height of rise can not be great (Brady, 1995). Water holding capacity was found similar in drain sediment (24.5-31.7 per cent) and pond sediment (29.5-36.3 per cent) and also between agricultural soil (36.1-43.7 per cent) and control soil (36.8-44.2 per cent) with much difference between soil and sediment. Yuandong *et al.* (2006) described the similar trend of water holding capacity on the basis of soil texture and chemical composition in the agricultural soil system. The pore space of a soil is that portion of the soil volume occupied by air and water. The amount of pore space is determined by the arrangement of the solid particles. If the particles lie close together, as in stands or compact sub-soils, the total porosity is low. If they are arranged in porous aggregates, as is often the case in medium textured soils high in organic matter, the pore space per unit volume will be high. Soil aeration may influence the availability of nutrients to plants (Miller and Turk, 2002). Bulk density relates to the combined volumes of the solids and pore spaces, soils with a high proportion of pore space to solids have lower bulk densities than those that are more compact and have less pore space. Consequently, any factor that influences soil pore space will affect bulk density. Mainly soil texture and compactness strongly influence the bulk density of soils (Brady, 1995). The bulk density was found similar between drain sediment (1.33-1.45 g/cm^3) and pond sediment (1.36-1.49 g/cm^3) and exhibited same between agricultural soil (1.15-1.24 g/cm^3) and control soil (1.10-1.22 g/cm^3) Bulk density was similarly described by Yuandong *et al.* (2006) and Krebs *et al.* (1999) in soil environment.

The soil structure influences its moisture content. A well granulated soil has more total pore space than one with poor granulation or one that has been compacted. The reduced pore space may be reflected in a lower water holding capacity. The compacted soil also has to higher proportion of small and medium sized pores, which tend to hold the water with greater tenacity than do the larger pores. Moisture content of drain sediment (13.3-20.8 per cent) and pond sediment (13.5-20.9 per cent), agricultural soil (3.5-10.7 per cent) and control soil (3.7-11.06 per cent) were found similar during the study. Rainwater leaching have made more efficient if antecedent moisture content of the soil profile was increased by saline water irrigation just before the onset of monsoons. On a wet soil, a rainfall event is expected to displace the salts to a greater depth. Conservation and proper utilization of rainwater for leaching salts is also utmost important.

The soil pH shows the acidity and basic nature of the soil. The soil is acidic if the pH is lower, while the higher pH means that soil is basic. The pH value was observed slightly high (7.11-7.63) in agriculture land, which shows that agriculture land is going to alkaline in nature which confirms that agriculture land again disturbed by its proper pH values whereas due to the decay of organic matter has produced the acidic properties of soil. If the decomposition of organic matter is there in soil, it will turn the pH low, other wise some foreign substances may alter the soil nature and properties towards alkaline (Nongkynrih *et al.*, 1996). Initial soil pH is an important property of the soil because of its influence on the infiltration rate of the soil (Gupta *et al.*, 1989). There are several causes of minor fluctuation in soil pH. For instance, the movements of salts into and out of soil moisture moves up and down the profile will influence pH. Similarly, the pH of mineral soils declines during the crop-growing season as a result of the acids produced by microorganisms and by the roots of higher plants. When soil temperatures are low, an increase in pH often is noted because biotic activities during these times are considerably slower. Soil pH significantly influences other soil chemical properties as well as biological organisms. High pH of textile industrial effluent can hampers plant crop growth (Sial *et al.*, 2006). The soil pH significantly affects the availability of most of the chemical element of importance to plants and microbes. Likewise, the tendency for toxic levels of element such as iron, manganese, aluminum to be established at low pH and deficiencies of iron and manganese at high pH have been mentioned. Continuous application of effluents with high pH will render the soils unfit for further cultivation (Sial *et al.*, 2006). Soil pH was found alkaline in drain sediment (7.8-8.3) and pond sediment (7.9-8.37), and slightly alkaline in agricultural soil (7.1-7.6) and control soil (7.2-7.4). Similar trend of pH fluctuation have been described by Kasem and Singh (1999) in agricultural soil environment.

Electric conductivity is the measure of current carrying capacity, which gives a clear idea of soluble salts present in soil. Conductivity generally depends upon the dilution of suspension. However, the salt composition is high in the water having high electric conductivity, but it may be beneficial for those soils having a lack or deficiency of some salts. The electric conductivity of drain sediment, pond sediment, agricultural soil and control soil was found in the range of 27.8-32.5, 27.3-31.9, 24.1-27.7 and 23.0-23.7 mho/cm respectively with some minor fluctuations. Electric conductivity has positive correlation with all heavy metals *viz.* Cd (0.3965) Cu (0.9454), Fe (0.4198), Mn (0.9076), Ni (0.8286), Pb (0.8596) and Zn (0.7586) in agricultural soil.

Chloride was found very high in effluent drain sediment and slightly low as with similar fluctuation in pond bed sediment, but in agricultural soil and soil near textile industries chloride was found comparatively low because of the high settlement of chloride in drain sediment from textile industrial effluent; while rest was reach in agricultural soil by various contacts. The chloride in drain sediment, pond sediment, agricultural soil and control soil was found 609.0-836.6, 588.5-827.0, 499.5-734.3 and 313.8-529.3 μg/g respectively with some minor fluctuations. Chloride has positive correlation with all heavy metals *viz.* Cd (0.3654), Cu (0.8479), Fe (0.3249), Mn (0.7254), Ni (0.6603), Pb (0.6972) and Zn (0.7744) in agricultural soil environment. Manchanda (1990) has also reported that pulse crops such as pea, mungbean, gram, lentil, and broadbean suffered more due to toxic accumulation of chloride, when irrigated with Cl ion then SO_4^{-2} dominant (70 per cent) water of more salinity.

The exchangeable potassium is the chief source of available potassium in soil, in most cases (Jha and Pandey, 1984). Potassium plays many essential roles in plants. It is an activator of dozens of enzymes responsible for such plants processes as energy metabolism, starch synthesis, nitrate reduction, and sugar degradation. Potassium is extremely mobile with in the plant and helps regulate the opening and closing of stomata in the leaves and the uptake of water by root cells. At any one time most of the potassium is in the primary mineral and non-exchangeable forms. To maintain the level of exchangeable and soil solution potassium high enough to encourage good crop production, chemical fertilizer are used. Potassium may deficient in alkaline soil and an excess of potassium may delay maturity of plant. In some crops it may tend to accumulate in the leaf and stem rather then in the grain (Miller and Turk, 2002). The potassium in drain sediment, pond sediment, agricultural soil and control soil was found almost in uniform pattern 27.3-33.5, 28.3-33.5, 31.0-36.5 and 30.3-34.5 μg/g respectively with some minor fluctuations. Potassium has slightly positive correlation with many heavy metals *viz.* Cd (0.6852), Cu (-0.1689), Fe (0.7116), Mn (0.0611), Ni (0.2623), Pb (0.2369) and Zn (-0.6269) in agricultural soil. Korfali and Davies (2004) and Soltan (1999) have been exhibited a clear idea about the fluctuations in potassium concentrations in soil system.

Sodium in alkali soils results in a deflocculation of the colloids and hence in a breaking down of the soil structural units. This puddle condition renders the soil more or less impervious and retards entrance of irrigation and rain water impedes drainage. In heavy textured soil, the penetration of roots may be restricted by the deflocculated zone. Aeration is also much reduced, setting up aeration conditions and resulting in the formation of reduced compounds which are toxic to plants (Miller and Turk, 2002). The highly alkaline soil was due to the presence of sodium carbonate. Large quantity of adsorbed sodium repressed the availability of several nutrients especially, iron, manganese and phosphorous (Miller and Turk, 2002).

The occurrence of sodium in pond sediment was found very high 143.6-556.0 µg/g with some minor fluctuations. Desalination of soil is usually accomplished by impounding of water at the surface of the alkaline soil. The relationship in between reduction of salts and exchangeable sodium from a given depth of soil and the amount of leaching water for the sandy loam soil (Minhas and Khosla, 1986). The amount of soluble salts in soil can be reduce to safe limits for agriculture crops by leaching (Rao *et al.*, 1987).

Usually the effects of one or more water quality parameters on the changes in physico-chemical characteristics of soils and crop productivity have been studied and evaluating the suitability of water for irrigation in most experiments, however, integration of the interactive and interdependent processes of various water quality indices with soils have required to obtain comprehensive knowledge of the whole system. It requires the development of computerized models for providing a quantitative description of the dynamic behavior of soils upon interaction with irrigation water quality and its relation with crop productivity. Modeling efforts should go a long way in making generalizations based on the experience gained under specific conditions and extrapolation of experimental findings to different locations and seasons in study area.

Chapter 5

Impact Assessment on Groundwater

5.1 Groundwater

Under groundwater is used for industries, domestic supply and agriculture irrigation in most part of the world as it is a replenishable resource and has inherent advantage over surface water. There has been a tremendous increase in the demand of fresh water due to growth in population, advanced agriculture practices and industrial usages (Bachmat, 1994). In the last two decades, the rapid growth of industrialization and urbanization has created negative impacts on the environment. Due to industrial, municipal and agricultural wastes containing pesticides, insecticides, fertilizers residues and heavy metals with water; groundwater has been polluted by leaching process of pollutants present in effluents (Sastry and Rathee, 1999). In Haryana, about 90 per cent of the population of rural areas utilizes groundwater for drinking, washing and other daily uses. But groundwater in Haryana is very saline and large variation in salinity can be observed over a short distance. Water table is also very high. Therefore, scarcity of drinking water is very big problem faced by population of this state. Panipat district of Haryana is well known for the textile industries. A large number of small-scale industries mainly textile printing and textile dying have been established in Panipat city. These industries discharge their effluent openly into pond or other water bodies. The seepage of effluents posed serious threat to the groundwater resources (Groff, 1993). So, groundwater must be managed and protected carefully for the most beneficial use in the future.

Clean groundwater is needed for more than drinking purposes. Agriculture depends highly on groundwater for irrigation; as a large number of groundwater pumps were used for crop irrigation in India. Poor or contaminated groundwater could jeopardize crops and threaten the health of livestock. Groundwater is connected to surface water in the hydrological cycle, and some aquifers actually feed area springs and rivers. Sometimes contamination occurs naturally by mixing the sallow aquifers with other contaminated sources of water table. Saline water from deeper aquifers may reach aquifers that provide water for human beings, livestock and others agricultural activities. Some groundwater may have naturally high levels of nitrates, metals, iron, sulfate, or chloride, all of which provides an odd odor, color, or taste in water. Groundwater contamination has become a major problem in recent decades due to exploring industrialization. Groundwater, as a wide resource for irrigation in most part of country, may contaminate the soil quality by repeated irrigation of agricultural fields. Bed quality of groundwater used in irrigation may alter the characteristics of agriculture soils of the area.

In the present study, the impact of textile effluents was found on pond environment and polluted pond may alter the ecological parameters of surroundings, so, groundwater may be polluted. This contaminated groundwater used in irrigation purposes may harmful for soils of agriculture fields and ultimately effects on productivity of agriculture crop plants, and soil macrobiotic community.

Different workers from India and abroad have been discussed various aspects of water pollution in term of characteristics of water bodies, affected agricultural soils and bottom sediments from time to time. This review includes some of important contribution made in the field of industrial pollution, with references to physical, chemical and biological parameters of groundwater, surface water, characterization of textile industrial effluents, contaminated agricultural irrigated soil and bottom sediment soils by different authors of country and abroad.

5.2 Historical Resume

Eaton (1950) described the significance of carbonates in waters mostly used for irrigation purposes in agriculture. Gardner (1957) suggested some steady state solutions of the unsaturated moisture flow equation with application to evaporation from a water table.

Kanwar (1961) assessed the quality of irrigation water as an index of suitability for irrigation purposes. Seeber (1962) highlighted the cation hydrological facies of groundwater in the English town Formation in New Jersey. Darra *et al.* (1964) evaluated the water quality of irrigation water used in Rajasthan. Hoston (1965) stated that water quality index is a rating of water quality parameters by a single numerical expression reflecting the composite influence of water quality parameters (physical, chemical and bacteriological) significant for a specific beneficial use.

Saksena *et al.* (1966) evaluated the quality of groundwaters for irrigation in Ahor developmental block, Jalore. Singh and Bhumbla (1968) analyzed the effect of quality of irrigation water on soil properties. Singh *et al.* (1969) reported the salinity problem in high water table areas of north India.

Singh and Sharma (1971) studied on the effects of saline irrigation waters on physico-chemical properties of some soils of Rajasthan.

Sharma and Parihar (1973) evaluated the effect of depth and salinity of groundwater on evaporation and soil salinization.

Sharma *et al.* (1975) presented the status of groundwaters quality of Jind district, Haryana with special reference to the chemical composition of waters used in irrigation in agricultural sector.

Sharma *et al.* (1976) described the status of groundwaters quality of Karnal district of Haryana with special reference to the chemical composition of waters used in irrigation in agricultural sector. Singh *et al.* (1977) highlighted the salinity and sodium hazards of underground irrigation waters of Bhatinda district of Punjab state. Shivakumar *et al.* (1977) made a geo-graphical approach for toxic trace element pollution in groundwater around Patancheru and Bollaram Industrial area, A.P., India.

Sharma (1978) studied on the testing of a model for prediction sodium hazards of irrigation waters.

Todd (1980) described the groundwater movement; groundwater properties and hydrogeology for explain the behaviour of groundwater pollution. Vates (1986) made a study on septic tank density and groundwater contamination by different composite effluents of textile industries.

Rhoades (1987) studied on the use of saline water for irrigation. Nriagu and Pacyna (1988) carried out a study on the quantitative assessment of worldwide contamination of air, water and soil with trace metals.

Zilliox (1989) determined the industrial impacts on the quality of groundwater in a large basin as a case of the Rhine aquifer in France and noticed it harmful for public consumption. Narayan and Suresh (1989) evaluated the chemical quality of groundwater of Mangalore city in state Karnataka.

Minhas and Gupta (1992) indicated the effects of wastewater irrigation in agriculture soils and found sodicity due to the poor water quality. Karle *et al.* (1992) studied groundwater quality in Pravaranagar area of Ahmednagar district and Kumar *et al.* (1992) studied the relationship between Fluoride, total hardness and total alkalinity in the groundwater of Bharmar District (Rajsthan). PPCB (1993) released a report on groundwater quality of Ludhiana city, Punjab based on the quality of the tube wells and hand pumps in shallow waters.

Kataria *et al.* (1995) studied organic contamination in Bore well waters of BHEL area of Bhopal and give a baseline data for further study on same problem. Sharma *et al.* (1995) assessed the quality of groundwater in municipal and fringe areas near Gwalior.

Singh *et al.* (1996) carried out a hydrological analysis on groundwater samples collected from 19 open wells in different parts of Varanasi city. Sharma and Patel (1996) determined copper in water of Raipur with the help of spectrophotometer. Kataria (1996) analysed the BOD and COD contents in bore well water of Bhopal (M.P.).

Prasad and Chandra (1997) studied well water quality status of Hyderabad and Khurshid and Shabeer (1997) observed water quality degradation due to heavy metal pollution.

Garg *et al.* (1998) observed the fluoride concentrations in undergroundwater of Hissar, Haryana. Mohan *et al.* (1998) observed heavy metals (Fe, Pb, Ca and Zn) in the groundwater of Naini industrial area; Dist. Allahabad, U.P. Kashem and Singh (1999) had also found the decreasing concentration of Cu, Mn, Pb and Zn in groundwater with increasing distance from disposal point.

Stetzenbach *et al.* (1999) evaluated the heavy metals in groundwater and used multivariate statistical analysis. Dutta (1999) highlighted the groundwater situation in Delhi. Rao and Rao (1999) identified the path lines of pollutants migration in groundwater of the Visakhpatanam urban area of south India using computer models. Particles path lines in the groundwater indicated a predominant northeast and southwest migration of groundwater pollution in the area.

Tyagi and Budhi (2000) postulated the degradation of groundwater quality due to heavy metals in industrial area of India. Reddy and Linn (2000) used catalysts against nitrate pollution in groundwater and observed it significantly suitable for abatement of groundwater pollution.

Kumar *et al.* (2001) studied on the impact of textile industry on groundwater quality of Sanagar, Jaipur and found the high amount of chemicals (e.g. sulphuric acid and acetic acid) and dyes (e.g. chrome and metal dyes) suspended solids and toxic metals in the waste which may enter in the groundwater and deteriorate its quality. Aurangabadkar *et al.* (2001) studied on groundwater quality at Chennai.

Srinivas *et al.* (2002) pointed out the groundwater pollution due to the industrial effluents in Kothur industrial area, Mahaboonagar, (A.P.). Siddiqui and Pathania (2002) made some studies on heavy metals in surface and groundwater of Jalandhar and Ludhiana districts of Punjab state. Vishwanath and Ananthmurthy (2002) assessed the groundwater quality around a solid waste dumping site and concluded that groundwater contains high electric conductivity; TDS, hardness and fluorides but the nitrates and chlorides were with in the permissible limits of potable drinking water standards.

Yadav *et al.* (2002) highlighted the post irrigation impact of domestic sewage effluent on composition of soils, crops and groundwater in Kurukshetra district. Mohan and Sharma (2002) and Yadav *et al.* (2003) determined the groundwater quality of different places of Rajasthan. Reshma and Prakasham (2003) studied the potability of Dug wells in Kerala.

Royee and Prakasham (2003) analysed the water characteristics of dug and tube wells of Kollam municipality in Tamilnadu. Yadav and Lata (2003) indicated the heavy metal pollution in groundwater in Bahadurgarh block of districts Jhajjar, Haryana. Mishra and Sahoo (2003) evaluated the groundwater quality in and around Deogarh. Mor *et al.* (2003) assessed the quality of groundwater of Jind city in Haryana state.

Malik *et al.* (2004) evaluated the accumulation of heavy metals in crop plants through irrigation of contaminated groundwater in Panipat region and emphasized on bioremediation processes for combat groundwater pollution. Mahesha and Prasad (2004) elaborated the physico-chemical characteristics of bore well water in Arsikara taluk, Hassan.

Reduction of the levels of toxic metals from industrial effluents is necessary before discharging into surface waters. The general methods of treating wastewater-containing metals are coagulation, ion exchange, reverse osmosis and adsorption. Adsorption is a highly effective physico-chemical treatment for removing heavy metals for effluents. Due to high cost of activated carbon, efforts are being directed towards finding low cost adsorbents. Recently much interest has been shown for removal of heavy metals using a large number of low cost materials. Coal fly ash, the solid waste of power plants is one of the cheapest and non-conventional adsorbents (De, 2005).

Dwivedi *et al.* (2005) indicated the contamination of groundwater due to the artificial groundwater recharging in Gengetic plain area. Kumar *et al.* (2005) described the status of water resources of India and highlighted the hydrological cycle, monsoon and precipitation, surface and groundwater resources, water requirements in domestic use, irrigation, hydroelectric power, industrial requirement and groundwater management in India. Dutta (2005) highlighted the ethics of groundwater for its sustainability and the vulnerability and depletion of groundwater due to rapid growth of population, urbanization, industrialization and competition for economic development and also stated the groundwater contamination, concepts of groundwater management, strategies for the groundwater protection and suggested use of contaminated groundwater.

Palanivelu *et al.* (2006) evidenced the poor groundwater quality in areas like Kottivakkam beach, Kuppam Oorurokot, Kuppam. Singh (2006) indicated the contamination of groundwater of Ganges-Brahmaputra river basin in north east India and found the groundwater unfit for direct human consumption and noticed the rice and vegetable fields irrigated by groundwater received a large amount of arsenic through shallow tube wells in the affected areas of north 24 *Parganas* and Murshidabad, West Bangal.

Malik *et al.* (2007) assessed the migration of heavy metals in the groundwater regime of Panipat industrial area on the behalf of different spatial variations the heavy metals were found in maximum quantities at point source of pollution.

Rizk *et al.* (2007) identified the sources of dissolved solids and water in Wadi Al Bih aquifer, Ras Al Khaimah Emirate in United Arab Emirates. Vijith and Satheesh (2007) completed the Geographical Information System (GIS) based assessment of spatiotemporal characteristics of groundwater quality of upland sub-watersheds of Meenachil River, parts of Western Ghats, Kottayam District of Kerala, India.

Obiri (2007) determined the heavy metals in water from boreholes in Dumasi in the Wassa West District of Western Region of Republic of Ghana.

Fransson (2007) studied on to verify the methods for estimating transmissivity distributions along boreholes. Mico *et al.* (2007) described the comparison of two digestion methods for the analysis of heavy metals by flame atomic absorption spectroscopy.

5.3 Observations

Experimental Site

Physico-chemical Parameters of Groundwater

Temperature (°C)

Temperature of groundwater was recorded maximum 31.63 ± 2.5 in monsoon 2006 and minimum 25.5 ± 0.71 in summer 2007 ranged between 25.0 to 34.5 °C. The highest value of temperature in groundwater was found 34.5 °C during July 2005 and lowest 25.0 °C in February 2007. The temperature variation during the study is given Table 5.1. Monthly variation in temperature is given in Figure 5.1, Figure 5.7 and Figure 5.13 for 40 ft, 80 ft and more than 120 ft depth respectively.

pH

The maximum pH value of groundwater was recorded 7.7 ± 0.1 in summer 2005 and minimum 7.43 ± 0.06 in monsoon 2005 ranged between 7.3 to 7.8 on pH scale. The highest value of pH in groundwater was found 7.8 during June 2006 and lowest 7.3 in October 2005. The variation in the pH during the study is given Table 5.1. Monthly variation in pH is given in Figure 5.1, Figure 5.7 and Figure 5.13 for 40 ft, 80 ft and more than 120 ft depth respectively.

Total Solids (mg/l)

The maximum total solids in groundwater were recorded 1464.67 ± 191.20 in summer 2005 and minimum 782.75 ± 121.72 in winter 2006-07 ranged between 641.0 to 1688.0 mg/l. The highest value of total solids in groundwater was found 1688 mg/l during June 2006 and lowest 641.0 mg/l in November 2005. The variation in the total solids during the study is given Table 5.1. Monthly variation in total solids is given in Figure 5.2, Figure 5.8 and Figure 5.14 for 40 ft, 80 ft and more than 120 ft depth respectively.

Total Dissolved Solids (mg/l)

Total dissolved solids in groundwater were recorded maximum 1331.67 ± 169.74 in summer 2005 and minimum 710.25 ± 110.09 in winter 2006-07 ranged between 612.0 to 1537.0 mg/l. The highest value of total dissolved solids in groundwater was found 1537.0 mg/l during June 2006 and lowest 612.0 mg/l in December 2005. The variation in the total dissolved solids during the study is given Table 5.1. Monthly variation in total dissolved solids is given in Figure 5.2, Figure 5.8 and Figure 5.14 for 40 ft, 80 ft and more than 120 ft depth respectively.

Total Suspended Solids (mg/l)

Total suspended solids in groundwater were recorded maximum 133.0 ± 21.52 in summer 2005 and minimum 72.50 ± 11.73 in winter 2006-07 ranged between 62.0 to 155.0 mg/l. The highest value of total suspended solids in groundwater was found 155.0 mg/l during June 2005 and lowest 62.0 mg/l in November 2006. The

Table 5.1: Seasonal Variation of Physico-chemical Characteristics of Groundwater at 40 ft Depth from Hand Pump in Village Binjhole

Para-meters	*Unit*	*Summer 2005*		*Monsoon 2005*		*Winter 2005-06*		*Summer 2006*		*Monsoon 2006*		*Winter 2006-07*		*Summer 2007*	
		Mean	*SD*	*Mean*	*SD*	*Mean*	*SD*	*Mean*	*SD*	*Mean*	*SD*	*Mean*	*SD*	*Mean*	*SD*
Temp.	°C	31.433	2.183	31.500	2.784	25.600	0.652	29.325	3.031	31.633	2.503	26.200	0.572	25.500	0.707
pH		7.700	0.100	7.433	0.058	7.460	0.114	7.700	0.082	7.500	0.100	7.475	0.096	7.550	0.071
TS	mg/l	1464.667	191.197	1229.000	154.088	797.200	165.915	1393.000	310.321	1280.667	166.572	782.750	121.722	1057.000	45.255
TSS	mg/l	133.000	21.517	120.667	16.921	79.000	14.816	125.500	27.887	118.333	13.503	72.500	11.733	95.000	4.243
TDS	mg/l	1331.667	169.742	1108.333	137.508	718.200	151.136	1267.500	283.580	1162.333	153.070	710.250	110.089	962.000	41.012
EC	mho/cm	2.081	0.265	1.732	0.215	1.122	0.236	1.980	0.443	1.816	0.239	1.110	0.172	1.503	0.064
DO	mg/l	2.567	0.208	3.967	0.681	3.800	0.604	2.925	0.206	4.000	0.436	3.950	0.420	3.650	0.212
Free CO_2	mg/l	0.933	0.153	0.167	0.289	0.220	0.303	0.700	0.183	0.167	0.289	0.100	0.200	0.250	0.354
BOD	mg/l	35.667	8.145	31.333	23.159	14.400	11.149	27.750	12.580	35.333	18.610	20.750	10.689	11.000	2.828
COD	mg/l	93.000	16.523	75.000	48.775	37.000	30.725	71.750	33.540	88.000	42.509	54.750	31.020	27.500	7.778
Total alkalinity	mg/l	557.667	59.744	584.667	23.714	493.600	31.061	579.500	38.631	596.667	48.911	502.500	22.128	564.500	28.991
Ca	mg/l	106.667	15.275	136.667	51.316	46.000	16.733	102.500	40.311	143.333	47.258	52.500	12.583	65.000	7.071
Mg	mg/l	92.667	25.423	88.000	32.604	32.400	16.227	82.500	59.360	97.667	35.233	39.000	14.024	22.000	4.243
K	mg/l	5.267	0.569	6.100	0.100	5.280	0.319	5.300	0.560	6.033	0.208	5.475	0.250	4.900	0.141
Cl	mg/l	104.333	34.269	145.000	33.151	72.200	35.252	66.838	62.522	151.333	17.954	83.750	39.778	47.500	3.536
Na	mg/l	88.333	4.163	95.000	11.790	51.000	13.454	75.500	24.365	97.667	11.240	59.000	12.780	51.500	4.950

Table-5.2: Seasonal Variation of Physico-chemical Characteristics of Groundwater at 80 ft Depth from Tube Well and Deep Hand Pump

Para-meters	*Unit*	*Summer 2005*		*Monsoon 2005*		*Winter 2005-06*		*Summer 2006*		*Monsoon 2006*		*Winter 2006-07*		*Summer 2007*	
		Mean	*SD*	*Mean*	*SD*	*Mean*	*SD*	*Mean*	*SD*	*Mean*	*SD*	*Mean*	*SD*	*Mean*	*SD*
Temp.	°C	30.833	2.021	31.100	2.651	26.340	0.422	29.250	2.327	31.333	2.255	26.550	0.666	26.900	0.141
pH		7.333	0.153	7.233	0.153	7.860	0.195	7.500	0.082	7.300	0.100	7.900	0.283	7.750	0.071
TS	mg/l	615.333	104.902	944.667	85.149	788.200	308.086	580.500	118.329	950.333	68.223	852.000	252.052	476.500	61.518
TSS	mg/l	59.333	7.371	85.000	11.790	75.000	31.217	52.000	9.695	87.333	5.686	80.000	25.100	45.000	4.243
TDS	mg/l	556.000	97.535	859.667	74.842	713.200	277.196	528.500	108.911	863.000	62.554	772.000	226.995	431.500	57.276
EC	mho/cm	0.869	0.152	1.343	0.117	1.114	0.433	0.826	0.170	1.348	0.098	1.206	0.355	0.674	0.089
DO	mg/l	2.633	0.513	3.900	0.361	4.000	0.381	2.850	0.351	3.900	0.265	3.950	0.300	3.950	0.636
Free CO_2	mg/l	0.900	0.100	0.167	0.289	0.100	0.224	0.700	0.183	0.167	0.289	0.100	0.200	0.250	0.354
BOD	mg/l	24.000	7.000	23.333	19.296	10.200	7.294	19.000	6.733	26.667	9.074	12.500	7.594	8.500	2.121
COD	mg/l	62.000	8.718	52.667	41.053	26.600	17.601	48.000	16.793	68.667	21.197	33.750	22.051	20.500	3.536
Total alkalinity	mg/l	337.333	50.063	296.000	27.785	261.600	5.225	327.750	47.183	303.000	34.117	262.000	10.985	283.500	7.778
Ca	mg/l	96.667	15.275	136.667	23.094	44.000	23.022	87.500	33.040	136.667	15.275	55.000	26.458	50.000	14.142
Mg	mg/l	82.333	22.368	89.000	28.160	29.400	14.622	71.500	40.739	97.333	23.159	38.000	16.145	23.000	2.828
K	mg/l	4.933	0.513	5.967	0.153	5.180	0.277	5.050	0.580	5.967	0.115	5.400	0.216	4.900	0.141
Cl	mg/l	4.700	0.954	18.333	6.028	12.940	11.812	5.025	1.459	19.667	8.083	15.125	11.229	4.800	0.424
Na	mg/l	57.333	5.508	62.333	11.240	24.800	10.592	46.500	20.486	69.333	12.503	31.250	10.210	23.000	2.828

Table 5.3: Seasonal Variation of Physico-chemical Characteristics of Groundwater (more than 120 ft depth) of Tube Well

Parameters	Unit	Summer 2005		Monsoon 2005		Winter 2005-06		Summer 2006		Monsoon 2006		Winter 2006-07		Summer 2007	
		Mean	SD	Mean	SD	Mean	SD	Mean	SD	Mean	SD	Mean	SD	Mean	SD
Temp.	°C	30.267	2.040	31.167	2.350	26.880	0.526	28.950	2.027	31.000	2.000	26.500	0.408	27.000	0.000
pH		7.167	0.058	7.167	0.115	7.620	0.179	7.300	0.082	7.200	0.100	7.600	0.183	7.500	0.000
TS	mg/l	285.667	38.004	494.333	70.002	620.800	201.268	290.500	38.144	505.667	71.038	687.250	139.801	370.000	98.995
TSS	mg/l	27.000	4.000	47.000	5.000	57.800	18.913	27.250	3.403	47.333	7.095	63.500	13.026	35.000	9.899
TDS	mg/l	258.667	34.005	447.333	65.003	563.000	182.391	263.250	34.760	458.333	64.010	623.750	126.790	335.000	89.095
EC	mho/cm	0.404	0.053	0.699	0.102	0.880	0.285	0.411	0.054	0.716	0.100	0.975	0.198	0.523	0.139
DO	mg/l	2.767	0.252	4.167	0.351	4.040	0.321	3.200	0.356	4.133	0.115	4.125	0.299	3.700	0.707
Free CO_2	mg/l	0.800	0.100	0.000	0.000	0.100	0.224	0.675	0.206	0.000	0.000	0.000	0.000	0.250	0.354
BOD	mg/l	14.000	5.292	12.000	14.422	3.600	5.128	10.500	7.937	16.333	8.505	6.250	7.228	2.000	2.828
COD	mg/l	38.000	14.799	28.000	33.422	8.400	11.845	25.750	18.464	42.333	23.798	17.250	20.123	4.500	6.364
Total alkalinity	mg/l	263.333	31.070	207.667	17.786	184.000	8.337	252.000	23.409	211.333	13.503	181.500	8.062	221.000	22.627
Ca	mg/l	80.000	20.000	116.667	15.275	34.000	11.402	75.000	35.119	116.667	15.275	50.000	21.602	40.000	14.142
Mg	mg/l	74.000	24.021	83.333	24.440	27.200	12.814	59.000	39.875	92.000	30.050	35.500	14.731	19.500	2.121
K	mg/l	4.633	0.513	5.700	0.200	4.960	0.305	4.550	0.480	5.467	0.252	5.150	0.332	4.350	0.212
Cl	mg/l	4.833	0.551	12.167	4.368	7.740	2.830	4.775	0.629	13.733	4.941	9.850	3.576	5.000	0.141
Na	mg/l	26.667	8.386	37.000	13.454	10.420	3.582	21.625	12.605	36.000	8.185	11.450	5.318	8.700	0.849

Table 5.4: Seasonal Variation of Physico-chemical Characteristics of Groundwater of 80 ft Depth at Control Site (C)

Para-meters	*Unit*	*Summer 2005*		*Monsoon 2005*		*Winter 2005-06*		*Summer 2006*		*Monsoon 2006*		*Winter 2006-07*		*Summer 2007*	
		Mean	*SD*	*Mean*	*SD*	*Mean*	*SD*	*Mean*	*SD*	*Mean*	*SD*	*Mean*	*SD*	*Mean*	*SD*
Temp.	°C	30.500	1.179	30.600	2.152	26.640	0.472	29.250	1.555	30.233	2.159	26.825	0.236	27.250	0.354
pH		7.133	0.058	7.133	0.058	7.420	0.164	7.250	0.129	7.200	0.100	7.425	0.171	7.450	0.071
TS	mg/l	451.333	87.512	630.000	50.269	726.600	151.085	446.750	76.164	651.000	50.507	780.500	110.301	485.500	103.945
TSS	mg/l	42.333	7.506	57.667	4.509	68.200	13.627	42.250	7.182	60.000	4.583	71.000	10.165	46.000	7.071
TDS	mg/l	409.000	80.019	572.333	45.764	658.400	137.467	404.500	69.019	591.000	46.033	709.500	100.142	439.500	96.874
EC	mho/cm	0.639	0.125	0.894	0.072	1.029	0.215	0.632	0.108	0.923	0.072	1.109	0.156	0.687	0.151
DO	mg/l	2.367	0.513	3.567	0.503	3.340	0.391	2.500	0.356	3.467	0.462	3.300	0.245	2.950	0.212
Free CO_2	mg/l	0.933	0.153	0.167	0.289	0.260	0.358	0.775	0.126	0.333	0.289	0.275	0.320	0.700	0.141
BOD	mg/l	15.000	3.000	13.667	7.371	6.600	5.683	13.250	4.787	15.667	6.110	7.500	5.802	8.500	0.707
COD	mg/l	38.000	7.000	35.667	18.903	16.000	13.172	33.250	12.816	40.667	15.044	18.000	13.880	21.500	4.950
Total alkalinity	mg/l	303.000	22.913	233.667	30.089	216.400	41.621	296.000	31.602	249.000	31.193	216.500	35.828	273.500	6.364
Ca	mg/l	86.667	15.275	110.000	10.000	42.000	8.367	80.000	31.623	106.667	5.774	55.000	17.321	40.000	14.142
Mg	mg/l	78.333	17.502	88.000	18.682	29.800	15.401	65.250	42.414	101.000	23.302	40.000	16.693	21.000	4.243
K	mg/l	4.833	0.666	5.800	0.100	5.060	0.404	4.925	0.506	5.733	0.058	5.250	0.311	4.650	0.212
Cl	mg/l	5.200	1.758	13.000	3.775	10.960	6.886	4.900	1.111	12.200	3.897	12.925	5.518	5.050	0.778
Na	mg/l	52.667	6.807	58.333	10.263	22.600	9.343	40.750	23.472	58.000	7.000	24.500	10.247	18.500	3.536

Table 5.5: Mean Values (±SD) of Physico-chemical Characteristics of Groundwater at Control and Experiment Sites

Parameters	Unit	2005-06				2006-07			
		Control Site	SD	Experimental sites	SD	Control Site	SD	Experimental sites	SD
Temperature	°C	28.583	2.323	28.661	0.134	28.500	1.942	28.583	2.630
pH		7.267	0.178	7.483	0.095	7.325	0.166	7.525	0.214
TS	mg/l	606.000	165.808	774.806	309.575	619.833	159.929	794.444	248.689
TSS	mg/l	56.583	15.163	73.528	30.262	57.250	13.877	72.778	22.893
TDS	mg/l	549.417	150.672	701.278	279.320	562.583	146.087	721.667	225.916
EC	mho/cm	0.858	0.235	1.096	0.436	0.879	0.228	1.128	0.353
DO	mg/l	3.083	0.644	3.583	0.118	3.083	0.497	3.692	0.581
Free-CO_2	mg/l	0.450	0.423	0.347	0.063	0.483	0.319	0.280	0.348
BOD	mg/l	10.583	6.244	15.944	7.682	11.583	5.869	17.861	10.414
COD	mg/l	26.667	15.848	39.917	19.398	29.250	15.149	45.917	26.573
Total alkalinity	mg/l	245.583	47.149	348.167	171.459	257.417	44.933	355.389	41.818
Ca	mg/l	70.833	31.467	77.500	9.610	74.167	31.176	84.433	41.800
Mg	mg/l	55.750	32.311	56.306	4.257	62.083	37.609	62.033	39.100
K	mg/l	5.142	0.571	5.231	0.226	5.225	0.465	5.306	0.479
Cl	mg/l	9.458	5.692	39.889	52.379	9.492	5.134	37.604	22.806
Na	mg/l	38.083	20.120	44.544	24.806	38.417	19.228	46.755	18.636

variation in the total suspended solids during the study is given Table 5.1. Monthly variation in total suspended solids is given in Figure 5.2, Figure 5.8 and Figure 5.14 for 40 ft, 80 ft and more than 120 ft depth respectively.

Electrical Conductivity (mho/cm)

Maximum electric conductivity in groundwater was recorded 2.08 ± 0.265 in summer 2005 and minimum 1.11 ± 0.17 in winter 2006-07 ranged between 0.90 and 2.40 mho/cm. The highest value of electric conductivity in groundwater was found 2.40 during June 2006 and lowest 0.90 in November 2005. The variation in the electric conductivity during the study is given Table 5.1. Monthly variation in electric conductivity is given in Figure 5.3, Figure 5.9 and Figure 5.15 for 40 ft, 80 ft and more than 120 ft depth respectively.

Dissolved Oxygen (mg/l)

Dissolved oxygen in groundwater was recorded maximum 4.0 ± 0.45 in monsoon 2006 and minimum 2.57 ± 0.21 in summer 2005 ranged between 2.4 to 4.8 mg/l. The highest value of dissolved oxygen in groundwater was found 4.8 mg/l during January 2006 and lowest 2.4 mg/l in April 2005. The variation in the dissolved oxygen content during the study is given Table 5.1. Monthly variation in dissolved oxygen is given in Figure 5.4, Figure 5.9 and Figure 5.15 for 40 ft, 80 ft and more than 120 ft depth respectively.

Free CO_2 (mg/l)

Free carbon dioxide in groundwater was recorded maximum 0.93 ± 0.15 in summer 2005 and minimum 0.10 ± 0.20 in winter 2006-07 ranged between 0.00 to 1.1 mg/l. The highest value of free carbon dioxide in groundwater was found 1.1 mg/l during April 2005 and lowest 0.0 mg/l in August 2005. The variation in the free carbon dioxide content during the study is given Table 5.1. Monthly variation in carbon dioxide is given in Figure 5.3, Figure 5.9 and Figure 5.15 for 40 ft, 80 ft and more than 120 ft depth respectively.

Biochemical Oxygen Demand (mg/l)

The biochemical oxygen demand in groundwater was recorded maximum 35.67 ± 8.15 in summer 2005 and minimum 11.00 ± 2.82 in summer 2007 ranged between 5.0 to 57.0 mg/l. The highest value of biochemical oxygen demand in groundwater was found 57.0 mg/l during July 2005 and lowest 5.0 mg/l in January 2006. The variation in biochemical oxygen demand during the study is given Table 5.1. Monthly variation in biochemical oxygen demand is given in Figure 5.4, Figure 5.10 and Figure 5.16 for 40 ft, 80 ft and more than 120 ft depth respectively.

Chemical Oxygen Demand (mg/l)

The chemical oxygen demand in groundwater was recorded maximum 93.00 ± 16.52 in summer 2005 and minimum 27.50 ± 7.78 in summer 2007 ranged between 12.0 to 130.0 mg/l. The highest value of chemical oxygen demand in groundwater was found 130.0 mg/l during July 2006 and lowest 12.0 mg/l in January 2006. The variation in chemical oxygen demand during the year 2005-06 is given Table 5.1.

Monthly variation in chemical oxygen demand is given in Figure 5.4, Figure 5.10 and Figure 5.16 for 40 ft, 80 ft and more than 120 ft depth respectively.

Total Alkalinity (mg/l)

Total alkalinity in groundwater was recorded maximum 596.67 ± 48.91 in monsoon 2006 and minimum 493.60 ± 31.1in winter 2005-06 ranged between 448.0 to 653.0 mg/l. The highest value of total alkalinity in groundwater was found 653.0 mg/l during July 2006 and lowest 448.0 mg/l in November 2005. The variation in total alkalinity during the study is given Table 5.1. Monthly variation in total alkalinity is given in Figure 5.6, Figure 5.12 and Figure 5.18 for 40 ft, 80 ft and more than 120 ft depth respectively.

Calcium (mg/l)

The concentration of calcium in groundwater was recorded maximum 143.3 ± 47.3 in monsoon 2006 and minimum 46.0 ± 16.7 in winter 2005-06 ranged between 20.0 to 180.0 mg/l. The highest value of calcium in groundwater was found 180.0 mg/l during July 2005 and lowest 20.0 mg/l in December 2005. The variation in calcium during the study is given Table 5.1. Monthly variation in calcium is given in Figure 5.5, Figure 5.11 and Figure 5.17 for 40 ft, 80 ft and more than 120 ft depth respectively.

Magnesium (mg/l)

The concentration of magnesium in groundwater was recorded maximum 97.7 ± 35.2 in monsoon 2006 and minimum 14.02 ± 22.0 in summer 2007 ranged between 15.0 to 141.0 mg/l. The highest value of magnesium in groundwater was found 141.0 mg/l during June 2006 and lowest 15.0 mg/l in January 2006. The variation in magnesium during the study is given Table 5.1. Monthly variation in magnesium is given in Figure 5.5, Figure 5.11 and Figure 5.17 for 40 ft, 80 ft and more than 120 ft depth respectively.

Potassium (mg/l)

The concentration of potassium in groundwater was recorded maximum 6.10 ± 0.10 in monsoon 2005 and minimum 4.90 ± 0.14 in summer 2007 ranged between 4.8 to 6.2 mg/l. The highest value of potassium in groundwater was found 6.2 mg/l during August 2006 and lowest 4.8 mg/l in April 2005. The variation in potassium during the study is given Table 5.1. Monthly variation in potassium is given in Figure 5.5, Figure 5.11and Figure 5.17 for 40 ft, 80 ft and more than 120 ft depth respectively.

Sodium (mg/l)

Sodium in groundwater was recorded maximum 97.67 ± 11.24 in monsoon 2006 and minimum 51.00 ± 13.45 in winter 2005-06 ranged between 36.0 to 110.0 mg/l. The highest value of Sodium in groundwater was found 110.0 mg/l during July 2006 and lowest 36.0 mg/l in December 2005. The variation in Sodium during the study is given Table 5.1. Monthly variation in Sodium is given in Figure 5.6, Figure 5.12 and Figure 5.18 for 40 ft, 80 ft and more than 120 ft depth respectively.

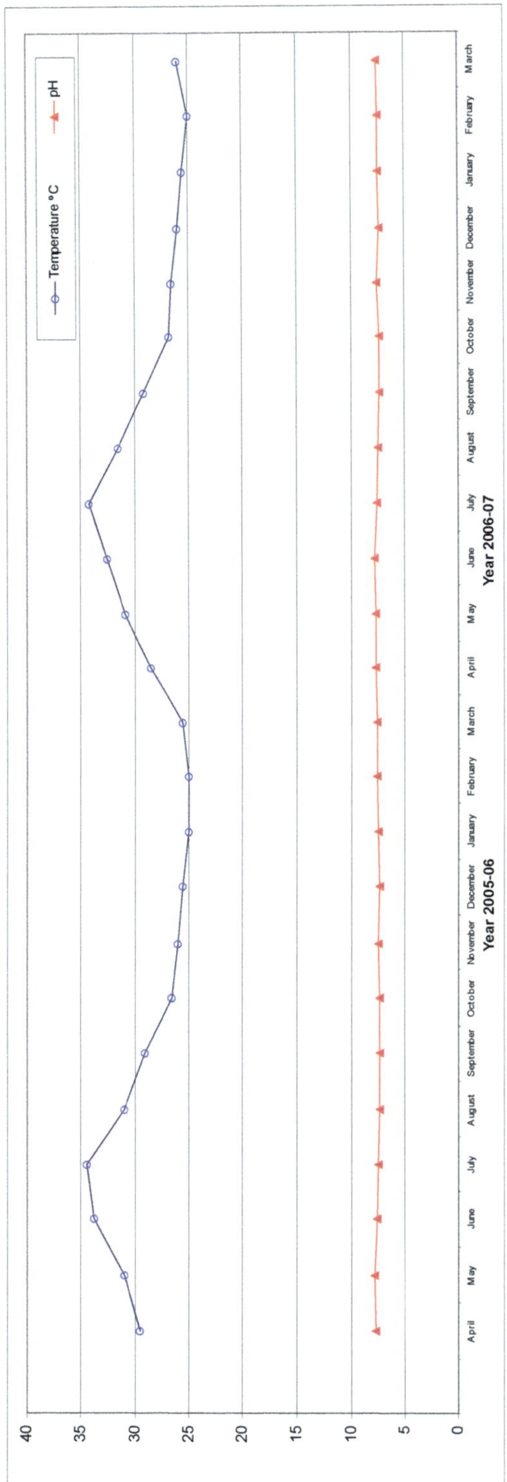

Figure 5.1: Monthly Variation of Temperature and pH in Groundwater at 40 ft Depth from Hand Pump in Village Binjhole

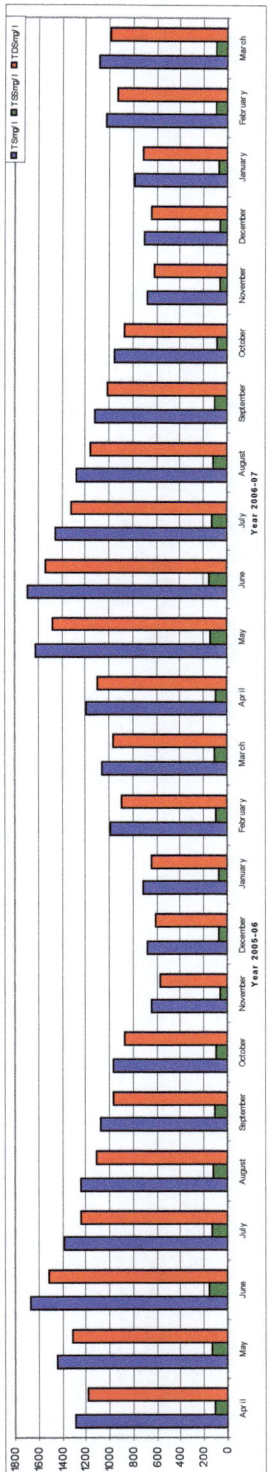

Figure 5.2: Monthly Variation of TS, TDS and TSS in Groundwater at 40 ft Depth from Hand Pump in Village Binjhole

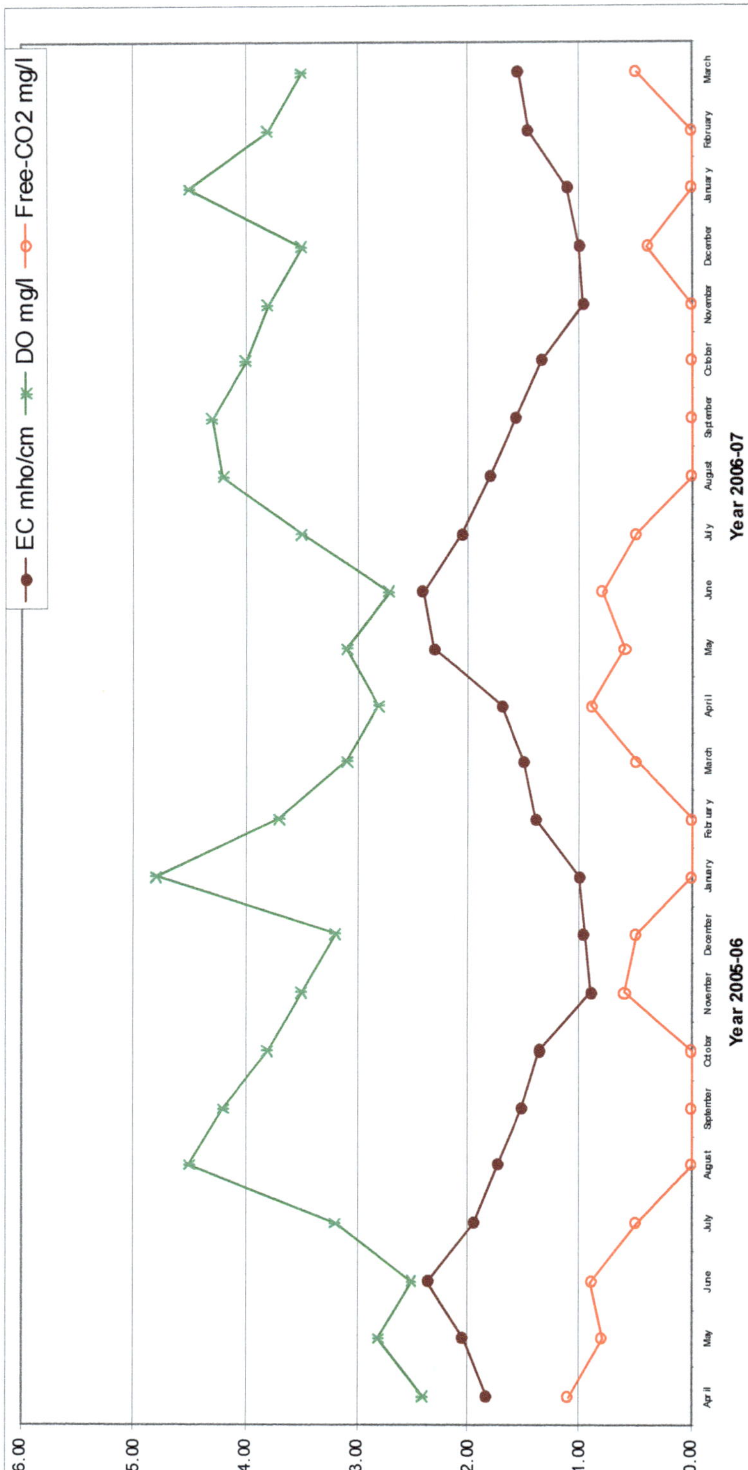

Figure 5.3: Monthly Variation of EC, DO and Free CO_2 in groundwater at 40 ft Depth from Hand Pump in Village Binjhole

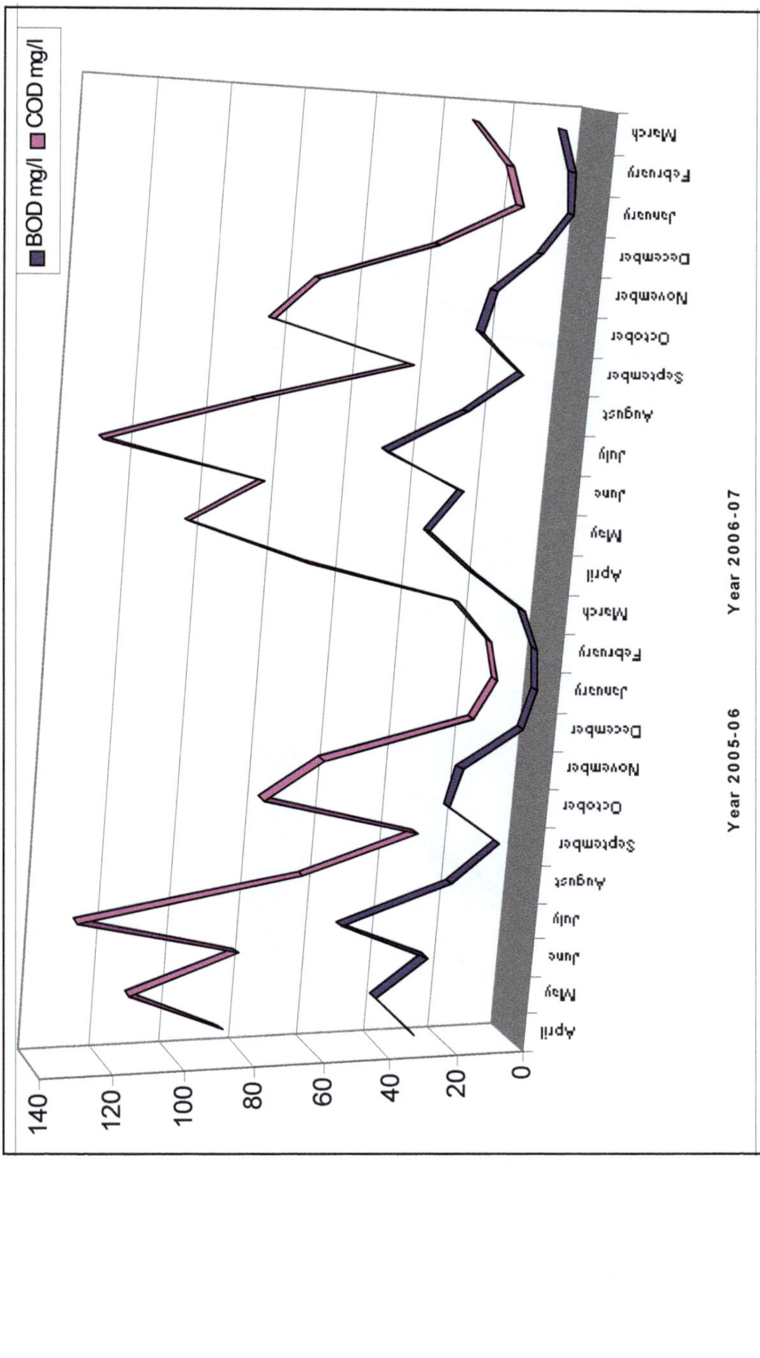

Figure 5.4: Monthly Variation of BOD and COD in Groundwater at 40 ft Depth from Hand Pump in Village Binjhole

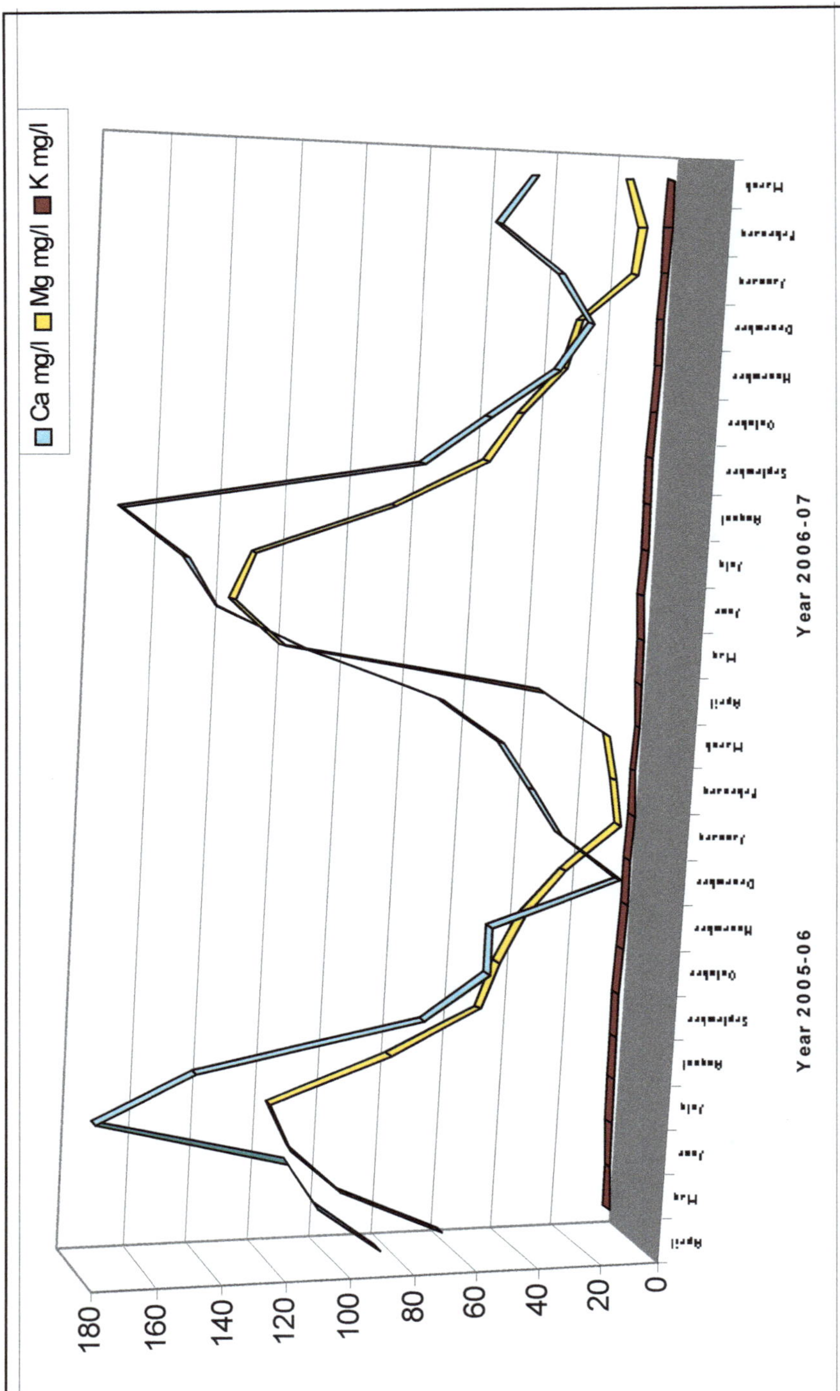

Figure 5.5: Monthly Variation of Calcium, Magnesium and Potassium in Groundwater at 40 ft Depth from Hand Pump in Village Binjhole

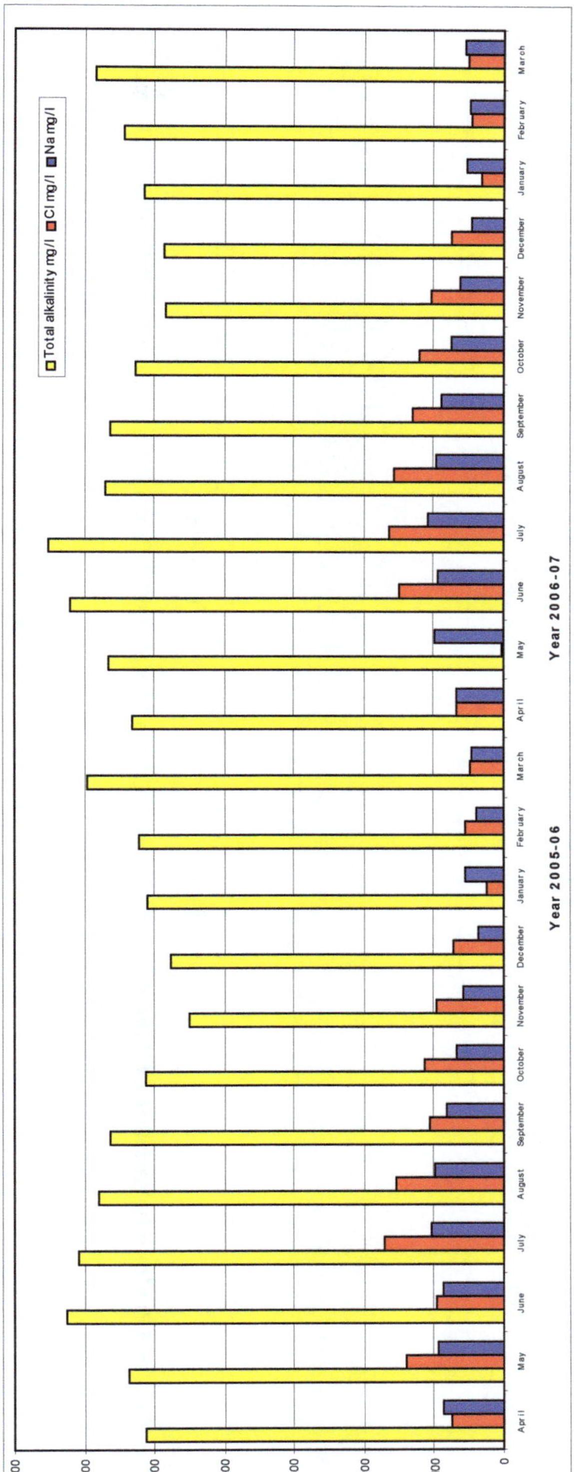

Figure 5.6: Monthly Variation of Chloride, Total Alkalinity and Sodium in Groundwater at 40 ft Depth from Hand Pump in Cillage Binjhole

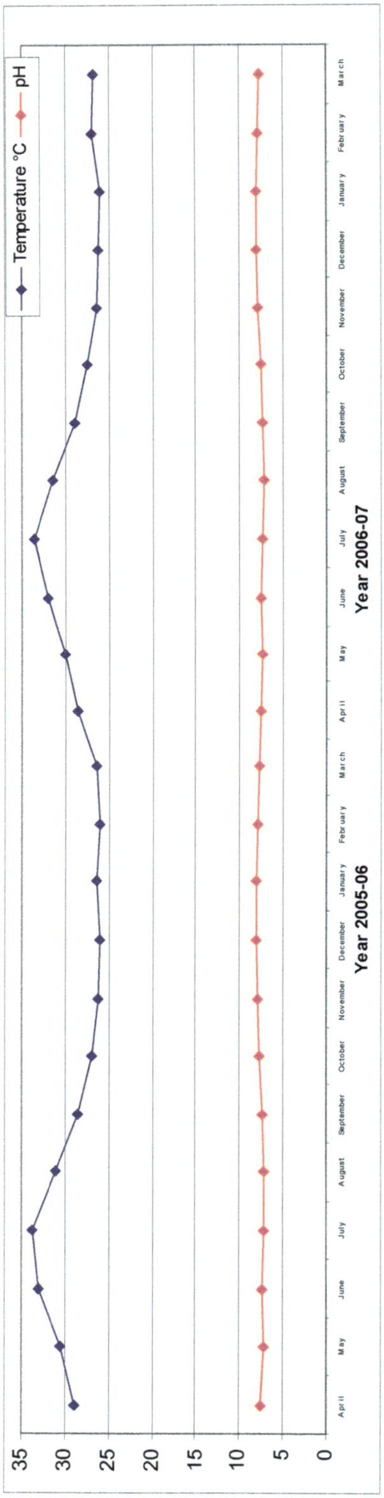

Figure 5.7: Monthly Variation of Temperature and pH in Groundwater at 80 ft Depth from Tubewell and Deep Hand Pump

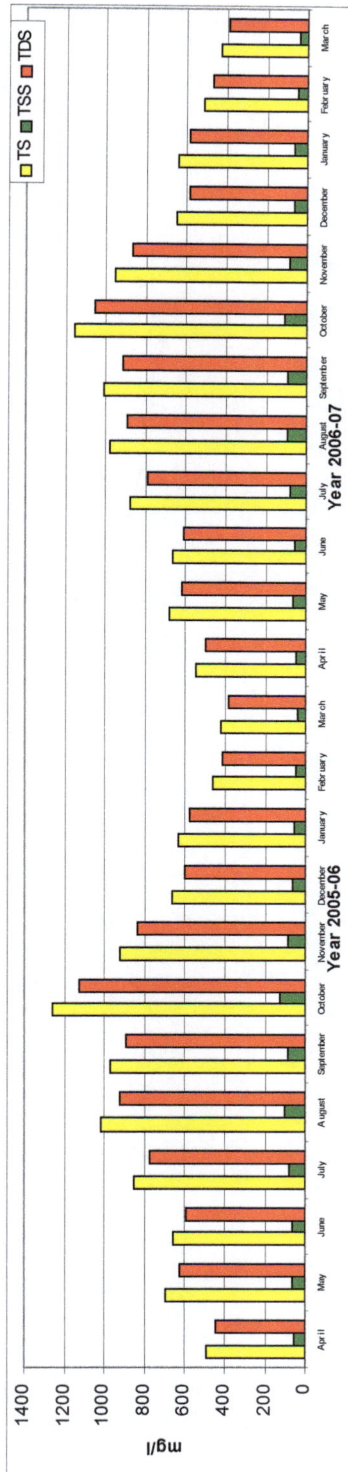

Figure 5.8: Monthly Variation of TS, TDS and TSS in Groundwater at 80 ft Depth from Tubewell and Deep Hand Pump

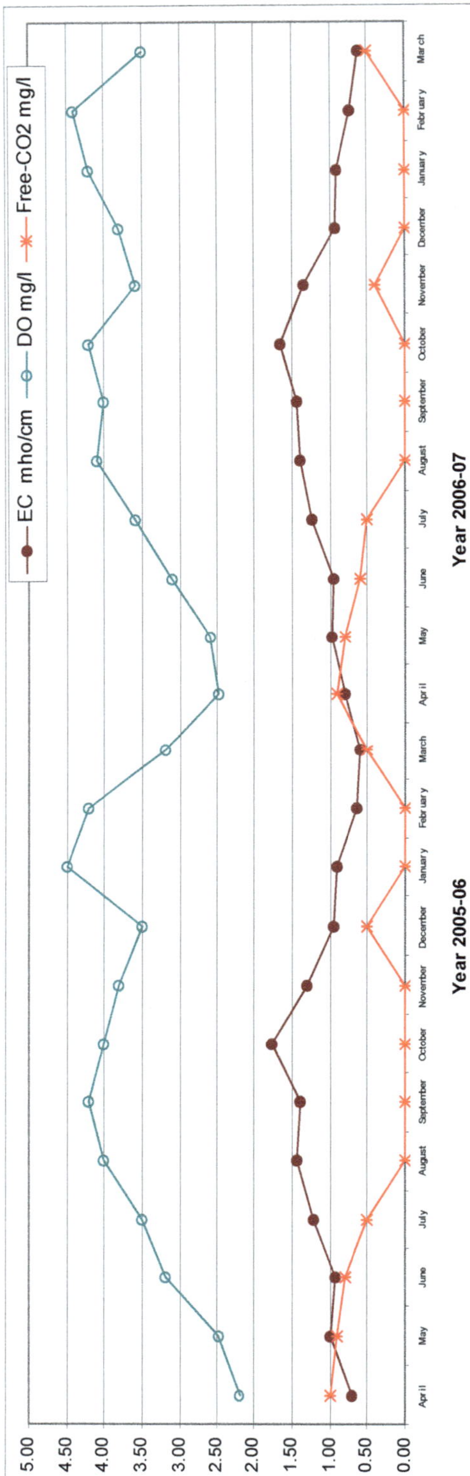

Figure 5.9: Monthly Variation of EC, DO and Free CO_2 in Groundwater at 80 ft Depth from Tubewell and Deep Hand Pump

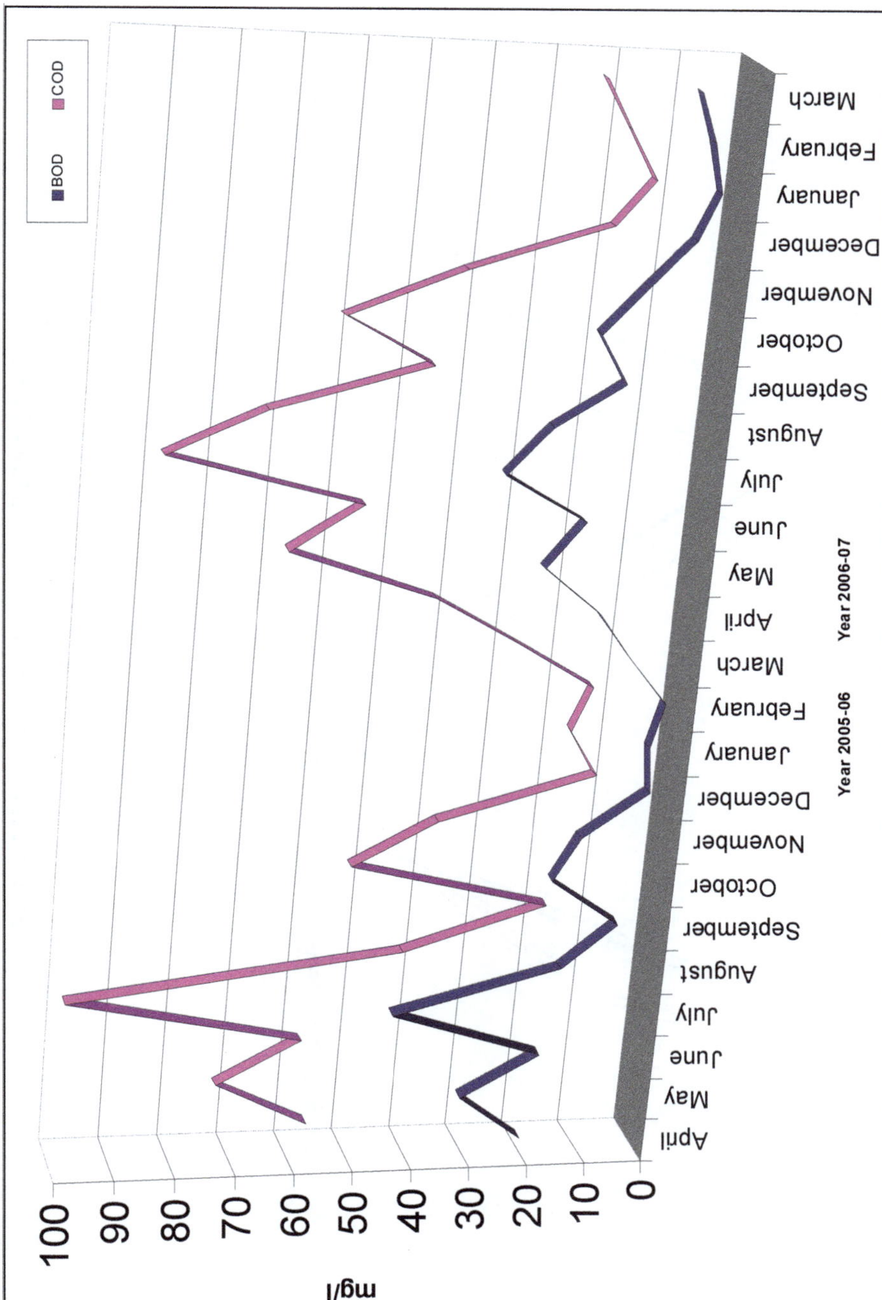

Figure 5.10: Monthly Variation of BOD and COD in Groundwater at 80 ft Depth from Tubewell and Deep Hand Pump

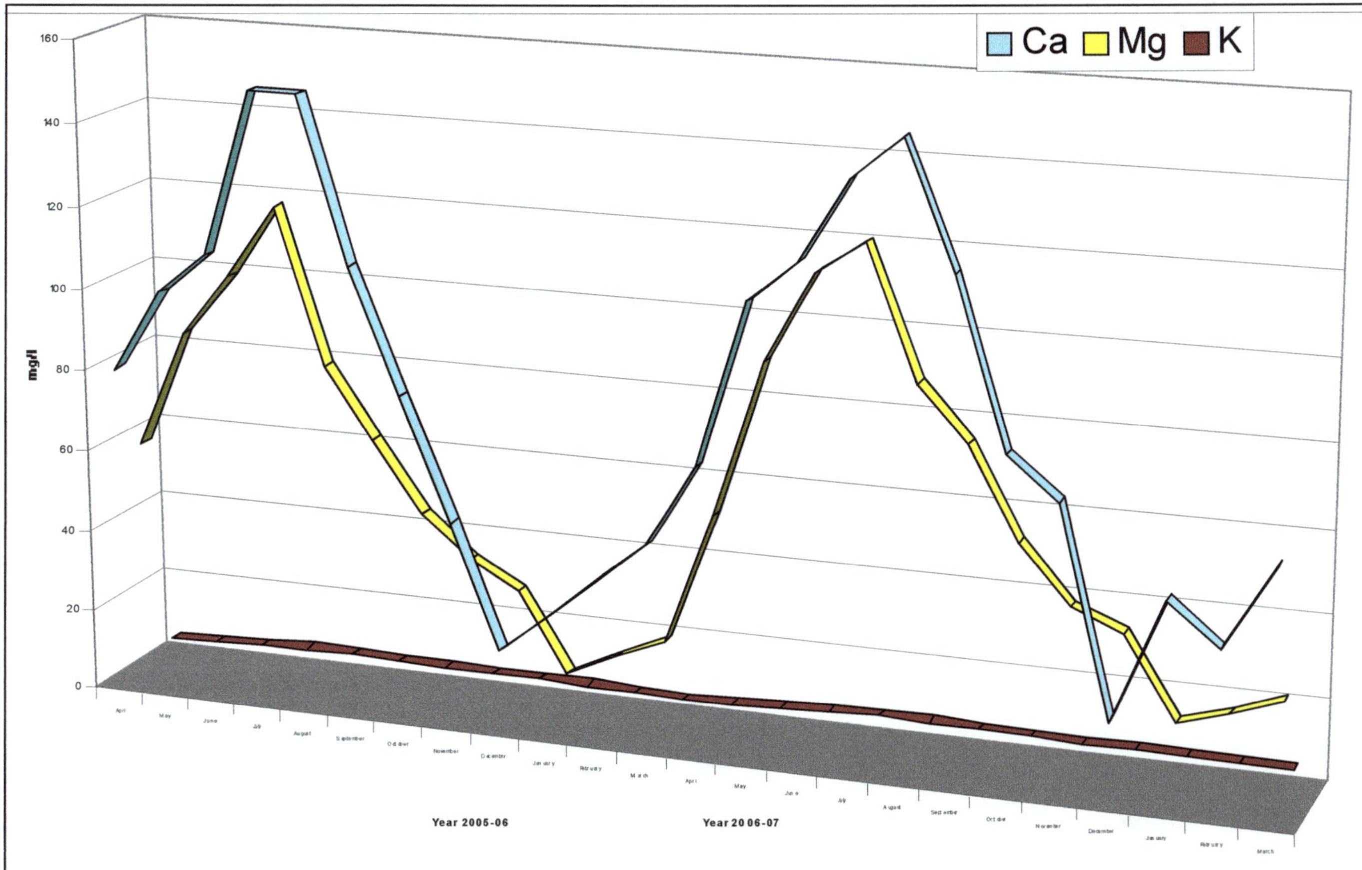

Figure 5.11: Monthly Variation of Calcium, Magnesium and Potassium in Groundwater at 80 ft Depth from Tubewell and Deep Hand Pump

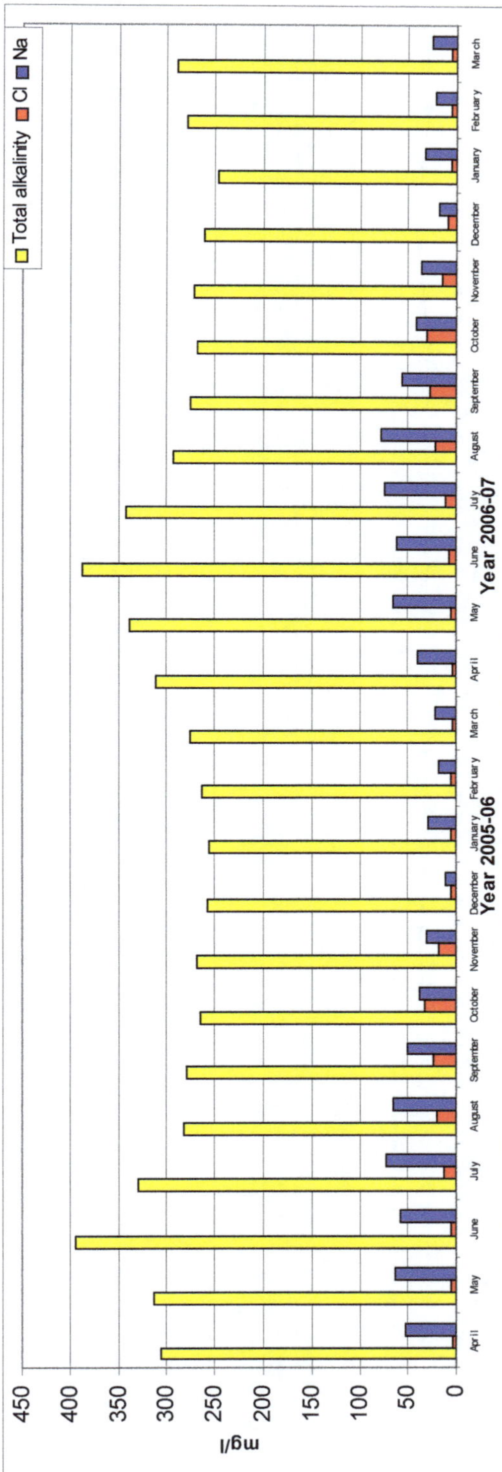

Figure 5.12: Monthly Variation of Chloride, Total Alkalinity and Sodium in Groundwater at 80 ft Depth from Tubewell and Deep Hand Pump

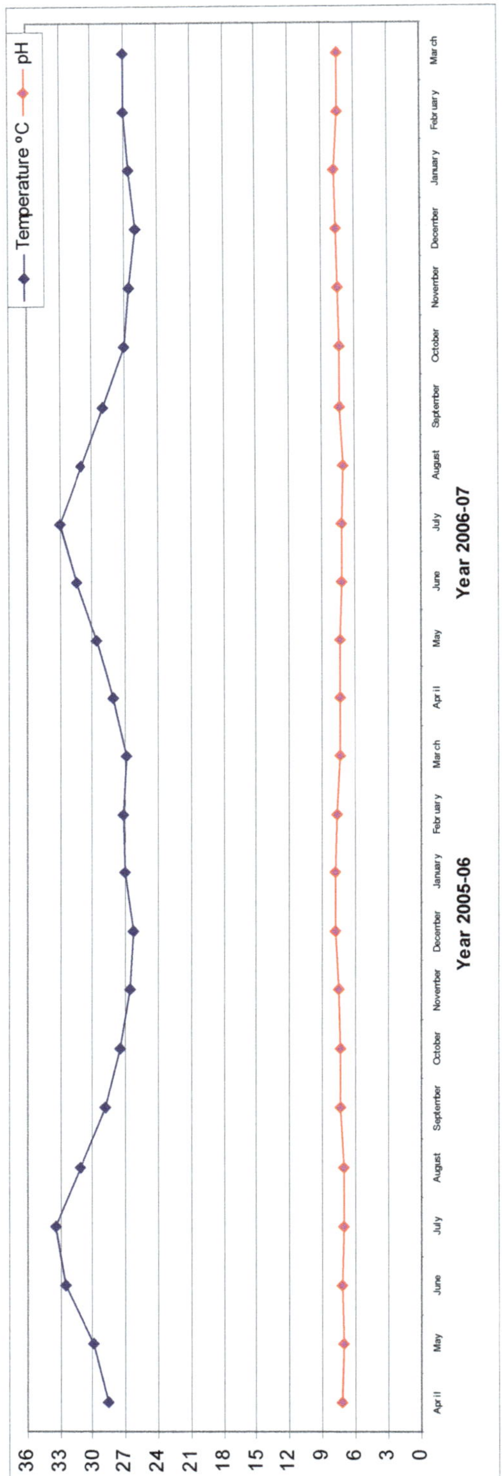

Figure 5.13: Monthly Variation of Temperature and pH in Deep Groundwater (120 ft depth) of Tube Well

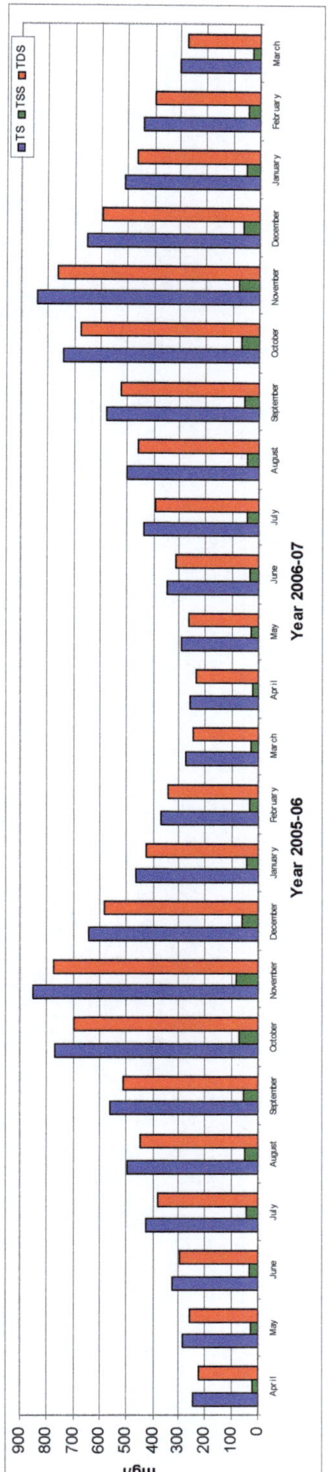

Figure 5.14: Monthly Variation of TS, TDS and TSS in Deep Groundwater (120 ft depth) of Tube Well

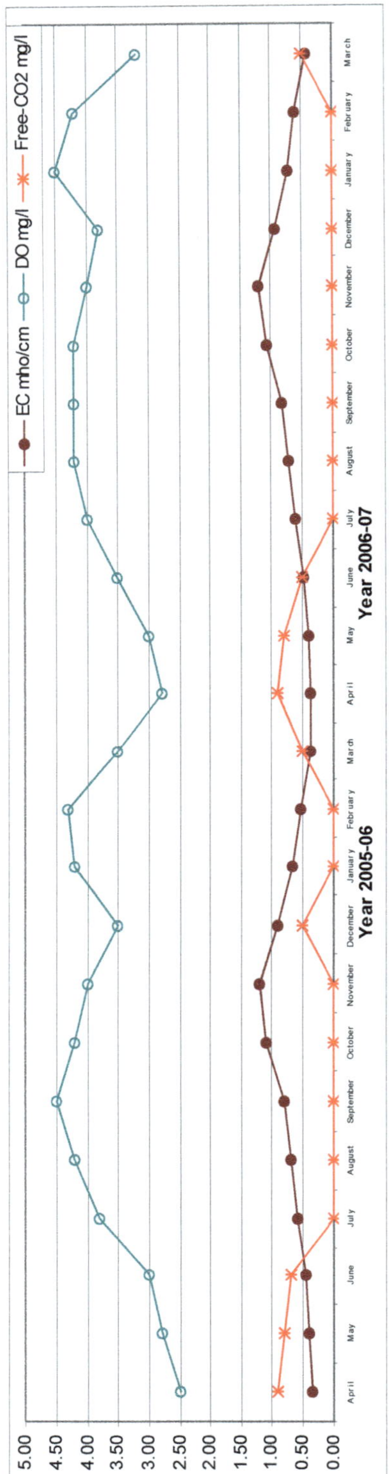

Figure 5.15: Monthly Variation of EC, DO and Free CO_2 in Deep Groundwater (120 ft depth) of Tube Well

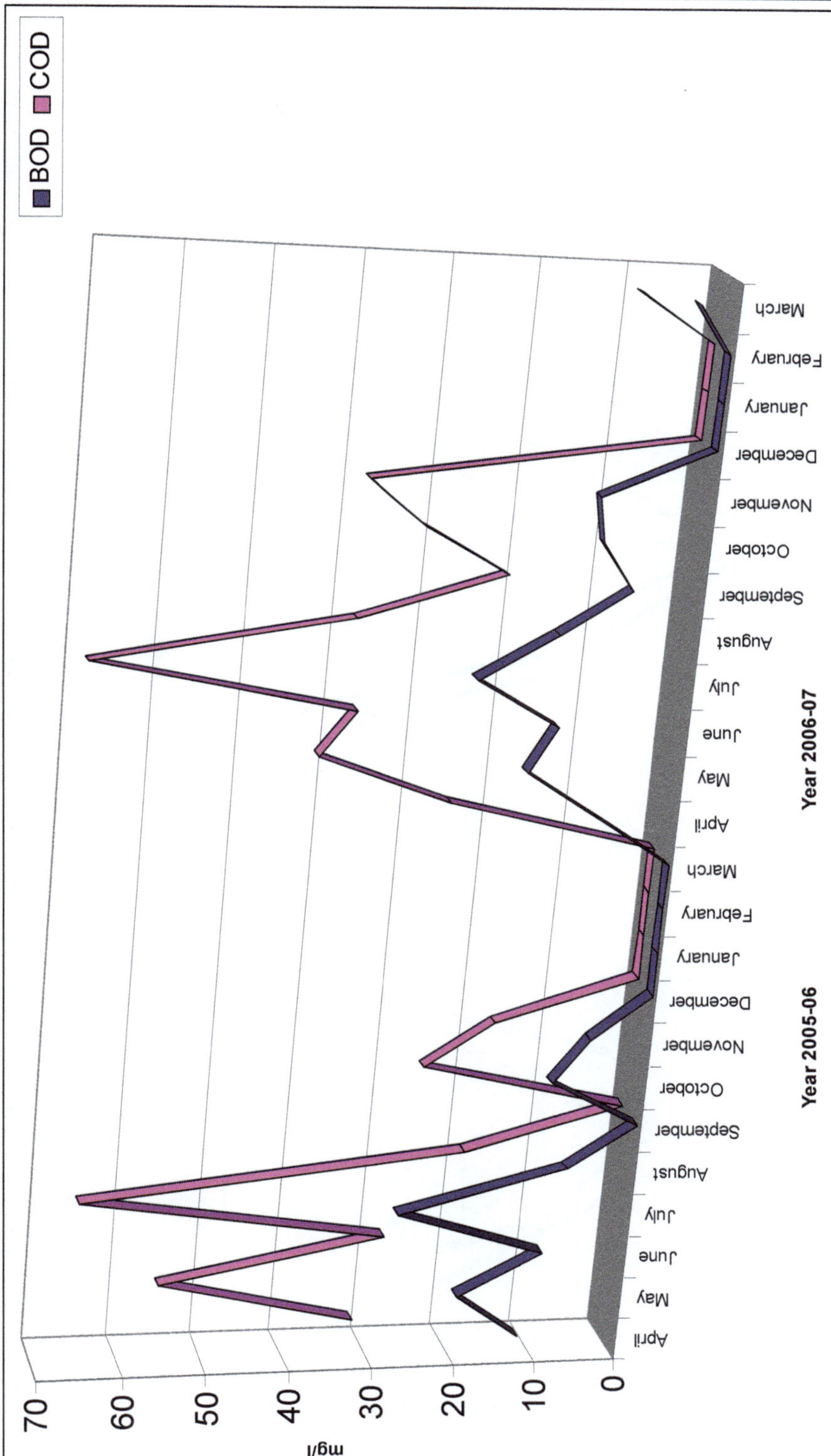

Figure 5.16: Monthly Variation of BOD and COD in Deep Groundwater (120 ft depth) of Tube Well

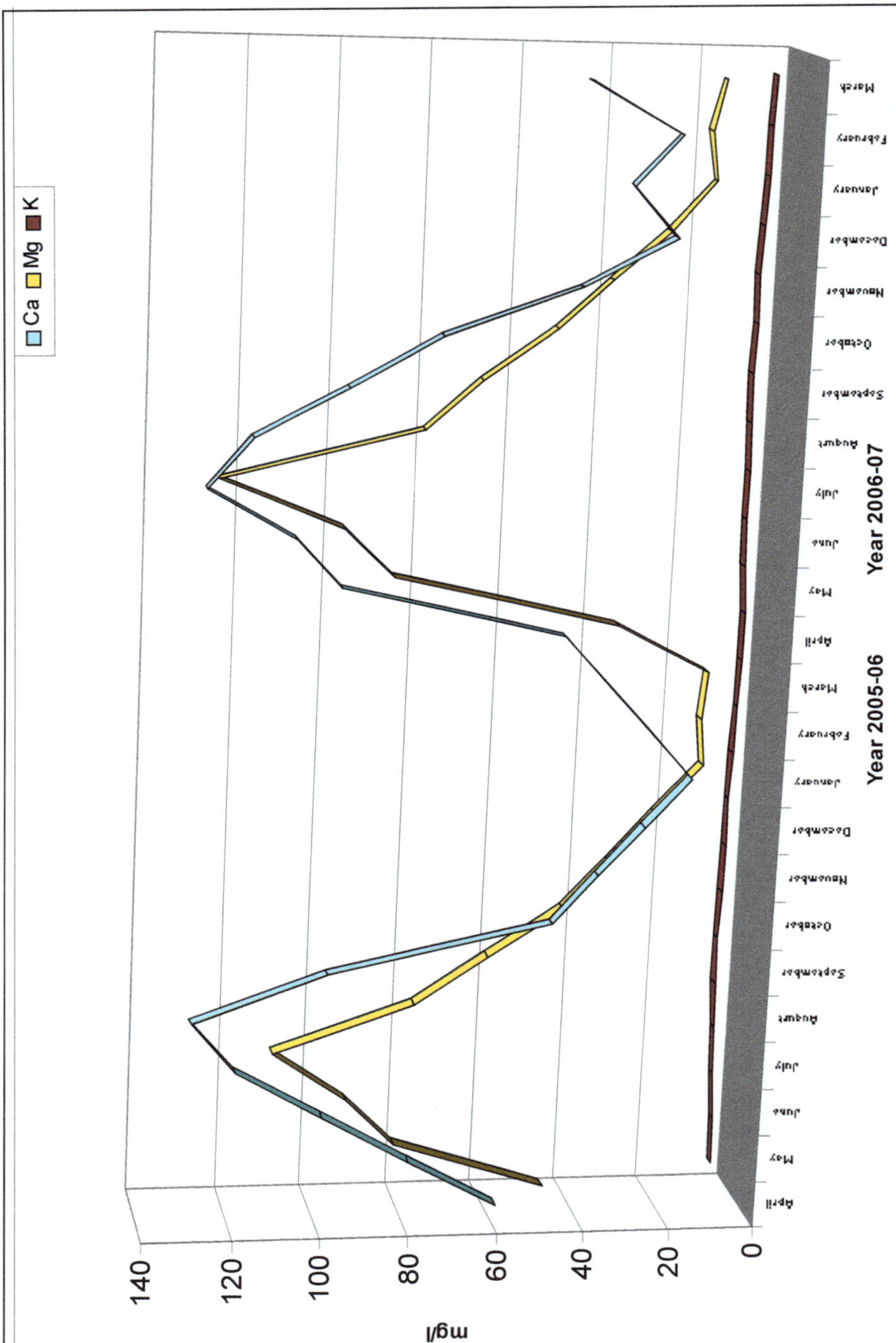

Figure 5.17: Monthly Variation of Calcium, Magnesium and Potassium in Deep Groundwater (120 ft depth) of Tube Well

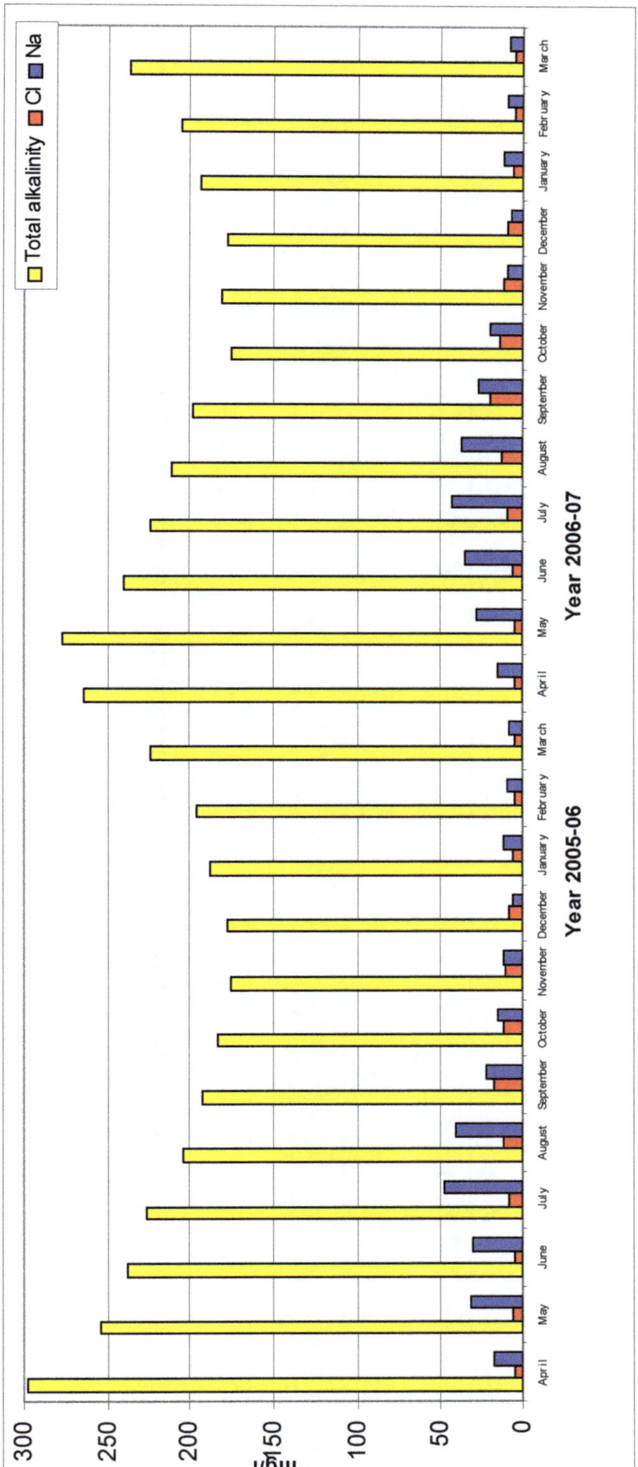

Figure 5.18: Monthly Variation of Chloride, Total Alkalinity and Sodium in Deep Groundwater (120 ft depth) of Tube Well

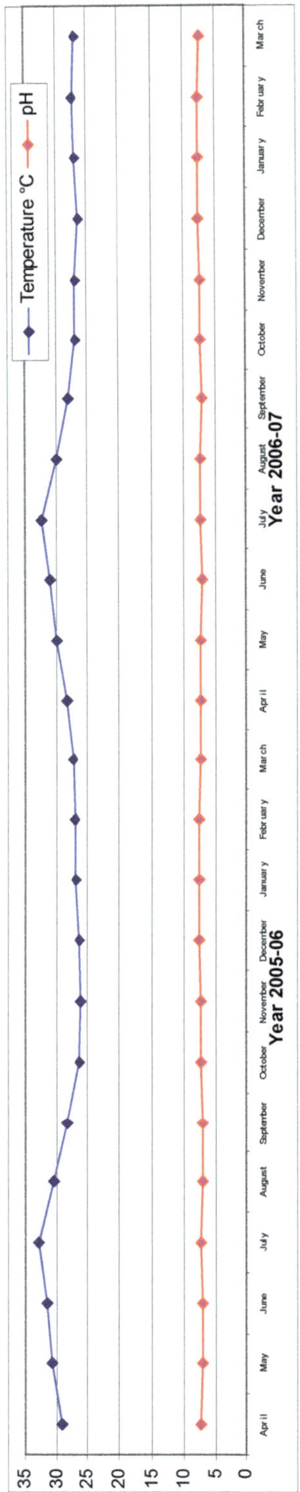

Figure 5.19: Temperature and pH Variation in Groundwater at Control Site

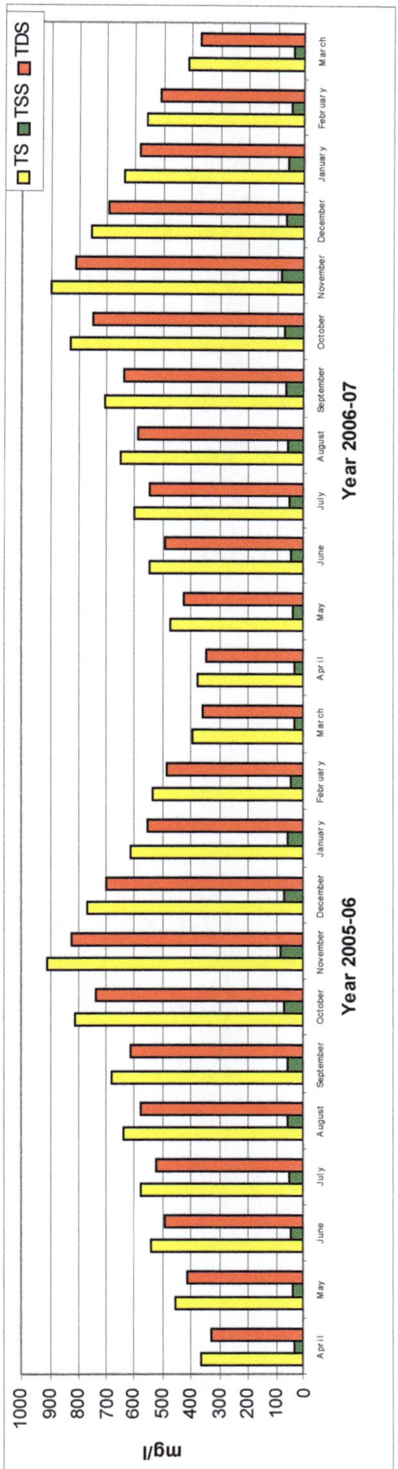

Figure 5.20.: TS, TDS and TSS Variation in Groundwater at Control Site

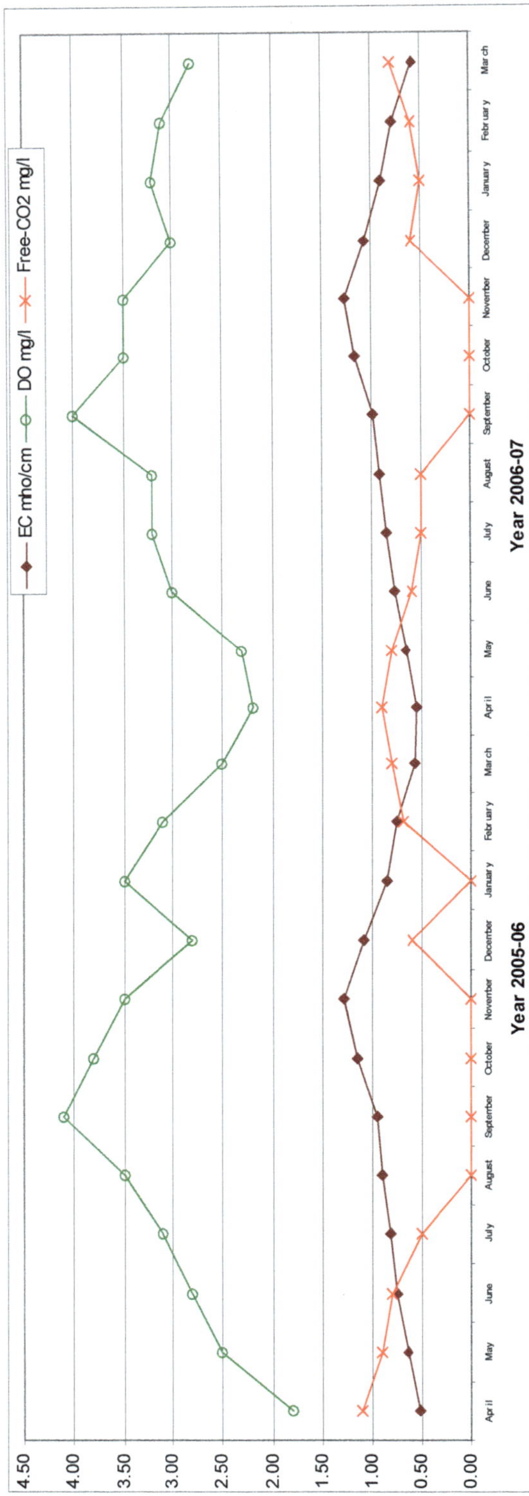

Figure 5.21: EC, DO and Free CO_2 Variation in Groundwater at Control Site

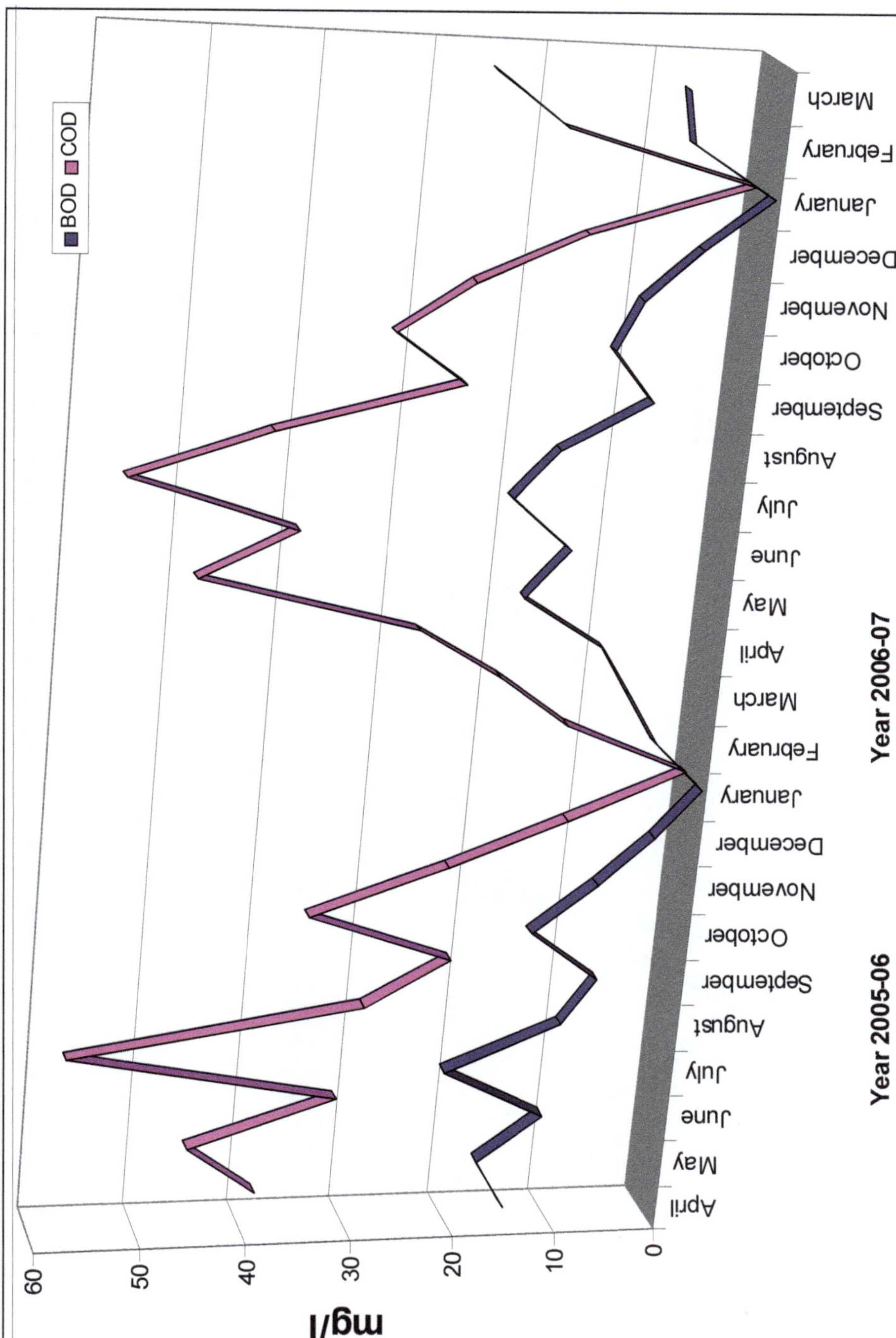

Figure 5.22: BOD and COD Variation in Groundwater at Control Site

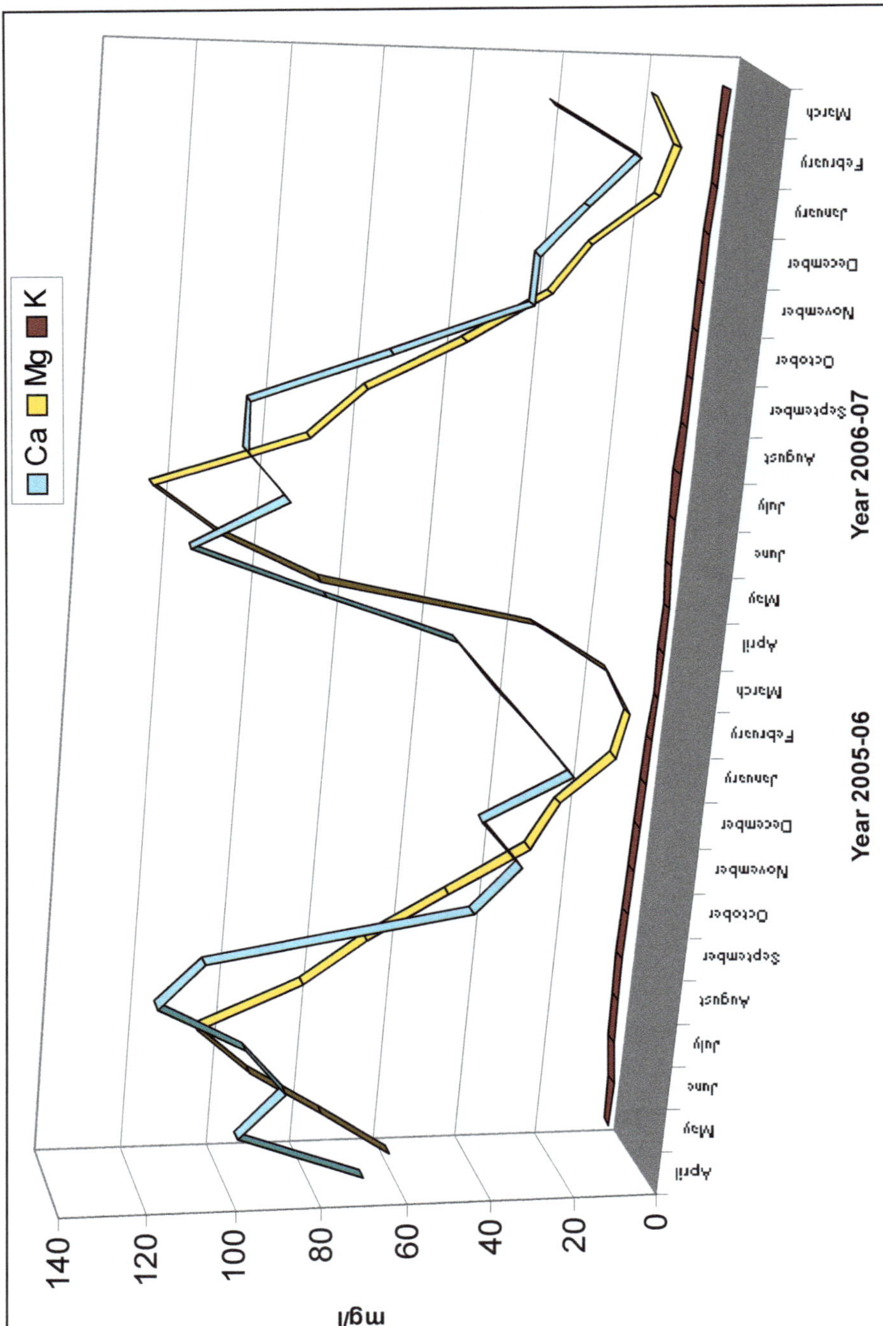

Figure 5.23: Calcium, Magnesium and Potassium Variation in Groundwater at Control Site

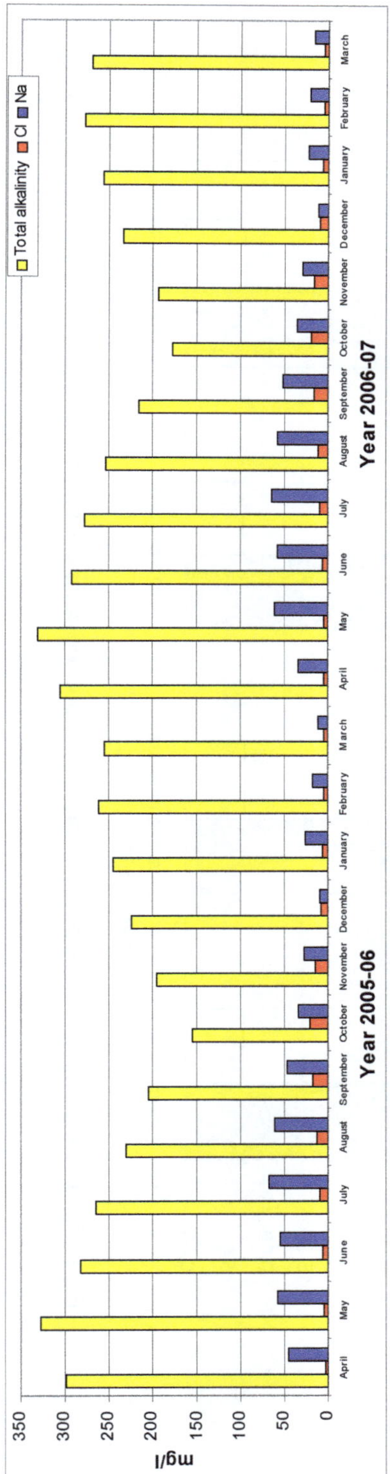

Figure 5.24: Chloride, Total Alkalinity and Sodium Variation in Groundwater at Control Site

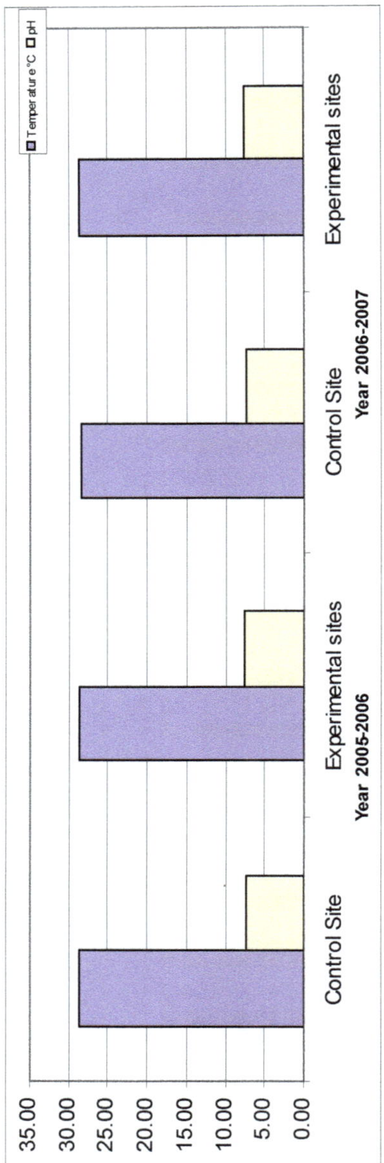

Figure 5.25: Mean Values of Temperature and pH in Groundwater at Control and Experimental Sites

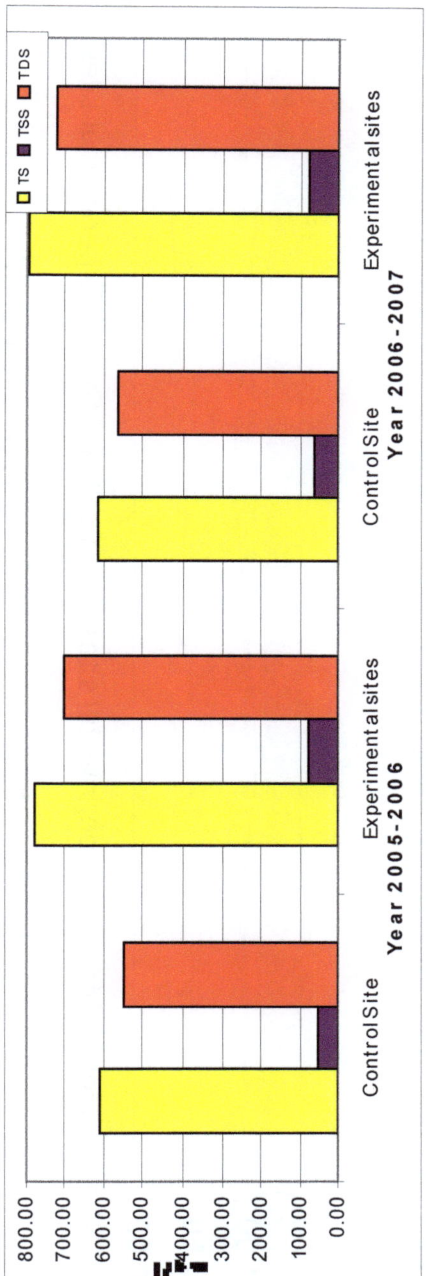

Figure 5.26.: Mean Values of TS, TDS and TSS in Groundwater at Control and Experimental Sites

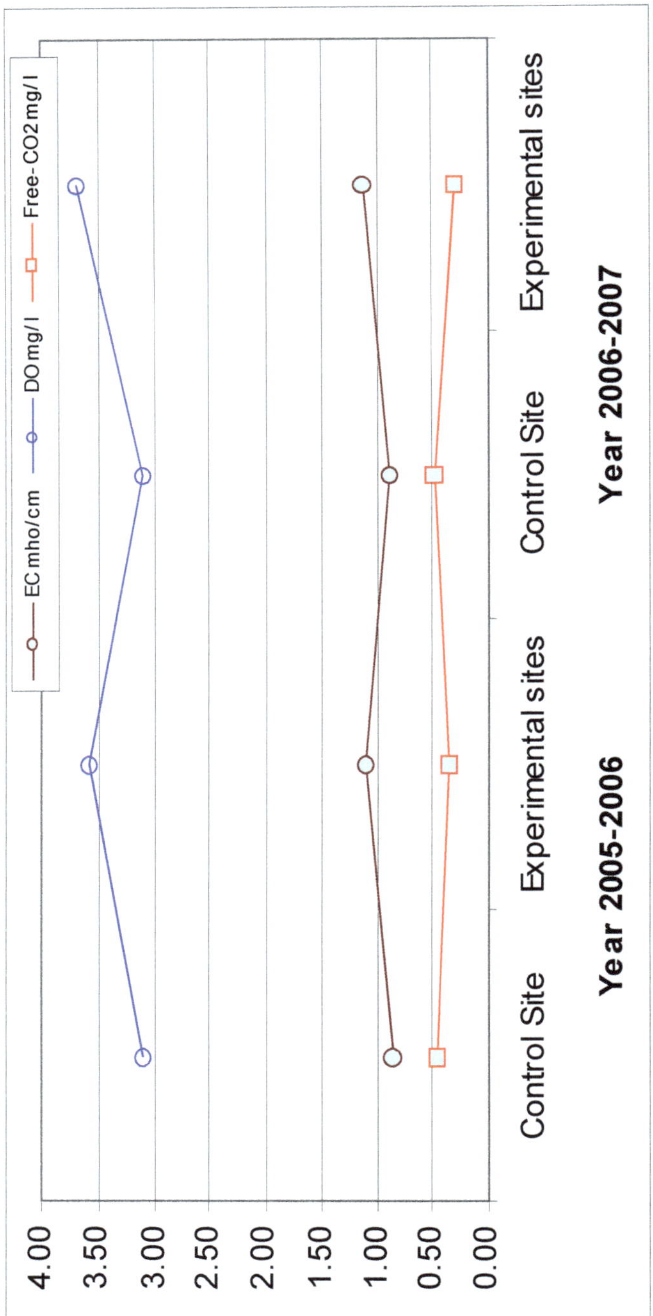

Figure 5.27: Mean Values of EC, DO and Free CO_2 in Groundwater at Control and Experimental Sites

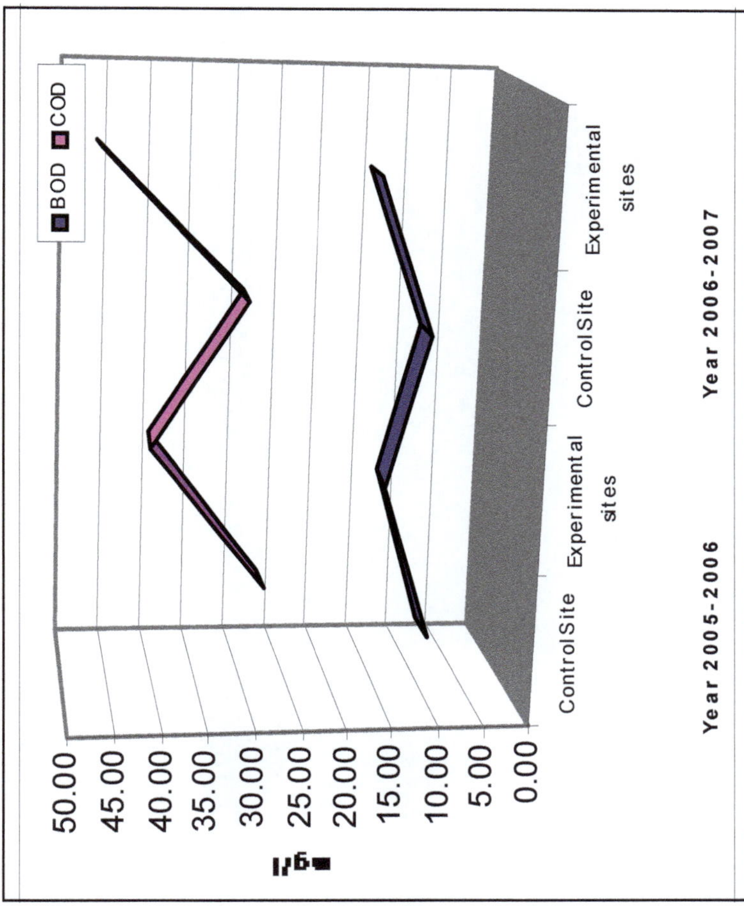

Figure 5.28: Mean Values of BOD and COD in Groundwater at Control and Experimental Sites

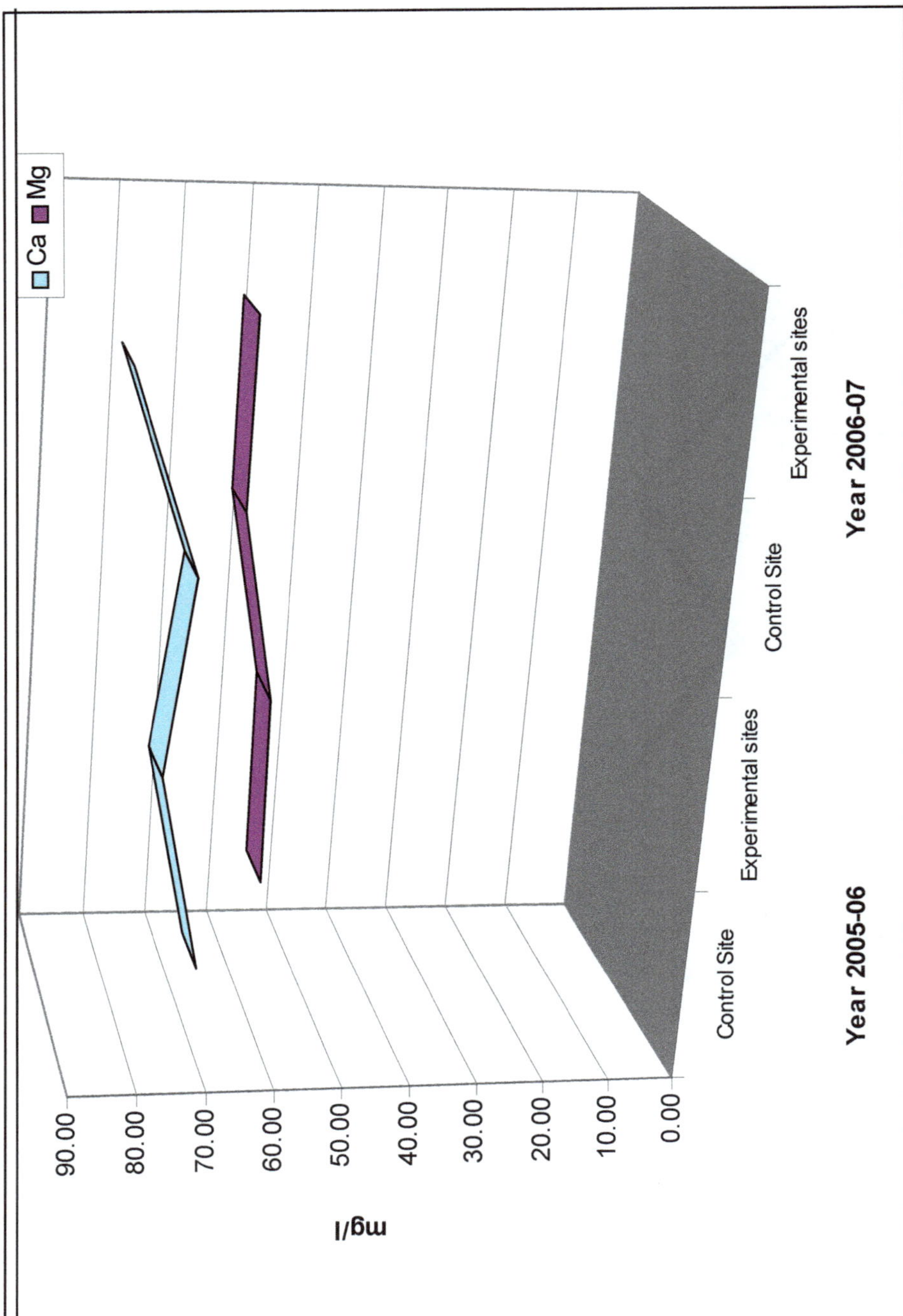

Figure 5.29: Mean Values of Calcium and Magnesium in Groundwater at Control and Experimental Sites

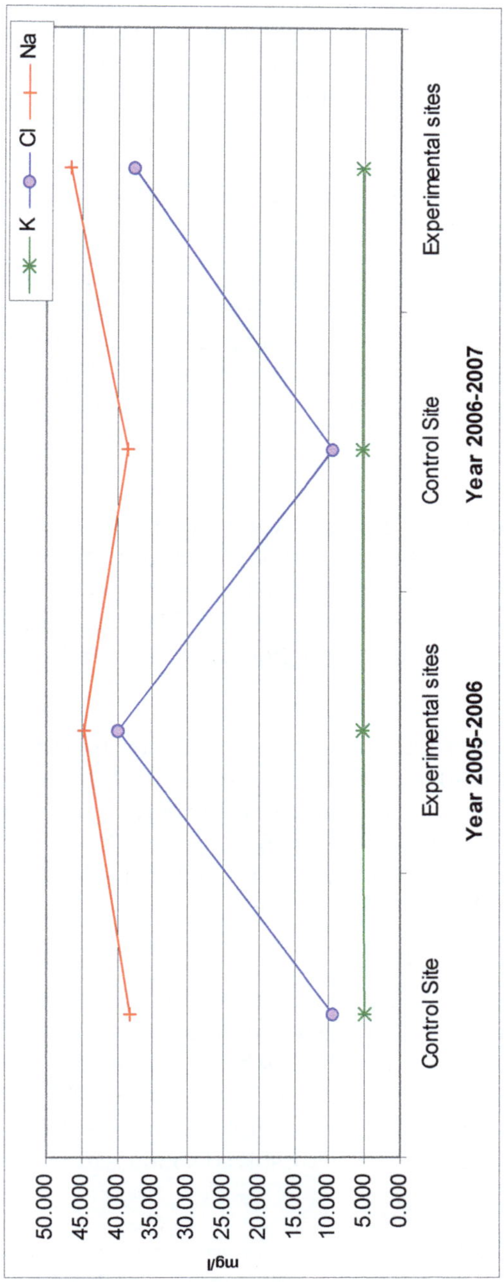

Figure 5.30: Mean Values of Cl, K and Na in Groundwater at Control and Experimental Sites

Chloride (mg/l)

Chloride in groundwater was recorded maximum 151.33 ± 17.95 in monsoon 2006 and minimum 47.50 ± 3.54 in summer 2007 ranged between 24.0 to 172.0 mg/l. The highest value of Chloride in groundwater was found 172.0 mg/l during July 2005 and lowest 24.0 mg/l in January 2006. The variation in Chloride during the study is given Table 5.1. Monthly variation in Chloride is given in Figure 5.6, Figure 5.12 and Figure 5.18 for 40 ft, 80 ft and more than 120 ft depth respectively.

Control Site

Physico-chemical Parameters of Groundwater

Temperature (°C)

Temperature of groundwater on control site was recorded maximum 30.60 ± 2.15 in monsoon 2005 and minimum 26.64 ± 0.47 in winter 2005-06 ranged between 26.0 to 32.8 °C. The highest value of temperature in groundwater was found 32.8 °C during July 2005 and lowest 26°C in November 2005. The temperature variation during the study is given in Table 5.4. Monthly variation in temperature is given in Figure 5.19.

pH

The maximum pH value of groundwater on control site was recorded 7.45 ± 0.07 in summer 2007 and minimum 7.13 ± 0.058 in summer 2005 ranged between 7.1 and 7.6 on pH scale. The highest value of pH in groundwater was found 7.6 during January 2007 and lowest 7.1 in September 2005. The variation in the pH during the study is given Table 5.4. Monthly variation in pH is given in Figure 5.19.

Total solids (mg/l)

The maximum total solids in groundwater on control site were recorded 780.50 ± 110.30 in winter 2006-07 and minimum 446.75 ± 76.16 in summer 2006 ranged between 363.0 to 907.0 mg/l. The highest value of total solids in groundwater was found 907.0 mg/l during November 2005 and lowest 363.0 mg/l in April 2005. The variation in the total solids during the study is given Table 5.4. Monthly variation in total solids is given in Figure 5.20.

Total Dissolved Solids (mg/l)

Total dissolved solids in groundwater on control site were recorded maximum 709.50 ± 100.14 in winter 2006-07 and minimum 404.50 ± 69.02 in summer 2006 ranged between 328.0 to 822.0 mg/l. The highest value of total dissolved solids in groundwater was found 822.0 mg/l during November 2005 and lowest 328.0 mg/l in April 2005. The variation in the total dissolved solids during the study is given Table 5.4. Monthly variation in total dissolved solids is given in Figure 5.20.

Total Suspended Solids (mg/l)

Total suspended solids in groundwater on control site were recorded maximum 71.00 ± 10.17 in winter 2006-07 and minimum 42.25 ± 7.18 in summer 2006 ranged between 35.0 to 85.0 mg/l. The highest value of total suspended solids in groundwater was found 85.0 mg/l during November 2005 and lowest 35.0 mg/l in April 2006. The variation in the total suspended solids during the study is given Table 5.4. Monthly variation in total suspended solids is given in Figure 5.20.

Electrical Conductivity (mho/cm)

Maximum electric conductivity in groundwater on control site was recorded 1.11 ± 0.16 in winter 2006-07 and minimum 0.632 ± 0.11 in summer 2006 ranged between 0.513 and 1.284 mho/cm. The highest value of electric conductivity in groundwater was found 1.284 mho/cm during November 2005, and lowest 0.513 mho/cm in April 2005. The variation in the electric conductivity during the study is given Table 5.4. Monthly variation in electric conductivity is given in Figure 5.21.

Dissolved Oxygen (mg/l)

Dissolved oxygen in groundwater on control site was recorded maximum 3.57 ± 0.50 in monsoon 2005 and minimum 2.37 ± 0.51 in summer 2005 ranged between 1.8 to 4.1 mg/l. The highest value of dissolved oxygen in groundwater was found 4.1 mg/l during September 2005 and lowest 1.8 mg/l in April 2005. The variation in the dissolved oxygen content during the study is given Table 5.4. Monthly variation in dissolved oxygen is given in Figure 5.21.

Free CO_2 (mg/l)

Free carbon dioxide in groundwater on control site was recorded maximum 0.93 ± 0.15 in summer 2005 and minimum 0.17 ± 0.29 in monsoon 2005 ranged between 0.00 to 1.1 mg/l. The highest value of free carbon dioxide in groundwater was found 1.1 mg/l during April 2005 and lowest 0.00 mg/l in October 2005. The variation in the free carbon dioxide content during the study is given Table 5.4. Monthly variation in carbon dioxide is given in Figure 5.21.

Biochemical Oxygen Demand (mg/l)

The biochemical oxygen demand in groundwater on control site was recorded maximum 15.67 ± 6.11 in monsoon 2006 and minimum 6.60 ± 5.69 in winter 2005-06 ranged between 0.00 to 22.0 mg/l. The highest value of biochemical oxygen demand in groundwater was found 22.0 mg/l during July 2005 and lowest 0.00 mg/l in January 2006. The variation in biochemical oxygen demand during the study is given Table 5.4. Monthly variation in biochemical oxygen demand is given in Figure 5.22.

Chemical Oxygen Demand (mg/l)

The chemical oxygen demand in groundwater on control site was recorded maximum 40.67 ± 15.04 in monsoon 2006 and minimum 16.00 ± 13.17 in winter 2005-06 ranged between 0.00 to 57.00 mg/l. The highest value of chemical oxygen demand in groundwater was found 57.0 mg/l during July 2005 and lowest 0.00 mg/l in January 2006. The variation in chemical oxygen demand during the study is given Table 5.4. Monthly variation in chemical oxygen demand is given in Figure 5.22.

Total Alkalinity (mg/l)

Total alkalinity in groundwater on control site was recorded maximum 303.00 ± 22.91 in summer 2005 and minimum 216.40 ± 41.62 in winter 2005-06 ranged between 156.0 to 331.0 mg/l. The highest value of total alkalinity in groundwater was found 331.0 mg/l during May 2006 and lowest 156.0 mg/l in October 2005. The variation in total alkalinity during the study is given Table 5.4. Monthly variation in total alkalinity is given in Figure 5.24.

Calcium (mg/l)

The concentration of calcium in groundwater on control site was recorded maximum 110.0 ± 10.0 in monsoon 2005 and minimum 40.0 ± 14.1 in summer 2005 ranged between 30.0 to 120.0 mg/l. The highest value of calcium in groundwater was found 120.0 mg/l during August 2005 and lowest 30.0 mg/l in February 2007. The variation in calcium during the study is given Table 5.4. Monthly variation in calcium is given in Figure 5.23.

Magnesium (mg/l)

The concentration of magnesium in groundwater on control site was recorded maximum 101.0 ± 23.3 in monsoon 2006 and minimum 21.0 ± 4.2 in summer 2007 ranged between 15.0 to 127.0 mg/l. The highest value of magnesium in groundwater was found 127.0 mg/l during July 2006 and lowest 15.0 mg/l in February 2006. The variation in magnesium during the study is given Table 5.4. Monthly variation in magnesium is given in Figure 5.23.

Potassium (mg/l)

The concentration of potassium in groundwater on control site was recorded maximum 5.80 ± 0.10 in monsoon 2005 and minimum 4.65 ± 0.21 in summer 2007 ranged between 4.4 to 5.9 mg/l. The highest value of potassium in groundwater was found 5.9 during August 2005 and lowest 4.4 in April 2005. The variation in potassium during the study is given Table 5.4. Monthly variation in potassium is given in Figure 5.23.

Sodium (mg/l)

Sodium in groundwater on control site was recorded maximum 58.33 ± 10.26 in monsoon 2005 and minimum 18.50 ± 3.54 in summer 2007 ranged between 9.0 to 67.0 mg/l. The highest value of Sodium in groundwater was found 67.0 mg/l during July 2005 and lowest 9.0 mg/l in December 2005. The variation in Sodium during the study is given Table 5.4. Monthly variation in Sodium is given in Figure 5.24.

Chloride (mg/l)

Chloride in groundwater on control site was recorded maximum 13.00 ± 3.78 in monsoon 2005 and minimum 4.90 ± 1.11 in summer 2006 ranged between 3.9 to 21.0 mg/l. The highest value of Chloride in groundwater was found 21.0 mg/l during October 2005 and lowest 3.9 mg/l in April 2005. The variation in Chloride during the study is given Table 5.4. Monthly variation in Chloride is given in Figure 5.24.

Correlation Coefficient

Correlations between the heavy metals of groundwater of 40 ft depth are given in Table 7.19b for each metal. Correlation between lead and manganese was found 0.9365, 0.8564 for nickel and lead, 0.8227 for nickel and zinc. Correlation value for iron and zinc was found 0.9363. Correlations between the heavy metals of groundwater of 80 ft depth are given in Table 7.19c for each metal. For manganese correlation was 0.9446, 0.9195 and 0.9207 for nickel, lead and zinc respectively. For copper correlation was found 0.9623, 0.9548, 0.8559 and 0.9000 for manganese, nickel, lead and zinc respectively. Correlations between the heavy metals of deep groundwater of more than 120 ft depth are given in Table 7.19d for each metal. Correlation for copper was found maximum 0.8289, 0.9427, 0.9122, 0.5557 and 0.9509 with iron, manganese, nickel, lead and zinc respectively. Correlation for iron was found 0.9104, 0.7884, 0.8446 and 0.8660 with manganese, nickel, lead and zinc respectively. Correlation for nickel was found 0.5299 and 0.8346 with lead and zinc respectively.

Correlation between heavy metals of groundwater and heavy metals of agricultural soil are given in Table 7.20a for each metal. For cadmium correlation was found maximum 0.7611 and 0.6466 with cadmium and lead respectively. Correlation for copper was found maximum 0.9733 and 0.9121 with iron and nickel respectively. Correlation for iron was found maximum 0.8074 and 0.8628 with manganese and lead respectively. Correlation for manganese was found maximum 0.9279 and 0.9548 with nickel and zinc respectively. Correlation for nickel was found maximum 0.9458 and 0.9475 with manganese and nickel respectively. Correlation for lead was found maximum 0.9482 and 0.9664 with copper and manganese respectively. Correlation for zinc was found maximum 0.8065 and 0.5301 with iron and nickel respectively.

5.4 Discussion

With the increase in industrialization our health problems are increasing day by day. Today the pollution especially of water resources is increasing steadily due to rapid population growth, industrial proliferation, urbanization increasing living standards and wide sphere of human activities. The greatest danger to the quality of environment is posed by human activities such as poorly organized and unregulated disposal of industrial, domestic waste and crude oil spillage are the major causes of the environmental deterioration. The industrial effluent over the years has caused serious damages of the surrounding soil, groundwater and agriculture crops (Chao *et al.*, 2005). Most of the dye houses of the Panipat region discharge their effluent directly into a common effluent drain, which carry the total of composite effluents towards the village pond at Binjhole adjoining to industrial area, ultimately has altered the water quality and biotic and nutrient dynamic of the pond ecosystem in a tremendous way and further may affect the groundwater quality. Due to post irrigation practices, the soil turns to barren or unfertile and affected agriculture productivity.

Generally, textile industry consumes substantial volume of water and chemicals in wet processing of textile materials. Wet processing of cloth or fibers dying requires huge amount of different chemical dyes and wash away them with the wastewater discharged through industrial outlet. Thus, the common effluent drain consist all types of dye residuals in a composite form. This composite effluent is very toxic in nature due to the presence of different synthetic dyes, which contain various toxic heavy metals. Further future of these heavy metals is not so managed because the common effluent drain ends up in to a pond near village Binjhole, Industrial area and some local surroundings of dye houses. The effluents discharged from dye houses are quite important for temporal monitoring. Because the effluents store pond is highly contaminated and is a depot of heavy metals due to the presence of chemical dye portions continuously since a long period of time. Heavy metals remain in water ecosystem, but these may travel through food web at various trophic levels and environmental components as well as. Metals moves pond sediments and reach in shallow water table by leaching processes and the metal contamination may disperse in a wide area of the region. Contamination of soils of close areas occurs when the effluents or polluted water comes under the contact of soil. The regular contact of polluted water with soil makes the sodicity and alters the physico-chemical characteristics, texture and profile of soil stratum.

Thus, the quality assessment of groundwater required to know the pollution load of heavy metals on hydrosphere and lithosphere and to determined the drinking water quality of Panipat industrial zone in term of suitability and possible impacts on human health. In the present study, water samples were collected and analyzed for determination of physico-chemical characteristics and distribution of heavy metals from April 2005 to March 2007 regularly. The results are depicted in the form of tables and graphs.

5.5 Physico-Chemical Parameters of Groundwater

In the present investigation of water resources, physico-chemical characteristics of water directly reflect its quality for irrigation, bathing, drinking and industrial uses. Physico-chemical parameter like total dissolved solids, pH, chloride, total alkalinity, biochemical oxygen demand, chemical oxygen demand, pH are the some important parameters for assessing the quality of water for irrigation and drinking purposes.

The distribution of groundwater is not uniform in a region due to the unequal utilization and exploitation rate of groundwater. In certain areas there are quite a large number of tube wells whereas in other areas there are practically no tube wells. Water consumption from groundwater is negligible in some areas due to the less exploitation of groundwater resources. So, for proper monitoring of water quality random sampling is the best option for analyzing the physico-chemical parameter of water.

Temperature mostly fluctuates according to climatic conditions of the region and directly depends upon seasonal air temperature. However, groundwater temperature remains comparatively lower (25.5-31.6 °C) due to the geothermal energy. The temperature of textile common effluents (25.7-32.8 °C) and pond water (25.5-32.4 °C) has fluctuated significantly due to the lentic behaviour of aquatic ecosystem. Temperature of composite effluent drops slightly while reaching in the pond (25.5-32.4 °C). Pathe *et al.* (2005) reported the temperature fluctuation trend from cluster of small cotton textile units from different sector, which follow permissible limit of B.I.S and WHO and remain always below 40°C during study. During the study period temperature of textile industrial effluent never cross the standards limits. Temperature slightly influences some parameters of water and affects the concentration of dissolved oxygen contents and can also influence the activity of bacteria and toxic chemicals in water. Textile industrial effluent may cause problems with high temperature (Groff, 1993). Temperature facilitates the oxidation process biochemical oxygen demand and chemical oxygen demand achieved maximum reduction at high temperature as compared to low temperature (Aslam *et al.*, 2004). Temperature of water depends upon the meteorological characteristics of existing environment; especially air temperature. Temperature fluctuated with air temperature with positive relation. However due to some processes in textile units, the temperature of textile effluents was slightly higher than pond water and groundwater. With the increasing of depth of water table temperature often decreases, but increased in winter season due to the geo-thermal energy. Temperature of agricultural soil fluctuated in the range of (17.4-27.8 °C), and comparatively low than the temperature of drain sediment (15.5-24.9 °C) and pond sediment (15.3-24.8 °C) in almost all seasons. Temperature has positive correlation with iron (0.9188) Mn (0.6670), Ni (0.7044), Cu (0.6860) in groundwater sample, while in soil and sediment having negative relation. In drain sediment, temperature has positive relation with lead (0.8233), Zn (0.673), Cd (0.6827); while temperature has not any significant correlation with heavy metal in pond sediment'. Similar observations were made by Mahesh and Prasad (2004), Mustafa *et al.* (1998) in groundwater samples.

In the present study, the pH of the textile effluents was recorded mostly alkaline. The pH of the composite effluents was highly alkaline, which slightly decreases in pond water system. Extreme high pH or rapid pH change can exert stresses conditions or kill aquatic life. Changes in pH can also affect aquatic system indirectly by altering other aspects of water chemistry. Low pH levels accelerate the release of metals from rocks or sediments in the water. Groff (1993) reported that textile wastewater may cause problem used as untreated effluent consist higher pH and temperature. pH has significant importance in all chemicals and biochemical reactions in an aquatic system (Basu *et al.*, 1973).

pH of textile industrial effluent was found mostly basic (9.6-10.7) due to the high salt load and high alkalinity. Pond water's pH was also alkaline (8.2-8.6) due to the directly mixing with effluent drains with pond. The pH of both, drain sediment (7.8-8.3) and pond sediments (7.9-8.4) were also noticed alkaline in nature. The pH of groundwater was found slightly alkaline (7.4-7.7) and decreasing alkalinity with increasing depth. The pH of deep groundwater and agricultural soil was found slightly alkaline (7.1-7.6) respectively and very near to 7 (neutral) during present study. Similar trend is in agreement with that reported by Sial *et al.* (2006), Pathe *et al.* (2005) in textile industrial effluent, Singh *et al.* (1994) in effluent drain and Mor *et al.* (2003), Yadav and Lata (2003) in groundwater.

Total solids were found very high in textile industrial effluent (2210.4-2643.7 mg/l) due to the high degree of pollution. Different dyes and fibres are present in the textile wastewater, which generally released dissolved solids into common effluent drain. The total solids in pond surface water were fluctuated in the range of 2033.2 to 2365.7 mg/l. Similar alterations in surface water quality of the pond were observed by Malik *et al.* (2006). The dissolved solids in wastewater are mainly inorganic salts and relatively unaffected by biological treatment processes and can accumulate in water recirculation system. Failure to remove them may lead to an increase in total solids content of surface water sources. The dissolved solids in wastewater may be harmful to vegetation and also preclude use in various irrigation processes. Total dissolved solids in the textile effluents were observed very high than the prescribed limits of C.P.C.B. and B.I.S values. In the pond, the values of total solids and total dissolved solids were noticed comparatively high, however, these were not found higher than textile industrial effluent. Dissolved solids vary qualitatively and quantitatively in different water depending upon the season, location and other factors. In the present study, the total dissolved solids values in effluents (2045.7-2219.7 mg/l) and pond water (1861.0-2037.5 mg/l) were not so high according to CPCB standards. Similar observations were also reported by Yusuf and Sonibare (2004), Garg and Kaushik (2006) in textile industrial effluent and Mor *et al.* (2003), Mustafa *et al.* (2003) in groundwater.

Total dissolved solids are the measure of the quantity of material dissolved in water that can pass through a filter (usually with a pore size of 0.45 micrometers). The material can include carbonate, bicarbonate, chloride, sulfate, phosphate, nitrate, calcium, magnesium, sodium, organic ions, and other ions. Similar to total suspended solids, high concentrations of total dissolved solids may also reduce water transparency, combine with toxic trace compounds and heavy metals, and lead to an increase in water temperature. Total dissolved solids are used to estimate the quality of drinking water, because it represents the amount of ions in the water. High dissolved solids concentrations can produce laxative effects and can give an unpleasant mineral taste to water. High dissolved solids concentrations in water are also unsuitable for water recycling processes and many industrial applications. The values of total dissolved solids concentrations recorded in the present study are comparable to those obtain by Pathe *et al.* (2005), Yusuf and Sonibare (2004), Garg and Kaushik (2006) in textile industrial effluent and Mustafa *et al.* (2003), Mor *et al.* (2003) and Sastry and Rathee (1999) in groundwater.

Total suspended solids include both organic and inorganic materials, which adversely affect the aquatic life, especially fishes and bottom fauna of surface water ecosystems. These produce carbon dioxide, hydrogen sulphide, methane gases, which are very noxious in nature. They increase the turbidity of water, reduce light penetration and impair photosynthetic activity of aquatic plants in surface water system. Total suspended solids in the effluents (141.8-570.7 mg/l) and pond water (96.8-487.7 mg/l) were found slightly high than recommended by C.P.C.B. and B.I.S. Chavan *et al.* (2005) reported the adverse impacts of total suspended solids on the surroundings environment of the industrial effluent near Sukhna river at Midc area, Aurangabad.

Because the suspended particles absorb heat from sunlight and changed dissolved oxygen levels to fall even further because warmer waters can hold less dissolved oxygen. High total suspended solids in a water body can often mean higher concentrations of bacteria, nutrients, pesticides, and metals in the water. These pollutants may attach to sediment particles on the land and be carried into water bodies with storm water. In the water, the pollutants may be released from the sediment or travel farther. The flow rate of the effluent drain is a primary factor in total suspended solids concentrations. Fast running water can carry more particles and larger-sized sediment. Heavy rains can pick up sand, silt, clay, organic particles and heavy rubbish particles from the land and carry it to surface water. A change in flow rate can also affect total suspended solids; if the speed or direction of the water current increases, particulate matter from bottom sediments may be re-suspended.

Total solids were found very high in textile industrial effluent due to the dirt release during the washing of textile materials. In pond water, total solids were slightly low due to the settlement of heavy particles on bottom sediment. In shallow groundwater, total solids were found slightly higher than those of in deep groundwater. Similarly, total dissolved solids and total suspended solids were also correlated to groundwater of various depth. The ratio of total dissolved solids and total suspended solids with total solids was found quite similar during the study. Such types of trends in determining the total suspended solids load have been reported by Garg and Kaushik (2006), Yusuf and Sonibare (2004) in textile industrial effluent and Mor *et al.* (2003), Mustafa *et al.* (2003) in groundwater. Liquid wastewater from textile industries are not fit for irrigation regarding electric conductivity and total suspended solids (Bohn *et al.*, 1976, Abbas, 1991).

Conductance is a measurement of an electrical current with increasing amount and mobility of ions. These ions, which come from the breakdown of compounds, conduct electricity because they are negatively or positively charged when dissolved in water. Therefore, conductance is an indirect measure of the presence of dissolved solids such as chloride, nitrate, sulfate, phosphate, sodium, magnesium, calcium, and iron, and can be used as an indicator of water pollution.

It is not necessary that the water having high electric conductivity is not fit for irrigation. Only one parameter can't ensure the suitability of the water quality. However, the salt composition is high in the water having high Electric conductivity, but it may be beneficial for those soils having a lack or deficiency of some salts. Electric conductivity has a positive correlation with total dissolved solids in textile industrial effluent and groundwater also. Generally, there are not regulatory levels for electric conductivity. Instead, the concentration of total dissolved solids is often regulated. However, electric conductivity is an easily-obtained parameter that is a good indicator of the amount of dissolved solids in water, and thus can be used to detect contaminants in water. Total dissolved solids have a direct positive relationship with electric conductivity. If, total dissolved solids of water are high, electric conductivity will be more definitely (Mor *et al.*, 2003).

Electric conductivity was recorded very highly positive correlated to total dissolved solids due to the direct relation of total dissolved solids with electric conductivity in textile industrial effluent (3.1-3.4 mho/cm), pond water (2.9-3.2 mho/cm) and groundwater (1.1-2.0 mho/cm), while minimum at control site (0.63-1.1 mho/cm). Fluctuation in electric conductivity chiefly depends upon the quantity of ions in groundwater in various seasons. Electric conductivity has highly positive relation with Manganese (0.8544), nickel (0.8021) and lead (0.8340) in groundwater samples, while positive relations with almost all metals in agricultural soil. Sial *et al.* (2006), Garg and Kaushik (2006) were made similar observations in textile industrial effluent, Singh *et al.* (1994) in drain, Mahesh and Prasad (2004), Mor *et al.* (2003), Mustafa *et al.* (2003) in groundwater; which support the result of present study.

The dissolved oxygen in water is often attributed to the fact the oxygen is dissolved more during the period of active photosynthesis. Hutchinson (1957) reported that the water has low temperature has greater capacity for holding gases. The value of dissolved oxygen in the all sampling sites was found out to be low dissolved oxygen. The DO was found low in textile effluent (0.87-3.5 mg/l) and pond water (1.2-3.4 mg/l), however, in some months dissolved oxygen was found nil in textile industrial effluent and very low in pond water due to the increasing rate of biodegradation of different compounds in water column. As the matter of fact that the effluents carried highly pollution load, which could not allow any flora and fauna to survive in unfavourable condition and there will be no photosynthesis process in polluted habitat. The value of dissolved oxygen was recorded even less values (2.56±0.21 mg/l) in the summer month. This trend was stated by Khanna *et al.* (1992) and Joshi (1996) in surface water system. The dissolved oxygen showed an inverse relationship with temperature as reported by Das and Srivastava (1956) and Dobriyal (1985) in aquatic system.

Groundwater usually has low concentrations of dissolved oxygen, but it is also often colder than stream water. Oxygen is more easily dissolved into water at low altitudes than at high altitudes, because of higher atmospheric pressure. Oxygen is more easily dissolved into water with low levels of dissolved or suspended solids. Very low dissolved oxygen concentrations in water can result in mobilization of trace metals. DO have mostly negative correlation (-0.4459, -0.4396, -0.6278, -0.6314, -0.4856, -0.6519, -0.4412) in groundwater and (-0.6152, -0.5741, -0.8010, 0.1536, -0.4807, -0.2132, 0.0685) in textile industrial effluent with almost heavy metals (Cd, Cu, Fe, Mn, Ni, Pb and Zn) respectively. The value of dissolved oxygen recorded in the present study are comparable to results obtained by Yusuf and Sonibare (2004), Garg and Kaushik (2006) in textile industrial effluent, Singh *et al.* (1994) in effluent drain, Mahesh and Prasad (2004) and Mustafa *et al.* (2003) in groundwater.

Free carbon dioxide in most textile industrial effluents was fluctuated in the range of 0.38 to 2.8 mg/l. However, the dissolved oxygen contents was also not found high in textile industrial effluent (0.87-3.5 mg/l) and pond water (2.57-4.00 mg/l) and remain low (2.5-4.0 mg/l) in groundwater samples. Free carbon dioxide mainly depends upon some physico-chemical parameter like dissolved oxygen, pH, temperature, etc. (ww.holar.is). Carbon dioxide has highly positive correlations (0.4060, 0.5115, 0.6729, 0.6306, 0.5276, 0.6839 and 0.5164) in groundwater and also (0.6627, 0.6112, 0.7573, -0.0984, 0.4410, 0.3252, -0.0395) in textile industrial effluents with almost all heavy metals (Cd, Cu, Fe, Mn, Ni, Pb and Zn) respectively.

Biochemical oxygen demand and chemical oxygen demand are the measurement of the oxygen consuming capabilities of organic matter. Textile effluent during their process of deposition exerts biochemical oxygen demand/chemical oxygen demand, which have a catastrophic effect on the ecosystem by depleting the oxygen supply. Decay process causes the production of noxious gases as hydrogen sulphide and methane. Water with high biochemical oxygen demand/chemical oxygen demand indicates presence of decomposing organic matter and subsequent high bacterial counts that degrade its quality and potential uses (Verma and Kharat, 2000). Biochemical oxygen demands of the effluents drain (138.6-170.0 mg/l) and pond water (99.0-136.0 mg/l) were found high as given in standards of C.P.C.B., B.I.S. and WHO. However, in groundwater of different depth biochemical oxygen demand was found to be low or sometimes nil. This trend is an agreement with that reported by Aslam *et al.* (2004), Yusuf and Sonibare (2004), Sial *et al.* (2006), Pathe *et al.* (2005) in textile industrial effluent, Singh *et al.* (1994) in drain and Mahesh and Prasad (2004), Mustafa *et al.* (2003) in groundwater.

Chemical oxygen demand is directly proportional to biochemical oxygen demand. Dilution of effluents for biochemical oxygen demand analysis always depends upon the value of chemical oxygen demand in a wastewater. In fact, chemical oxygen demand explains the load of pollutants and degree of pollution in effluent drain. Effluents of dye houses having a tremendous value of COD in the range of 351.5-445.0 mg/l, while it was observed 274.5-359.7 mg/l in pond water due to the high rate of decomposition of organic material in polluted and stored water. COD shows a positive relationship with BOD and negative with DO (Khanna, 1993) and mostly remains in the ratio of approximately $1:3 \pm 0.5$ in the case of wastewater. Sevimli *et al.* (2005) observed the increasing trends of biochemical oxygen demand and chemical oxygen demand together while performing an experiment on decolorization of textile effluent and dye solution by ozonation. Senthinathan and Azeez (1999) and Groff (1993) reported that textile wastewater may cause problem used as untreated effluent consist higher COD, pH, temperature, Phosphorous and absorbable organic halogen. COD of textile wastewater may be reduced by using some appropriate technology Bhat *et al.* (2005). Aslam *et al.* (2004), Yusuf and Sonibare (2004), Sial *et al.* (2006), Pathe *et al.* (2005) were also reported the similar values in textile industrial effluent, and Mahesh and Prasad (2004), Mustafa *et al.* (2003) in groundwater. Waste containing high biochemical oxygen demand and chemical oxygen demand are responsible for a heavy depletion of oxygen levels in the particular sector of water body or soil (Sial *et al.*, 2006).

Water alkalinity is caused by strong bases and the salts of strong alkalis and weak acids. Carbonates are added to a water system if the water passes through soil and rock that contain carbonate minerals, such as calcite ($CaCO_3$). Where limestone and sedimentary rocks and carbonate-rich soils are predominant, waters will often have high alkalinity. Where igneous rocks (such as granite) and carbonate-poor soils are predominant waters will have low alkalinity. Alkalinity is a measure of the buffering capacity of water, or the capacity of bases to neutralize acids. Waters with low alkalinity are very susceptible to changes in pH. Waters with high alkalinity are able to resist major shifts in pH. As increasing amounts of acid are added to a water body, the pH of the water decreases, and the buffering capacity of the water is consumed. Because alkalinity and pH are so closely related, changes in pH can also affect alkalinity, especially in a poorly buffered system. Above pH 8.3, alkalinity is mostly in the form of carbonate (CO_3^{2-}); below 8.3, alkalinity is present mostly as bicarbonate (HCO_3^-).

The values of alkalinity in the effluents (633.6-889.0 mg/l) and in the pond water (563.4-732.3 mg/l) were found very high, while in the groundwater of various depths it was quite low according to depths as it was found 493.6-596.7 in 40 ft, 261.6-337.3 in 80 ft and 181.5-263.3 in 120 ft water samples. Bhat *et al.* (2005) observed the increasing trend in alkalinity of various textile effluents according to discharge points. Alkalinity not only helps regulate the pH of a water body, but also the metal content. Alkalinity has found with positive correlations with Cd (0.0237), Cu (0.809), Fe (0.8896), Mn (0.7664), Ni (0.7918), Pb (0.5841) and Zn (0.7069) Bicarbonate and carbonate ions in water can remove toxic metals (such as lead, arsenic, and cadmium) by precipitating the metals out of solution. Pathe *et al.* (2005), Garg and Kaushik (2006) have been made some observations in textile industrial effluent and Sastry and Rathee (1999), Mahesh and Prasad (2004), Mor *et al.* (2003), Yadav and Lata (2003) in groundwater; which support the results of present study. In irrigated area, a major portion receives water through exploitation of groundwater aquifers. The quality of groundwater is poor on account of excessive salt contents, high sodium or high toxic elements. The interactions of these parametric factors of water quality with soils influence their physico-chemical properties (Minhas and Gupta, 1992).

Hardness generally represents the concentration of calcium and magnesium ions, because these are the most common polyvalent cations. Other ions, such as iron and manganese, may also contribute to the hardness of water, but are generally present in much lower concentrations. Calcium, magnesium, and other polyvalent cations such as iron and manganese may be added to a natural water system as it passes through soil and rock containing large amounts of these elements in mineral deposits. Hardness mainly depends on the amount of calcium and magnesium salts or both present in the groundwater, water hardness is not health hazard but its values should remain below permissible level to restore the taste to water (WHO, 1984).

Some textile industrial processes may also produce significant amounts of calcium and manganese that are later discharged as industrial effluents. Calcium was found 10.58-27.4 mg/l in textile industrial effluent, 10.4-23.6 mg/l in pond water, 4.6-14.3 mg/l in groundwater of 40 ft and 3.4-11.7 mg/l in groundwater of 120 ft depth. Similar conditions were also reported by Garg and Kaushik (2006), Yusuf and Sonibare (2004) and Helena *et al.* (1999) in textile industrial effluent, Mustafa *et al.* (1998), Senthilnathan and Azeez (1999), Mor *et al.* (2003) and Yadav and Lata (2003) in groundwater.

When hardness equals alkalinity, the only cations present in significant concentrations in the water are calcium and magnesium. When hardness is greater than alkalinity, the waters may contain considerable amounts of other cations. Magnesium was found low in textile industrial effluent (4.7-16.1 mg/l), pond water (4.2-14.1 mg/l), groundwater of 40 ft (2.2-9.7 mg/l) and minimum in groundwater of 120 ft depth (1.9-9.2 mg/l). The Mg concentrations levels were comparable with those found by Yusuf and Sonibare (2004) and Helena *et al.* (1999) in textile effluents, Garg *et al.* (1998), Mor *et al.* (2003), Yadav and Lata (2003), Senthilnathan and Azeez (1999) and Mustafa *et al.* (1998) in groundwater.

Potassium is one of the important element among ten chemical element required in rather large quantities fir growth of all crops. Plants growth may be limited by a deficient available supply of potassium but its high some quantity may alter the soil composition. High K in water used in irrigation may alter the soil quality (Miller and Turk, 2002). Potassium was found low in textile industrial effluent (7.4-8.8 mg/l), pond water (7.3-8.4 mg/l), groundwater of 40 ft (4.9-6.1 mg/l) and groundwater at 120 ft (4.3-5.7 mg/l). Similar observation were made by Pathe *et al.* (2005) and Garg and Kaushik (2006) in textile industrial effluent, Das and Sahoo (2003), Mor *et al.* (2003), Yadav and Lata (2003) and Senthilnathan and Azeez (1999) in groundwater.

Sodium salts in the water are much more objectionable than salts of calcium and magnesium because of the tendency of sodium to cause deflocculation of the colloidal fraction of the soil and developed an undesirable structure water contain sufficient heavy metals to prove toxic to continue use by plants. As sodium chloride used in textile industry so the concentration of sodium and chloride will be high in the wastewater (Aslam *et al.*, 2004). Sodium was found very high in groundwater of industrial area, because sodium should not be higher than 20 mg/l in drinking water (WHO, 2006). Sodium values were found very high in textile industrial effluent (145.0-177.0 mg/l), pond water (128.5-158.7 mg/l), groundwater of 40 ft (51.0-97.7 mg/l) and groundwater at 120 ft (8.7-37.0 mg/l). Helena *et al.* (1999) and Garg and Kaushik (2006) in textile industrial effluent and Das and Sahoo (2003), Mor *et al.* (2003), Mahesh and Prasad (2004) and Senthilnathan and Azeez (1999) in groundwater were observed the similar observations.

Chloride is one of the most conservative constituents in wastewater, being affected very little either by vegetative growth or other biological processes, or by precipitation or anion exchange reaction. Chloride concentrations in the effluents (588.5-825.7 mg/l) and pond water (576.5-770.7 mg/l) were also expected to be high by the permissible limit of C.P.C.B., while the occurrence of chloride have reported in GW40 (47.5-151.3 mg/l), GW80 (4.7-19.6 mg/l) and GW120 (4.7-13.7 mg/l) in very low concentrations. A work conducting on the effect of natural flocculants on textile wastewater, chloride concentration is increased in wastewater used in textile industries and delayed later (Aboulhassan *et al.*, 2005).

Chloride does not have a health-based standard for drinking water. Chloride has been suggested as a standard value for drinking purposes (EPA, 1989). The Maximum Contaminant Level (MCL) of 250000 μg/L is based on poor taste chloride imparts to water. Chloride is not affected by adsorption or degradation processes and is therefore a conservative chemical in soil and groundwater. Chloride concentrations begin to increase at some depth due to dissolution reactions within the aquifer. Chloride is therefore a good indicator of human impacts in the environment (MPCA, 2001). In the wastewater the concentration of chloride will be high due to the use of sodium chloride in textile industry. Chloride concentration in irrigation water should be less then 75 ppm, otherwise it may be harmful for soil structure. Textile effluents are high in toxicity not only for human beings of surroundings but also a serious threat to groundwater and surface water resources (Aslam *et al.*, 2004). The chloride concentrations levels were comparable with those found by Helena *et al.* (1999), Garg and Kaushik (2006) in textile industrial effluent, Senthilnathan and Azeez (1999), Yadav and Lata (2003), Mor *et al.* (2003), Mahesh and Prasad (2004), in groundwater.

Chapter 6

Impact Assessment on Agricultural Soil

6.1 Agricultural Soil

Agricultural soil requires sufficient irrigation for high production in agricultural fields. Irrigation water with poor quality can result in the alteration of physico-chemical characteristics of irrigated agricultural soil. Several minerals such as chloride, sulphate, bicarbonate, sodium, calcium and magnesium in irrigation water may change the soil quality with its continuous application on soil. These minerals contained in the irrigation water can build up the soil's trophic level and ultimately influence the agriculture productivity (Malik and Bharti, 2007).

6.2 Historical Resume

The contaminated groundwater used in irrigation purposes may harmful for soils of agriculture fields and ultimately effects on productivity of agriculture crop plants, and soil macrobiotic community also.

Many workers from India and abroad have been discussed various aspects of agricultural soil pollution in term of characteristics of affected agricultural soils and bottom sediments from time to time. Some of important contribution made in the field of industrial pollution, with references to physical, chemical and biological parameters of contaminated agricultural irrigated soil and bottom sediment soils by different authors of country and abroad.

Aggarwal and Mehrotra (1952) surveyed the soil and soil water affected by effluent in some selected regions of Uttar Pradesh. Doneen (1954) reported the stalinization of soils of salts in irrigation waters. Richards (1954) evaluated the quality of saline and alkali soils and gave a technique for the improvement of these sick soil systems in the agricultural sector for good performance of agriculture productivity.

Singh and Bhumbla (1968) analyzed the effect of quality of irrigation water on soil properties. Singh *et al.* (1969) reported the salinity problem in soil of high water table areas of north India.

Singh and Sharma (1971) studied on the effects of saline irrigation waters on physico-chemical properties of some soils of Rajasthan.

Bhargava *et al.* (1972) evaluated the characteristics of some typical saline sodic soils occurring in Karnal district of Haryana. Williums (1972) described the characteristics and various properties of Metals. Sharma and Parihar (1973) evaluated the effect of depth and salinity of groundwater on evaporation and soil salinization.

Dowedy and Larson (1975) reported the availability of some metals in various vegetable crops in a polluted region in the vicinity of an industrial area. Dulka and Risby (1976) stated ultra trace metals in some environmental and biological systems in the vicinity of textile industrial sector.

Ajmal and Khan (1985) analysed the effects of textile waste on physico-chemical properties of soil and seed germination. The BOD, COD and total solids, Cl, SO_4, Na, Ca and Mg concentrations in the effluents were found to be very high.

Nriagu and Pacyna (1988) carried out a study on the quantitative assessment of worldwide contamination of air, water and soil with trace metals.

Minhas and Gupta (1992) indicated the effects of wastewater irrigation in agriculture soils and found sodicity due to the poor water quality. Singh *et al.* (1994) highlighted the degradation of water and soil quality of Parwanoo area with respect to heavy metals. Srinivasan and Murali (1996) carried out some studies on effect of industrial pollutants on properties of soils.

Pujari and Sinha (1999) studied on the water and soil quality of some villages of Attabira area irrigated by bargarg main canal originated from Hirakund reservoir of Orissa.

Yadav *et al.* (2002) highlighted the post irrigation impact of domestic sewage effluent on composition of soils, crops and groundwater in Kurukshetra district. Shrivastava *et al.* (2003) collected soil samples from surface of the soil from different agriculture fields in the Khandesh region and Tapi River sediment samples were also collected from five different stations which were 7-8 km away from each other and detected the concentrations of heavy metals have been determined by ICP-AES.

Chao *et al.* (2005) reported the heavy metal contaminated site in Taiwan and described the status and distribution of heavy metals in environment. Zheng *et al.* (2006) assessed the heavy metals pollution of agricultural soil in Guanzhong district of China. Lokeshwari and Chandrappa (2006) evaluated the impact of heavy metals contamination of a Lake on soil and cultivated vegetation near Bangalore and found the presence of some heavy metals in rice and vegetables, beyond the limit of Indian Standards and calculated metal transfer factors from soil to vegetation are found significant for Zn, Cu, Pb and Cd. Hedge *et al.* (2006) revealed the provenance of heavy minerals with special reference to ilmenite of the Honnevar beach in central west coast of India and observed some trace heavy elements like Co, Cr, V and Ni in the sandy soil of the beach.

Malik and Bharti (2007) studied on soil quality of irrigated agricultural fields in a textile industrial area of Panipat city. Walker *et al.* (2007) identified the influence of heavy metals and mineral nutrient supply on *Bituminaria bituminosa*. Gharaibeh *et al.* (2007) found the impact of field application of treated wastewater on hydraulic properties of vertisols.

Golia *et al.* (2007) measured the accumulation of metals on tobacco leaves (primings) grown in an agricultural area in relation to soil. Trevors and Saier (2007) described the regulation of pollution in various environmental components *viz.* water, air and soil. Mico *et al.* (2007) described the comparison of two digestion methods for the analysis of heavy metals by flame atomic absorption spectrophotometer.

6.3 Observations

Experimental Site

Physico-chemical Parameters of Agriculture Soils

Temperature (°C)

Temperature of agricultural soil was recorded maximum 27.87 ± 3.56 in summer 2005 and minimum 17.40 ± 3.92 in winter 2005-06 ranged between 11.3 to 31.5 °C. The highest value of temperature in agricultural soil was found 31.5 °C during June 2006 and lowest 11.3 °C in January 2006. The temperature variation during the study is given Table 6.2. Monthly variation in temperature is given in Figure 6.4.

Water Holding Capacity (Per cent)

Water holding capacity of agricultural soil was recorded maximum 43.75 ± 0.49 in summer 2005 and minimum 36.13 ± 4.92 in monsoon 2005 ranged between 31.58 to 44.35 per cent . The highest value of water holding capacity in agricultural soil was found 44.35 per cent during June 2006 and lowest 31.58 per cent in September 2005. The water holding capacity variation during the study is given Table 6.2. Monthly variation in water holding capacity is given in Figure 6.4.

Bulk Density (g/cm^3)

The bulk density of agricultural soil was recorded maximum 1.24 ± 0.04 in monsoon 2006 and minimum 1.15 ± 0.03 in summer 2005 ranged between 1.08 to 1.31 g/cm^3. The highest value of bulk density in agricultural soil was found 1.31 g/cm^3 during October 2005 and lowest 1.08 g/cm^3 in January 2006. The bulk density variation during the study is given Table 6.2. Monthly variation in bulk density is given in Figure 6.5.

Soil Moisture (per cent)

The soil moisture of agricultural soil was recorded maximum 10.76 ± 2.26 in monsoon 2006 and minimum 3.52 ± 1.04 in summer 2005 ranged between 2.75 to 12.2 per cent . The highest value of soil moisture in agricultural soil was found 12.2 per cent during August 2005 and lowest 2.75 per cent in May 2005. The soil moisture

Table 6.1: Seasonal Variation of Physico-Chemical Characteristics of Soil Near Textile Industries

Parameters	Unit	Summer 2005		Monsoon 2005		Winter 2005-06		Summer 2006		Monsoon 2006		Winter 2006-07		Summer 2007	
		Mean	*SD*	*Mean*	*SD*	*Mean*	*SD*	*Mean*	*SD*	*Mean*	*SD*	*Mean*	*SD*	*Mean*	*SD*
Temp.	°C	27.667	3.456	25.833	2.517	17.160	3.866	26.050	4.063	26.133	2.003	18.075	3.969	19.750	2.475
WHC	per cent	42.590	0.476	34.997	4.926	35.800	2.814	42.203	1.288	35.693	4.836	35.000	2.386	38.930	1.018
Bulk density	mg/cm^3	1.203	0.021	1.273	0.021	1.274	0.048	1.213	0.015	1.280	0.026	1.295	0.030	1.225	0.021
Soil moisture	Per cent	3.377	1.027	10.600	2.360	5.210	1.166	3.670	0.853	10.760	2.249	5.513	1.603	4.605	0.856
pH		7.700	0.100	7.800	0.100	7.920	0.084	7.800	0.082	7.833	0.058	7.850	0.058	8.050	0.071
EC	mho/cm	30.833	1.041	29.667	1.457	26.120	0.944	30.075	1.680	30.267	1.201	26.750	1.012	27.050	1.061
Cl	µg/g	769.333	140.479	721.000	86.000	551.600	35.211	716.500	172.055	739.333	96.862	567.250	32.253	552.000	8.485
Na	µg/g	424.000	117.656	100.333	86.489	264.600	159.054	461.750	94.132	133.667	73.921	228.000	125.748	503.500	13.435
K	µg/g	33.333	1.528	30.333	1.528	33.000	1.581	34.500	1.291	31.000	1.000	32.750	0.957	33.500	2.121

Table 6.2: Seasonal Variation of Physico-Chemical Characteristics of Agricultural Irrigated Soil

Parameters	Unit	Summer 2005		Monsoon 2005		Winter 2005-06		Summer 2006		Monsoon 2006		Winter 2006-07		Summer 2007	
		Mean	SD	Mean	SD	Mean	SD	Mean	SD	Mean	SD	Mean	SD	Mean	SD
Temp.	°C	27.867	3.556	26.200	2.551	17.400	3.922	26.375	4.328	26.567	2.601	18.325	4.105	19.750	3.182
WHC	Per cent	43.747	0.485	36.127	4.920	36.942	2.828	43.188	1.220	36.767	4.868	36.443	2.857	41.190	0.905
Bulk density	mg/cm^3	1.147	0.025	1.217	0.031	1.194	0.086	1.148	0.034	1.240	0.040	1.218	0.085	1.175	0.035
Soil moisture	Per cent	3.517	1.040	10.643	2.424	5.272	1.178	3.668	0.820	10.763	2.264	5.550	1.429	4.645	1.181
pH		7.300	0.200	7.563	0.185	7.324	0.319	7.330	0.215	7.633	0.153	7.453	0.301	7.110	0.014
EC	mho/cm	27.667	2.082	26.500	1.323	24.180	1.117	27.450	1.873	27.733	1.662	24.538	0.912	25.600	0.566
Cl	µg/g	734.333	112.010	641.667	69.292	514.800	71.521	683.500	148.586	673.000	71.631	535.500	69.150	499.500	37.477
Na	µg/g	537.333	125.085	192.000	91.995	363.400	157.267	566.750	101.385	200.000	72.062	329.000	124.887	599.000	36.770
K	µg/g	35.000	2.000	31.000	1.000	34.400	1.517	36.000	1.414	31.000	1.000	33.750	1.258	36.500	0.707

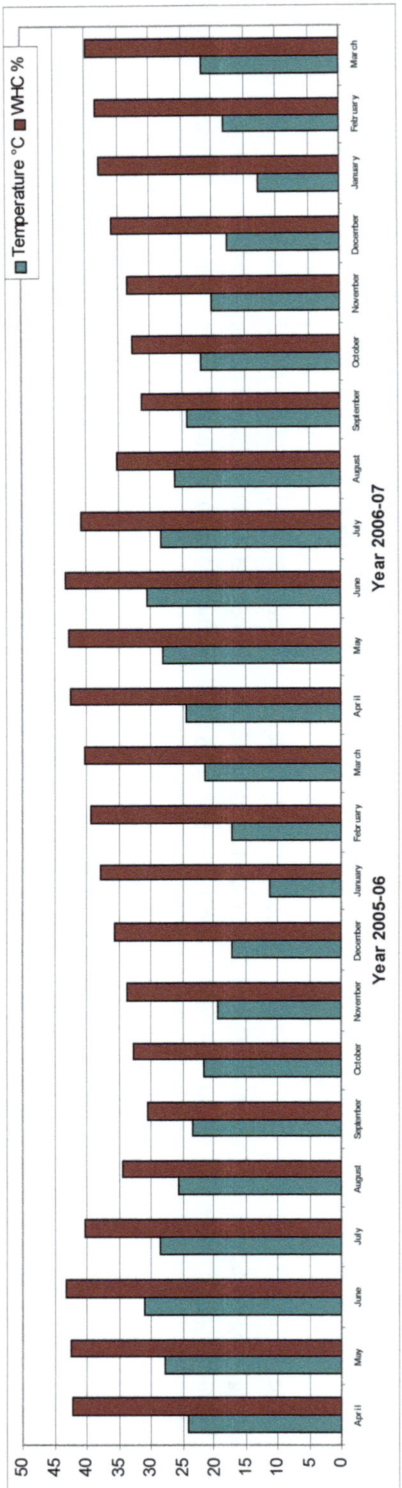

Figure 6.1: Monthly Variation of Temperature and WHC in Soil Near Textile Industries

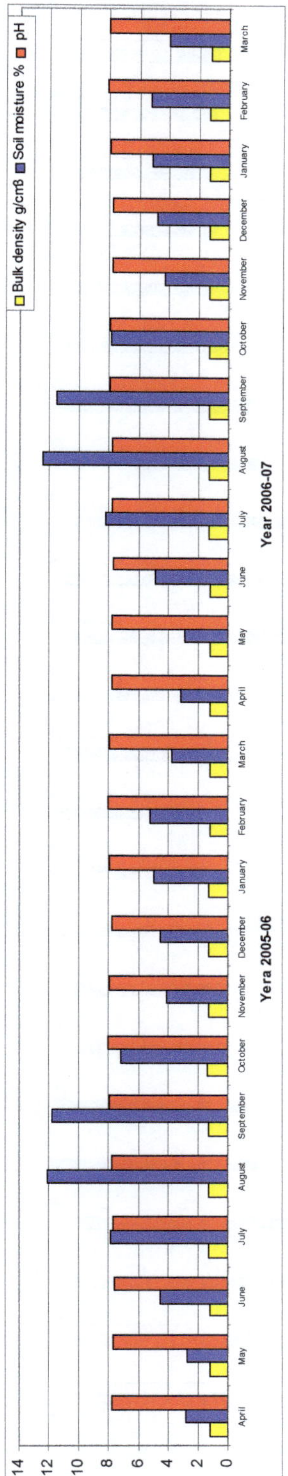

Figure 6.2: Monthly Variation of Bulk Density, Soil Moisture and pH in Soil Near Textile Industries

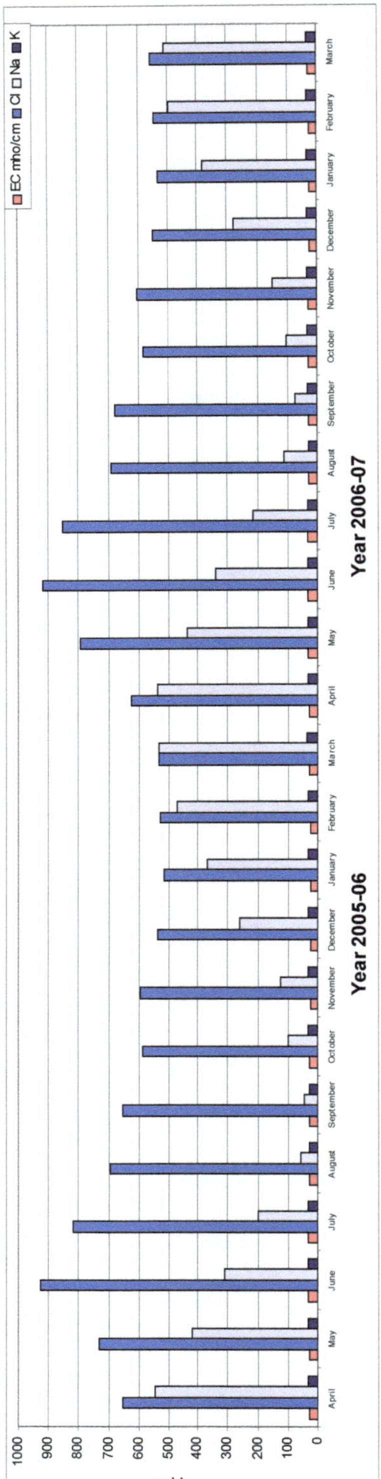

Figure 6.3: Monthly Variation of EC, Chloride, Sodium and Potassium in Soil Near Textile Industries

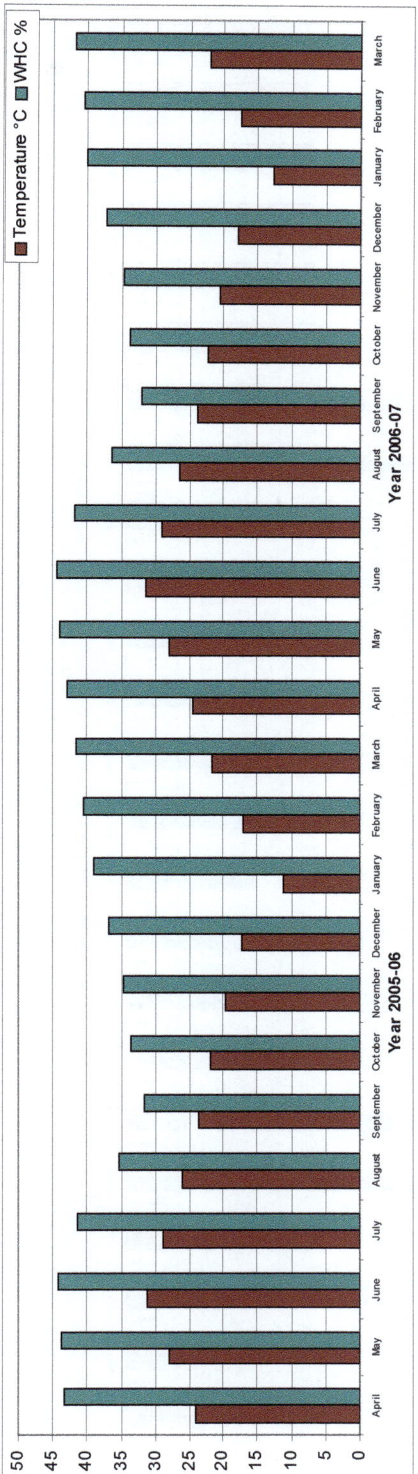

Figure 6.4: Monthly Variation of Temperature and WHC in Agricultural Soil

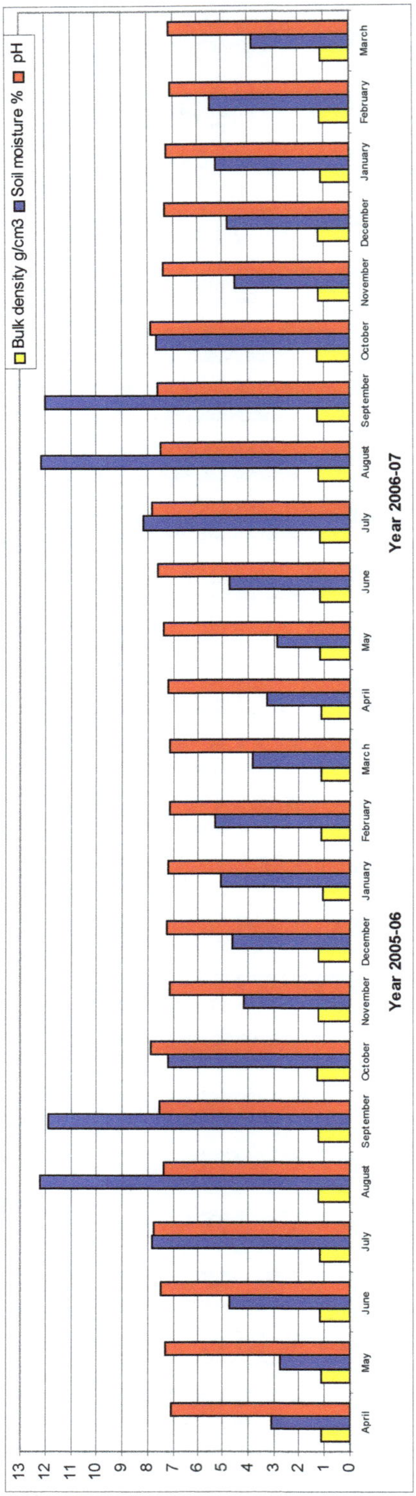

Figure 6.5: Monthly Variation of Bulk Density, Soil Moisture and pH in Agricultural Soil

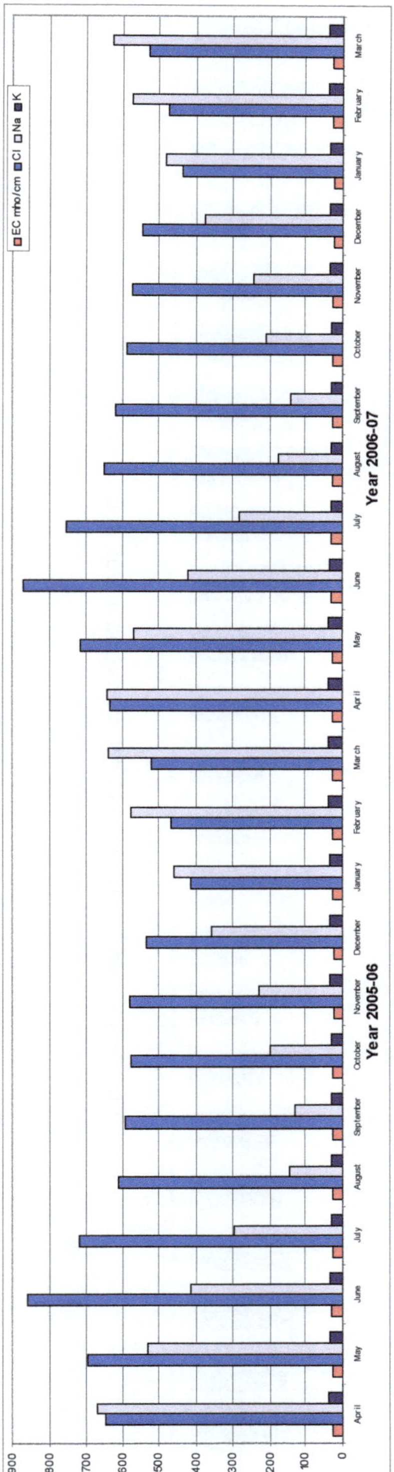

Figure 6.6: Monthly Variation of EC, Chloride, Sodium and Potassium in Agricultural Soil

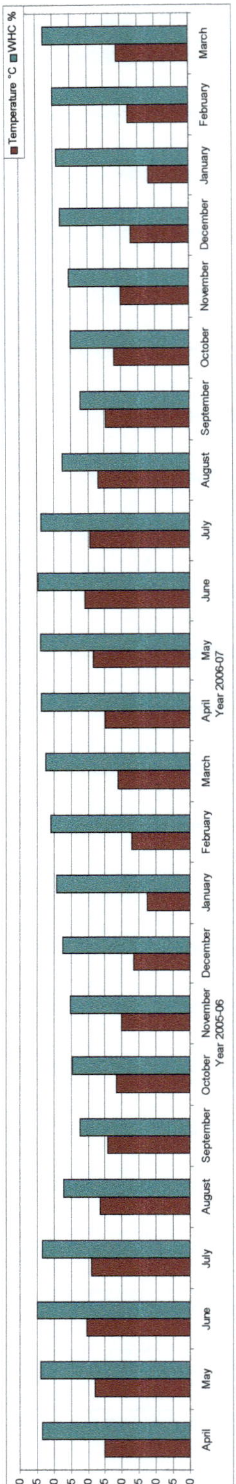

Figure 6.7: Temperature and WHC Variation in Soil at Control Site

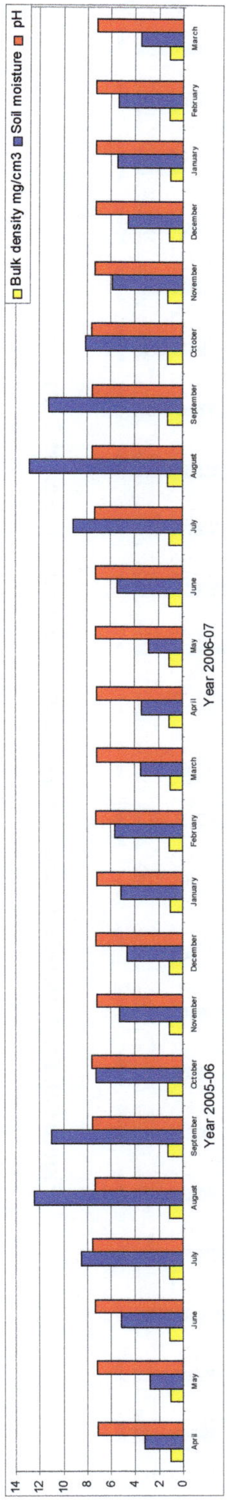

Figure 6.8: Bulk Density, Soil Moisture and pH Variation in Soil at Control Site

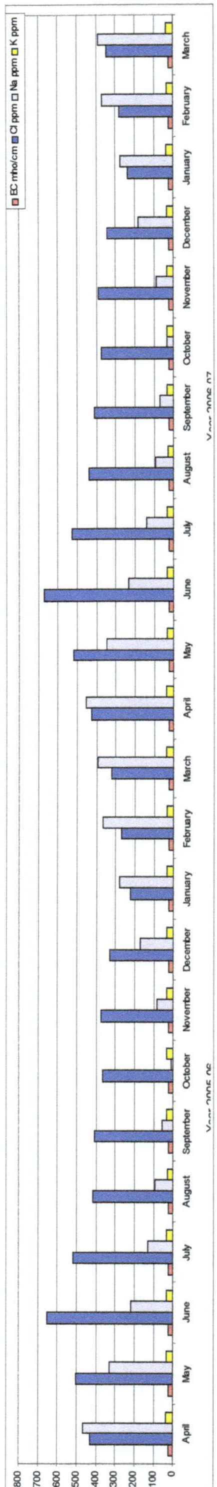

Figure 6.9: EC, Chloride, Sodium and Potassium Variation in Soil at Control Site

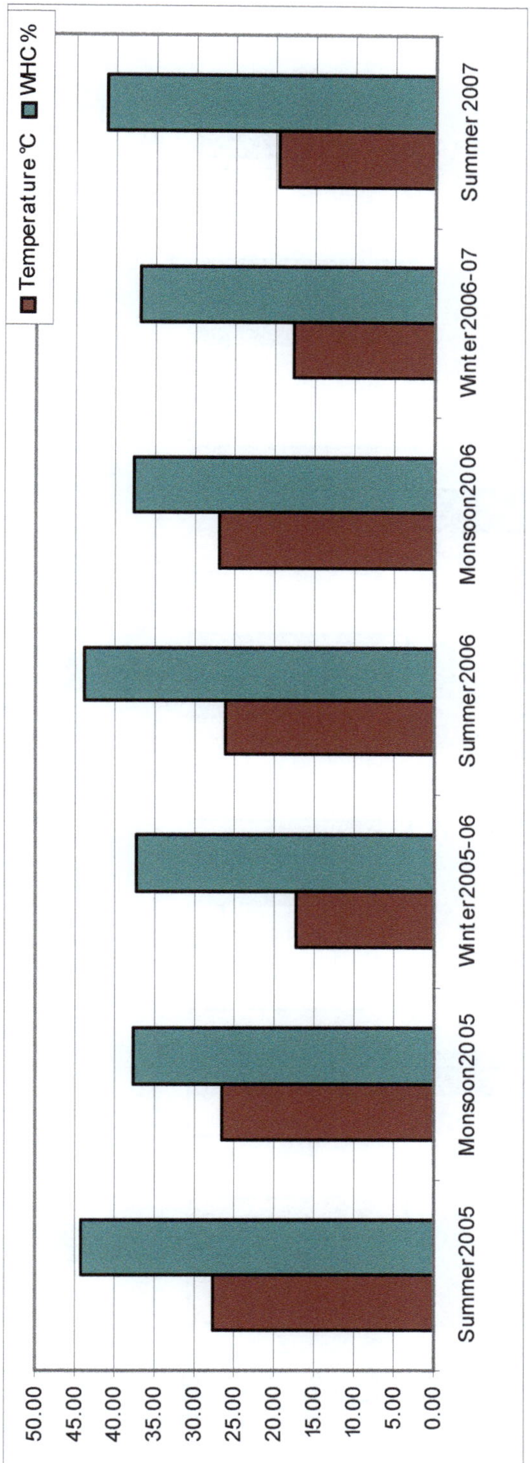

Figure 6.10: Seasonal Fluctuation in Temperature and WHC in Soil at Control Site

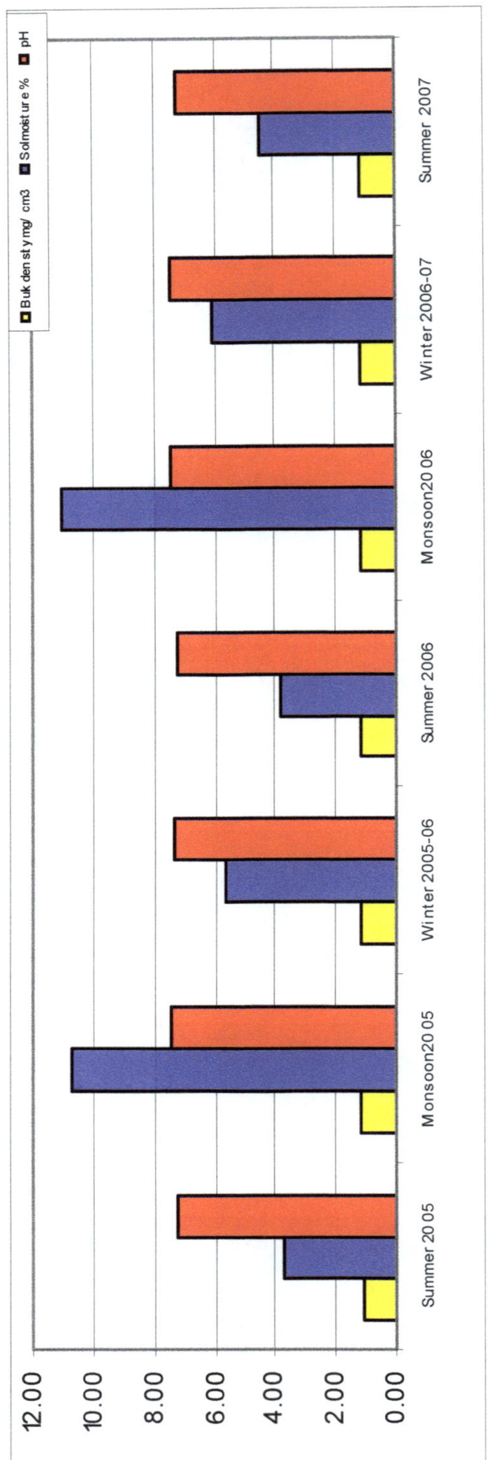

Figure 6.11: Seasonal Fluctuation in Bulk Density, Soil Moisture and pH in Soil at Control Site

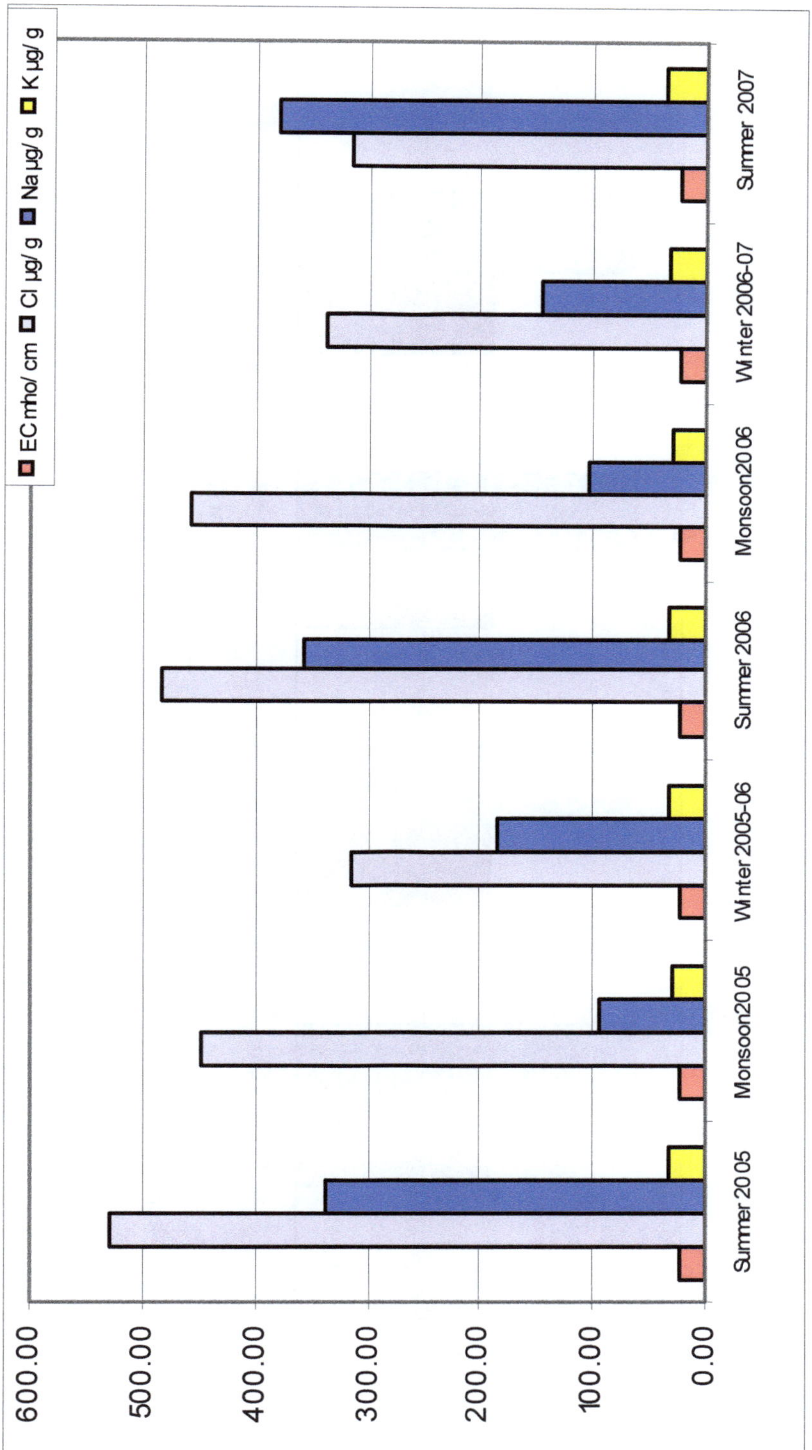

Figure 6.12: Seasonal Fluctuation in EC, Chloride, Sodium and Potassium in Soil at Control Site

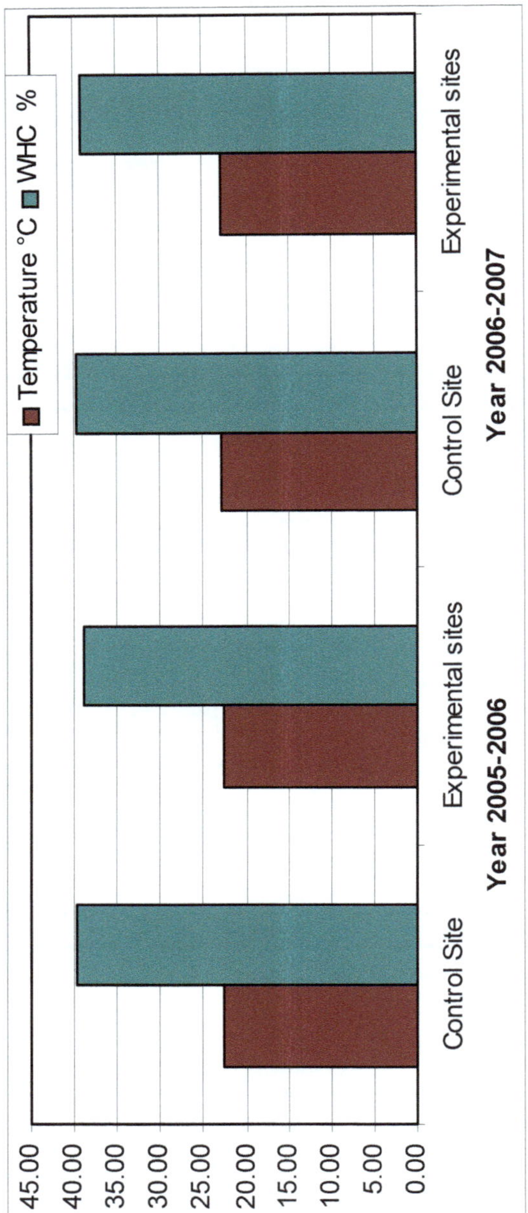

Figure 6.13: Mean Values of Temperature and WHC in Soil at Control and Experimental Site

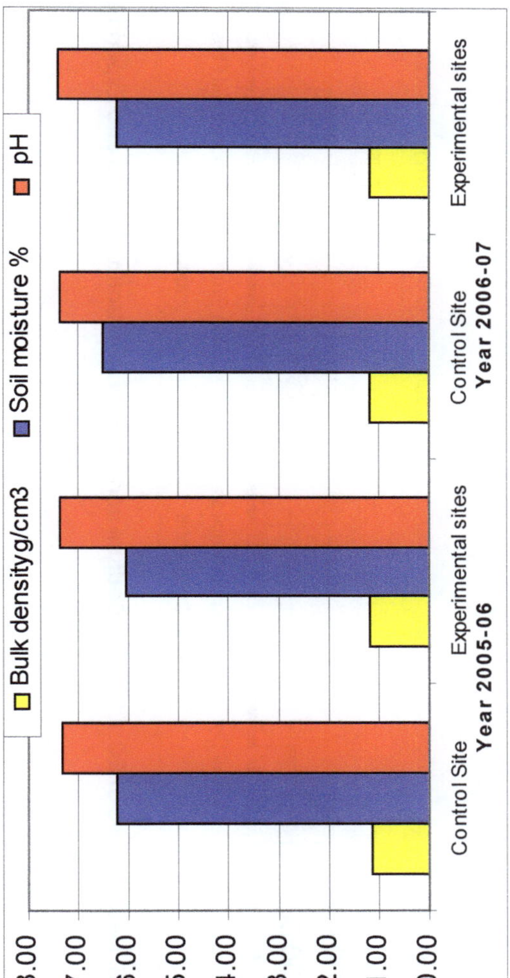

Figure 6.14: Mean Values of Bulk Density, Soil Moisture and pH in Soil at Control and Experimental Site

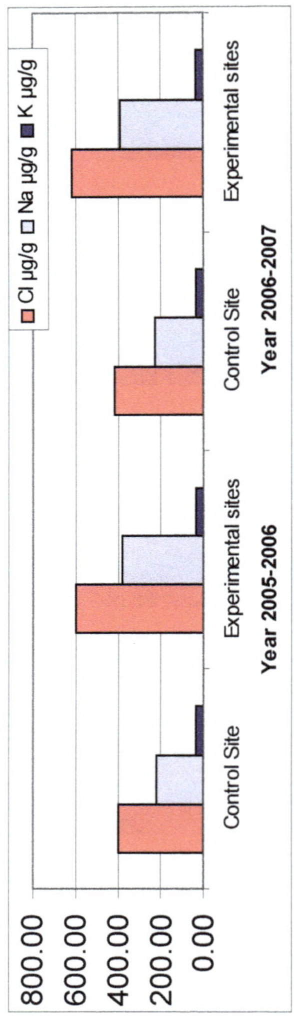

Figure 6.15: Mean Values of EC, Chloride, Sodium and Potassium in Soil at Control and Experimental Site

variation during the study is given Table 6.2. Monthly variation in soil moisture is given in Figure 6.5.

pH

pH of agricultural soil on pH scale was recorded maximum 7.63 ± 0.15 in monsoon 2006 and minimum 7.11 ± 0.01 in summer 2007 ranged between 7.1 to 7.9. The highest value of pH in agricultural soil was found 7.9 during October 2006 and lowest 7.1 in April 2005. The pH variation during the study is given Table 6.2. Monthly variation in pH is given in Figure 6.5.

Electrical Conductivity (mho/cm)

Electric conductivity of agricultural soil was recorded maximum 27.73 ± 1.66 in monsoon 2006 and minimum 24.18 ± 1.12 in winter 2005-06 ranged between 22.3 and 30.0 mho/cm. The highest value of Electric conductivity in agricultural soil was found 30.0 mho/cm during June 2005 and lowest 22.3 mho/cm in December 2005. The Electric conductivity variation during the study is given Table 6.2. Monthly variation in Electric conductivity is given in Figure 6.6.

Chloride (µg/g)

Chloride concentration in agricultural soil was recorded maximum 734.33 ± 112.01 in summer 2005 and minimum 499.50 ± 37.47 in summer 2007 ranged between 416.0 to 871.0 µg/g. The highest value of chloride in agricultural soil was found 871.0 µg/g during June 2006 and lowest 416.0 µg/g in January 2006. The chloride variation during the study is given Table 6.2. Monthly variation in chloride is given in Figure 6.6.

Potassium (µg/g)

Potassium concentration in agricultural soil was recorded maximum 36.50 ± 0.71 in summer 2007 and minimum 31.00 ± 1.00 in monsoon 2005 ranged between 30.0 to 37.0 µg/g. The highest value of potassium in agricultural soil was found 37.0 µg/g during March 2007 and lowest 30.0 µg/g in August 2005. The potassium variation during the study is given Table 6.2. Monthly variation in potassium is given in Figure 6.6.

Sodium (µg/g)

Sodium concentration in agricultural soil was recorded maximum 599.0 ± 36.77 in summer 2007 and minimum 192.0 ± 91.99 in monsoon 2005 ranged between 133.0 to 665.0 µg/g. The highest value of sodium in agricultural soil was found 665.0 µg/g during April 2005 and lowest 133.0 µg/g in September 2005. The sodium variation during the study is given Table 6.2. Monthly variation in sodium is given in Figure 6.6.

Control Site

Physico-chemical Parameter of Agriculture Soil

Temperature (°C)

Temperature of soil at control site was recorded maximum 27.83 ± 2.70 in summer 2005 and minimum 17.50 ± 3.48 in winter 2005-06 ranged between 11.8 to 31.0 °C.

The highest value of temperature in soil was found 31.0 °C during June 2006 and lowest 11.8 °C in January 2007. The temperature variation during the study is given Table 6.3. Monthly variation in temperature is given in Figure 6.7.

Water Holding Capacity (Per cent)

Water holding capacity of soil at control site was recorded maximum 44.18 ± 0.74 in summer 2005 and minimum 36.81 ± 1.89 in winter 2006-07 ranged between 32.17 to 45.0 per cent . The highest value of water holding capacity in soil was found 45.0 per cent during June 2005 and lowest 32.17 per cent in September 2006. The water holding capacity variation during the study is given Table 6.3. Monthly variation in water holding capacity is given in Figure 6.7.

Bulk Density (g/cm^3)

The bulk density of soil at control site was recorded maximum 1.22 ± 0.01 in monsoon 2006 and minimum 1.11 ± 0.051 in summer 2006 ranged between 1.05 to 1.25 g/cm^3. The highest value of bulk density in soil was found 1.25 g/cm^3 during October 2006 and lowest 1.05 g/cm^3 in January 2006. The bulk density variation during the study is given Table 6.3. Monthly variation in bulk density is given in Figure 6.8.

Soil Moisture (Per cent)

The soil moisture of soil at control site was recorded maximum 11.06 ± 1.86 in monsoon 2006 and minimum 3.73 ± 1.29 in summer 2005 ranged between 2.8 to 12.82 per cent . The highest value of soil moisture in soil was found 12.82 per cent during August 2006 and lowest 2.8 per cent in May 2005. The soil moisture variation during the study is given Table 6.3. Monthly variation in soil moisture is given in Figure 6.8.

pH

pH of soil at control site was recorded maximum 7.47 ± 0.058 in monsoon 2005 and minimum 7.23 ± 0.15 in summer 2005 ranged between 7.1 to 7.6. The highest value of pH in soil was found 7.6 during October 2006 and lowest 7.1 in April 2005. The pH variation during the study is given Table 6.3. Monthly variation in pH is given in Figure 6.8.

Electrical Conductivity (mho/cm)

Electric conductivity of soil at control site was recorded maximum 23.75 ± 0.27 in summer 2006 and minimum 23.00 ± 0.163 in winter 2006-07 ranged between 22.5 to 24.1 mho/cm. The highest value of Electric conductivity in soil was found 24.1mho/cm during April 2006 and lowest 22.5 mho/cm in September 2005. The Electric conductivity variation during the study is given Table 6.3. Monthly variation in Electric conductivity is given in Figure 6.9.

Chloride (µg/g)

Chloride concentration in soil at control site was recorded maximum 529.33 ± 114.47 in summer 2005 and minimum 313.80 ± 64.32 in winter 2005-06 ranged between 225.0 to 671.0 µg/g. The highest value of chloride in soil was found 671.0

Table 6.3: Seasonal Variation of Physico-Chemical Characteristics of Soil at Control Site

Para-meters	Unit	Summer 2005		Monsoon 2005		Winter 2005-06		Summer 2006		Monsoon 2006		Winter 2006-07		Summer 2007	
		Mean	SD	Mean	SD	Mean	SD	Mean	SD	Mean	SD	Mean	SD	Mean	SD
Temp.	°C	27.833	2.701	26.500	2.500	17.500	3.482	26.325	4.369	26.933	2.503	17.800	4.537	19.500	2.404
WHC	Per cent	44.183	0.742	37.733	5.519	37.500	2.670	43.720	0.916	37.843	5.826	36.810	1.890	41.295	2.029
Bulk density	mg/cm^3	1.107	0.051	1.207	0.012	1.168	0.076	1.125	0.017	1.220	0.010	1.183	0.070	1.180	0.042
Soil moisture	Per cent	3.733	1.286	10.683	1.994	5.630	0.972	3.825	1.147	11.063	1.857	6.010	1.532	4.440	1.329
pH		7.233	0.153	7.467	0.058	7.310	0.167	7.250	0.058	7.467	0.058	7.400	0.141	7.250	0.071
EC	mho/cm	23.633	0.153	23.100	0.557	23.040	0.182	23.750	0.265	23.300	0.436	23.000	0.163	23.650	0.212
Cl	µg/g	529.333	114.474	446.667	62.067	313.800	64.317	483.000	148.272	458.333	58.969	336.750	69.014	314.500	47.376
Na	µg/g	338.000	124.864	95.333	37.005	183.800	143.796	356.250	91.940	104.000	35.384	145.500	105.583	380.000	12.728
K	µg/g	33.000	2.646	30.333	2.517	33.200	0.837	34.000	1.414	30.667	2.082	33.250	1.708	34.500	0.707

Table 6.4: Mean Values (±SD) of Physico-chemical Characteristics of Soil at Control and Experiment Sites

Parameters	*Unit*	*2005-06*				*2006-07*			
		Control Site	*SD*	*Experimental sites*	*SD*	*Control Site*	*SD*	*Experimental sites*	*SD*
Temperature	°C	22.625	5.527	22.558	5.707	22.942	5.694	23.042	5.397
WHC	per cent	39.646	4.205	38.825	4.271	39.645	4.165	39.136	4.169
Bulk density	mg/cm^3	1.157	0.066	1.180	0.066	1.179	0.051	1.203	0.059
Soil moisture	per cent	6.242	3.052	6.054	3.163	6.493	3.179	6.220	3.144
pH		7.321	0.156	7.361	0.262	7.358	0.124	7.428	0.258
EC	mho/cm	23.242	0.380	25.742	1.960	23.392	0.425	26.404	1.976
Cl	μg/g	401.583	116.923	601.667	119.654	413.417	116.613	614.667	121.975
Na	μg/g	217.667	148.153	386.667	189.092	223.833	141.224	395.500	182.030
K	μg/g	32.583	2.193	33.917	2.314	32.917	2.021	34.000	2.335

µg/g during June 2006 and lowest 225.0 µg/g in January 2006. The chloride variation during the study is given Table 6.3. Monthly variation in chloride is given in Figure 6.9.

Potassium (µg/g)

Potassium concentration in soil at control site was recorded maximum 34.50 ± 0.71 in monsoon 2005 and minimum 30.33 ± 2.52 in winter 2006-07 ranged between 28.0 to 36.0 µg/g. The highest value of potassium in soil was found 36.0 µg/g during April 2005 and lowest 28.0 µg/g in August 2005. The potassium variation during the study is given Table 6.3. Monthly variation in potassium is given in Figure 6.9.

Sodium (µg/g)

Sodium concentration in soil at control site was recorded maximum 380.00 ± 12.73 in monsoon 2005 and minimum 95.35 ± 37.01 in winter 2006-07 ranged between 11.0 to 468.0 µg/g. The highest value of sodium in soil was found 468.0 µg/g during April 2005 and lowest 11.0 µg/g in October 2005. The sodium variation during the study is given Table 6.3. Monthly variation in sodium is given in Figure 6.9.

Correlation Coefficient

Correlation between physico-chemical parameters and heavy metals of agricultural soil are given in Table 7.14 for each metal. For cadmium correlation was found maximum 0.8421 and 0.7499 with water holding capacity and sodium respectively. Correlation for copper was found maximum 0.9134 and 0.9454 with temperature and Electric conductivity respectively. Correlation for iron was found maximum 0.8736 and 0.8436 with water holding capacity and sodium respectively. Correlation for manganese was found maximum 0.8162 and 0.9076 with temperature and Electric conductivity respectively. Correlation for nickel was found maximum 0.7798 and 0.8286 with water holding capacity and Electric conductivity respectively. Correlation for lead was found maximum 0.8483 and 0.8596 with water holding capacity and Electric conductivity respectively. Correlation for zinc was found maximum 0.8513 and 0.7586 with temperature and Electric conductivity respectively.

Correlation between physico-chemical parameters and heavy metals of soil at control site are given in Table 7.16 for each metal. For cadmium correlation was found maximum 0.8706 and 0.7532 with water holding capacity and Electric conductivity respectively. Correlation for copper was found maximum 0.9142 and 0.8290 with temperature and chloride respectively. Correlation for iron was found maximum 0.8397 and 0.8795 with water holding capacity and sodium respectively. Correlation for manganese was found maximum 0.8814 and 0.7925 with temperature and chloride respectively. Correlation for nickel was found maximum 0.7267 and 0.6699 with temperature and water holding capacity respectively. Correlation for lead was found maximum 0.8608 and 0.8055 with water holding capacity and Electric conductivity respectively. Correlation for zinc was found maximum 0.8704 and 0.7613 with temperature and chloride respectively.

Correlation between heavy metals of groundwater and heavy metals of agricultural soil are given in Table 7.20a for each metal. For cadmium correlation was found maximum 0.7611 and 0.6466 with cadmium and lead respectively. Correlation for copper was found maximum 0.9733 and 0.9121 with iron and nickel respectively. Correlation for iron was found maximum 0.8074 and 0.8628 with manganese and lead respectively. Correlation for manganese was found maximum 0.9279 and 0.9548 with nickel and zinc respectively. Correlation for nickel was found maximum 0.9458 and 0.9475 with manganese and nickel respectively. Correlation for lead was found maximum 0.9482 and 0.9664 with copper and manganese respectively. Correlation for zinc was found maximum 0.8065 and 0.5301 with iron and nickel respectively.

Correlation between heavy metals of agricultural soil and heavy metals of effluent drain sediment are given in Table 7.21 for each metal. For cadmium correlation was found maximum 0.7650 and 0.7646 with copper and manganese respectively. Correlation for copper was found maximum 0.2703 and 0.2542 with cadmium and iron respectively. Correlation for iron was found maximum 0.2989 with cadmium. Correlation for manganese was found maximum 0.7224 and 0.5619 with cadmium and iron respectively. Correlation for nickel was found maximum 0.8617 and 0.8799 with cadmium and iron respectively. Correlation for lead was found maximum 0.8432 and 0.7390 with copper and manganese respectively. Correlation for zinc was found maximum 0.7216 and 0.8553 with manganese and lead respectively.

6.4 Discussion

Soil's physico-chemical parameters may demonstrate the pollution status in the industrial region, so, physical and chemical characteristics are important for the assessment of contamination and pollution load.

6.5 Physico-chemical Parameters of Agriculture Soil

The temperature of soil greatly affects the physical, biological and chemical processes occurring in soil system. Biological decomposition can come to a near stand still, thereby limiting the rate at which nutrients such as nitrogen, phosphorus, sulphur and calcium are made available. Absorption and transport of water and nutrient ions by higher plants are adversely affected by low temperature (Brady, 1995). The temperature of field soils is not subject to radical human regulation. However, two kinds of management practices have significant effects on soil temperature, those that keep some type of cover or mulch on the soil and those that reduce excess soil moisture. There is not any significant variation in the temperature of agricultural soil (17.4-27.8 °C) and control soil (17.5-27.8 °C). The chemical processes and activities of microorganism which convert plant nutrients into available forms are also materially influenced by temperature. Removal of excess water from soil may facilitate changes in soil temperature. Humus content, good tillage, dark color of soil are the some other factors which influence the soil temperature (Miller and Turk, 2002).

The phenomenon of capillarity is a common one, the classic example being the movement of water up a wick when the lower end is immersed in water. Capillary forces are at work in all moist soils. However, the rate of movement and the rise in height are less than one would expect on the basis of soil pore size. One reason is that soil pores are filled with air, which may be entrapped, showing down or preventing the movement of water by capillary. Usually the height of the rise resulting from capillarity is greater with fine textured soils if sufficient time is allowed and the pores are not too small. This is readily explained on the basis of the capillary size and the continuity of the pores. With sandy soils the adjustment is rapid but so many of the pores are non-capillary that the height of rise can not be great (Brady, 1995). Water holding capacity is a property as to water quantity is absorbed or retained by given amount of soil (Ramani and Srivastava, 1989). Water held in the soil pores with varying degrees of tenacity depending on the amount of water present and the size of pores in soils. Depending on the soil, as much as one forth to two thirds of the moisture may remain in the soil after the plants have wilted or died due to the lack of soil water. Water holding capacity was found similar in agricultural soil (36.1-43.7 per cent) and control soil (36.8-44.2 per cent) with much difference between soil and sediment. Yuandong *et al.* (2006) described the similar trend of water holding capacity on the basis of soil texture and chemical composition in the agricultural soil system.

Bulk density relates to the combined volumes of the solids and pore spaces, soils with a high proportion of pore space to solids have lower bulk densities than those that are more compact and have less pore space. Consequently, any factor that influences soil pore space will affect bulk density. Mainly soil texture and compactness strongly influence the bulk density of soils (Brady, 1995). The pore space of a soil is that portion of the soil volume occupied by air and water. The amount of pore space is determined by the arrangement of the solid particles. If the particles lie close together, as in stands or compact sub-soils, the total porosity is low. If they are arranged in porous aggregates, as is often the case in medium textured soils high in organic matter, the pore space per unit volume will be high. Soil aeration may influence the availability of nutrients to plants (Miller and Turk, 2002). The bulk density was found similar between agricultural soil (1.15-1.24 g/cm^3) and control soil (1.10-1.22 g/cm^3) Bulk density was similarly described by Yuandong *et al.* (2006) and Krebs *et al.* (1999) in soil environment.

The structure of the soil influences its moisture content. A well granulated soil has more total pore space than one with poor granulation or one that has been compacted. The reduced pore space may be reflected in a lower water holding capacity. The compacted soil also has to higher proportion of small and medium sized pores, which tend to hold the water with greater tenacity than do the larger pores. Moisture content of agricultural soil (3.5-10.7 per cent) and control soil (3.7-11.06 per cent) were found similar during the study. Rainwater leaching have made more efficient if antecedent moisture content of the soil profile was increased by saline water irrigation just before the onset of monsoons. On a wet soil, a rainfall event is expected to displace the salts to a greater depth. Conservation and proper utilization of rainwater for leaching salts is also utmost important.

The measurement of soil pH shows the acidity and basic nature of the soil. The soil is acidic if the pH is lower, while the higher pH means that soil is basic. The pH value was observed slightly high (7.11-7.63) in agriculture land, which shows that agriculture land is going to alkaline in nature which confirms that agriculture land again disturbed by its proper pH values whereas due to the decay of organic matter has produced the acidic properties of soil. If the decomposition of organic matter is there in soil, it will turn the pH low, other wise some foreign substances may alter the soil nature and properties towards alkaline (Nongkynrih *et al.*, 1996). Initial soil pH is an important property of the soil because of its influence on the infiltration rate of the soil (Gupta *et al.*, 1989). There are several causes of minor fluctuation in soil pH. For instance, the movements of salts into and out of soil moisture moves up and down the profile will influence pH. Similarly, the pH of mineral soils declines during the crop-growing season as a result of the acids produced by microorganisms and by the roots of higher plants. When soil temperatures are low, an increase in pH often is noted because biotic activities during these times are considerably slower.

pH of soil significantly influences other soil chemical properties as well as biological organisms. High pH of textile industrial effluent can hampers plant crop growth (Sial *et al.*, 2006). The soil pH significantly affects the availability of most of the chemical element of importance to plants and microbes. Likewise, the tendency for toxic levels of element such as iron, manganese, aluminum to be established at low pH and deficiencies of iron and manganese at high pH have been mentioned. Continuous application of effluents with high pH will render the soils unfit for further cultivation (Sial *et al.*, 2006). Soil pH was found slightly alkaline in agricultural soil (7.1-7.6) and control soil (7.2-7.4). Similar trend of pH fluctuation have been described by Kasem and Singh (1999) in agricultural soil environment.

Electric conductivity is the measure of current carrying capacity, which gives a clear idea of soluble salts present in soil. Conductivity generally depends upon the dilution of suspension. However, the salt composition is high in the water having high electric conductivity, but it may be beneficial for those soils having a lack or deficiency of some salts. The electric conductivity of agricultural soil and control soil was found in the range of 24.1-27.7 and 23.0-23.7 mho/cm respectively with some minor fluctuations. Electric conductivity has positive correlation with all heavy metals *viz.* Cd (0.3965), Cu (0.9454), Fe (0.4198), Mn (0.9076), Ni (0.8286), Pb (0.8596) and Zn (0.7586) in agricultural soil.

Chloride was found very high in effluent drain sediment and slightly low as with similar fluctuation in pond bed sediment, but in agricultural soil and soil near textile industries chloride was found comparatively low because of the high settlement of chloride in drain sediment from textile industrial effluent; while rest was reach in agricultural soil by various contacts. The chloride in agricultural soil and control soil was found 499.5-734.3 and 313.8-529.3 µg/g respectively with some minor fluctuations. Chloride has positive correlation with all heavy metals *viz.* Cd (0.3654) Cu (0.8479), Fe (0.3249), Mn (0.7254), Ni (0.6603), Pb (0.6972) and Zn (0.7744) in agricultural soil environment. Manchanda (1990) has also reported that pulse crops such as pea, mungbean, gram, lentil, and broadbean suffered more due to toxic accumulation of chloride, when irrigated with Cl ion then SO_4^{-2} dominant (70 per cent) water of more salinity.

Potassium is extremely mobile with in the plant and helps regulate the opening and closing of stomata in the leaves and the uptake of water by root cells. At any one time most of the potassium is in the primary mineral and non-exchangeable forms. The exchangeable potassium is the chief source of available potassium in soil (Jha and Pandey, 1984). Potassium plays many essential roles in plants. It is an activator of dozens of enzymes responsible for such plants processes as energy metabolism, starch synthesis, nitrate reduction, and sugar degradation. To maintain the level of exchangeable and soil solution potassium high enough to encourage good crop production, chemical fertilizer are used. Potassium may deficient in alkaline soil and an excess of potassium may delay maturity of plant. In some crops it may tend to accumulate in the leaf and stem rather then in the grain (Miller and Turk, 2002). The potassium in agricultural soil and control soil was found almost in uniform pattern 31.0-36.5 and 30.3-34.5 µg/g respectively with some minor fluctuations. Potassium has slightly positive correlation with many heavy metals *viz.* Cd (0.6852), Cu (-0.1689), Fe (0.7116), Mn (0.0611), Ni (0.2623), Pb (0.2369) and Zn (-0.6269) in agricultural soil. Korfali and Davies (2004) and Soltan (1999) have been exhibited a clear idea about the fluctuations in potassium concentrations in soil system.

Sodium in alkali soils results in a deflocculation of the colloids and hence in a breaking down of the soil structural units. This puddle condition renders the soil more or less impervious and retards entrance of irrigation and rain water impedes drainage. In heavy textured soil, the penetration of roots may be restricted by the deflocculated zone. Aeration is also much reduced, setting up aeration conditions and resulting in the formation of reduced compounds which are toxic to plants (Miller and Turk, 2002). The highly alkaline soil was due to the presence of sodium carbonate. Large quantity of adsorbed sodium repressed the availability of several nutrients especially, iron, manganese and phosphorous (Miller and Turk, 2002). The occurrence of sodium in agricultural soil and control soil was found very high 192.0-599.0 and 95.33-380.0 µg/g respectively with some minor fluctuations. Sodium has slightly positive correlation with many heavy metals *viz.* Cd (0.7499), Cu (0.0853), Fe (0.8436), Mn (0.3169), Ni (0.4813), Pb (0.4876) and Zn (-0.4581) in agricultural soil. Similar observations were made by Soltan (1999) and Korfali and Davies (2004) in agricultural soil system. Desalination of soil is usually accomplished by impounding of water at the surface of the alkaline soil. The relationship in between reduction of salts and exchangeable sodium from a given depth of soil and the amount of leaching water for the sandy loam soil (Minhas and Khosla, 1986). The amount of soluble salts in soil can be reduce to safe limits for agriculture crops by leaching (Rao *et al.*, 1987).

The effects of water quality parameters on the changes in physico-chemical characteristics of soils and crop productivity have been studied and evaluating the suitability of water for irrigation, however, integration of the interactive and interdependent processes of various water quality indices with soils have required to obtain comprehensive knowledge of the whole system. It requires the development of computerized models for providing a quantitative description of the dynamic behavior of soils upon interaction with irrigation water quality and its relation with crop productivity. Modeling efforts should go a long way in making generalizations based on the experience gained under specific conditions and extrapolation of experimental findings to different locations and seasons in the study area.

Chapter 7

Distribution of Heavy Metals

7.1 Heavy Metals

Metals that are naturally introduced into the waterbody come primarily from such sources as rock weathering, soil erosion, or the dissolution of water-soluble salts. Naturally occurring metals move through aquatic environments independently of human activities, usually without any detrimental effects. Humans consume metallic elements through both water and food. Some metals such as sodium, potassium, magnesium, calcium, and iron are found in living tissue and are essential to human life-biological anomalies arise when they are depleted or removed. Probably less well known is that currently no less than six other heavy metals including molybdenum, manganese, cobalt, copper, and zinc, have been linked to human growth, development, achievement, and reproduction (Vahrenkamp, 1979; Friberg *et al.*, 1979). Even these metals, however, can become toxic or aesthetically undesirable when their concentrations are too great. Several heavy metals, like cadmium, lead, and mercury, are highly toxic at relatively low concentrations, can accumulate in body tissues over long periods of time, and are nonessential for human health.

The toxic heavy metals entering the ecosystem may lead to geo-accumulation, bioaccumulation and biomagnifications. Heavy metals like Fe, Cu, Zn, Ni and other trace elements are important for proper functioning of biological systems and their deficiency or excess could lead to a number of disorders. Food chain contamination by heavy metals has become a burning issue in recent years because of their potential accumulation in biosystems through contaminated water, soil and air. Therefore, a better understanding of heavy metal sources, their accumulation in the soil and the effect of their presence in water and soil on plant systems seem to be particularly important issues of present-day research on risk assessments (Lokeshwari and Chandrappa, 2006).

7.2 Historical Resume

Different researchers from India and abroad have been discussed various aspects of heavy metals pollution of water and soil in term of characteristics of water bodies, affected agricultural soils and bottom sediments from time to time. This review includes some of important contribution made in the field of industrial pollution, with references to heavy metals parameters of groundwater, surface water, characterization of textile industrial effluents, contaminated agricultural irrigated soil and bottom sediment soils by different authors of country and abroad.

Aggarwal and Mehrotra (1952) surveyed the soil and soil water affected by effluent in some selected regions of Uttar Pradesh. Richards (1954) evaluated the quality of saline and alkali soils and gave a technique for the improvement of these sick soil systems in the agricultural sector for good performance of agriculture productivity.

Kanwar (1961) assessed the quality of irrigation water as an index of suitability for irrigation purposes. Darra *et al.* (1964) evaluated the water quality of irrigation water used in Rajasthan. Saksena *et al.* (1966) evaluated the quality of groundwaters for irrigation in Ahor developmental block, Jalore. Bhakuni and Bopardikar (1967) stated the method of recovery Zinc from the spinning bath waste of Viscos reyon factory by using ion exchange method and found it significant useful in metal removal process.

Williams (1972) described the characteristics and various properties of Metals. Katz (1975) highlighted the effect of heavy metals on some environmental components and biotic organisms of aquatic environment in New York. Dowedy and Larson (1975) reported the availability of some metals in various vegetable crops in a polluted region in the vicinity of an industrial area. Dulka and Risby (1976) stated ultra trace metals in some environmental and biological systems in the vicinity of textile industrial sector. Shivakumar *et al.* (1977) made a geo-graphical approach for toxic trace element pollution in groundwater around Patancheru and Bollaram Industrial area, A.P., India.

Martin and Cougherty (1982) stated the basic monitoring techniques for detection of heavy metals pollution through the biological agents. Ireland (1983) described the heavy metals uptake and their distribution in tissues of organisms. Numberg (1984) studied on the volumetric approach in trace metal chemistry of natural waters and atmospheric precipitation.

Nriagu and Pacyna (1988) carried out a study on the quantitative assessment of worldwide contamination of air, water and soil with trace metals. Mishra *et al.* (1990) reported the occurrence of heavy metals in crop plants of Sanagar more than the permissible limits presence of which may cause the adverse effect on the plant/ human health and may cause high mortality of fish population in affected aquatic environment.

Singh *et al.* (1994) highlighted the degradation of water and soil quality of Parwanoo area with respect to heavy metals. Sharma and Patel (1996) determined copper in water of Raipur with the help of spectrophotometer. Prasad and Chandra

(1997) studied well water quality status of Hyderabad and Khurshid and Shabeer (1997) observed water quality degradation due to heavy metal pollution.

Mohan *et al.* (1998) observed heavy metals (Fe, Pb, Ca and Zn) in the groundwater of Naini industrial area; Dist. Allahabad, U.P. Kashem and Singh (1999) had also found the decreasing concentration of Cu, Mn, Pb and Zn with increasing distance from disposal point. Stetzenbach *et al.* (1999) evaluated the heavy metals in groundwater and used multivariate statistical analysis.

Tyagi and Budhi (2000) postulated the degradation of groundwater quality due to heavy metals in industrial area of India. Kumar *et al.* (2001) studied on the impact of textile industry on groundwater quality of Sanagar, Jaipur and found the high amount of chemicals (e.g. sulphuric acid and acetic acid) and dyes (e.g. chrome and metal dyes) suspended solids and toxic metals in the waste which may enter in the groundwater and deteriorate its quality.

Siddiqui and Pathania (2002) made some studies on heavy metals in surface and groundwater of Jalandhar and Ludhiana districts of Punjab state. Shrivastava *et al.* (2003) collected soil samples from surface of the soil from different agriculture fields in the Khandesh region and Tapi River sediment samples were also collected from five different stations which were 7-8 km away from each other and detected the concentrations of heavy metals have been determined by ICP-AES.

Khan *et al.* (2003) harvested some *Solanum melongena* plant in a laboratory experiment using five different levels of textile wastewater and observed the contamination of heavy metals as 1.417 mg/g of Zn, 1.003 mg/g of Cu, 0.378 mg/g of Ni, 0.378 mg/g of Cd, 0.773 mg/g of Cr, 1.139 mg/g of Pb and 0.427 mg/g of Co in the soil of pots treated with highest ratio of distilled water and wastewater. Yadav and Sumanlata (2003) indicated the heavy metal pollution in groundwater in Bahadurgarh block of districts Jhajjar, Haryana.

Malik *et al.* (2004) evaluated the accumulation of heavy metals in crop plants through irrigation of contaminated groundwater in Panipat region and emphasized on bioremediation processes for combat groundwater pollution. Chao *et al.* (2005) reported the heavy metal contaminated site in Taiwan and described the status and distribution of heavy metals in environment.

It is necessary to reduce the levels of toxic metals from industrial effluents before discharging into surface waters. The general methods of treating wastewater-containing metals are coagulation, ion exchange, reverse osmosis and adsorption. Adsorption is a highly effective physico-chemical treatment for removing heavy metals for effluents. Due to high cost of activated carbon, efforts are being directed towards finding low cost adsorbents. Recently much interest has been shown for removal of heavy metals using a large number of low cost materials. Coal fly ash, the solid waste of power plants is one of the cheapest and non-conventional adsorbents (De, 2005).

Zheng *et al.* (2006) assessed the heavy metals pollution of agricultural soil in Guanzhong district of China. Lokeshwari and Chandrappa (2006) evaluated the impact of heavy metals contamination of a Lake on soil and cultivated vegetation near Bangalore and found the presence of some heavy metals in rice and vegetables, beyond the limit of Indian Standards and calculated metal transfer factors from soil to vegetation are found significant for Zn, Cu, Pb and Cd. Hedge *et al.* (2006) revealed the provenance of heavy minerals with special reference to ilmenite of the Honnevar beach in central west coast of India and observed some trace heavy elements like Co, Cr, V and Ni in the sandy soil of the beach. Singh (2006) indicated the contamination of groundwater of Ganges-Brahmaputra river basin in north east India and found the groundwater unfit for direct human consumption and noticed the rice and vegetable fields irrigated by groundwater received a large amount of arsenic through shallow tube wells in the affected areas of north 24 *Parganas* and Murshidabad, West Bangal.

Malik *et al.* (2007) assessed the migration of heavy metals in the groundwater regime of Panipat industrial area on the behalf of different spatial variations the heavy metals were found in maximum quantities at point source of pollution. Malik and Bharti (2007) studied on soil quality of irrigated agricultural fields in a textile industrial area of Panipat city. Va'zquez, *et al.* (2007) monitored the heavy metal pollution in San Antonio Bay, Rio Negro, Argentina. Obiri (2007) determined the heavy metals in water from boreholes in Dumasi in the Wassa West District of Western Region of Republic of Ghana. Huang *et al.* (2007) indicated the speciation and mobility of heavy metals in mud in coastal reclamation areas in Shenzhen in China. Walker *et al.* (2007) identified the influence of heavy metals and mineral nutrient supply on *Bituminaria bituminosa.*

Golia *et al.* (2007) measured the accumulation of metals on tobacco leaves (primings) grown in an agricultural area in relation to soil. Mico *et al.* (2007) described the comparison of two digestion methods for the analysis of heavy metals by flame atomic absorption spectroscopy. Padmavathiamma and Li (2007) explained the phytoremediation technology through the hyper-accumulation metals in plants and different aspects of phytoremediation technology and the biological mechanisms underlying phytoremediation.

Analysis of Metals in Groundwater, Pond Water, Textile Effluents, Agriculture Soil, Effluent Drain Sediment and Pond Sediment

Collected, preserved, diluted water samples were prepared for the evaluation of metal concentrations with the help of AAS 4129. In the case of soil samples, a suspension (1:5) should be prepared in the distilled water; rest processes are like surface water and groundwater samples.

Metals enter in the surface, groundwater and soil/sediments through the disposing of toxic textile effluents of dye houses. For every metal a standard was required and concentration may be detected by AAS. Diluted and completely digested samples were required for metal detection by AAS 4129. Metal elements were determined with the help of Atomic absorption spectrophotometer (Model AAS 4129). Alan Walsh first developed the technique of atomic absorption spectrophotometry in 1955, and since then it has emerged as a powerful tool in quantitative analysis of more than 70 metals. The major advantages of the method are that it is free from any kind of interferences and very small concentrations are possible to be measured (De, 2002). The AAS is based on the principle that atoms of elements, which normally remain in ground state under flame condition, absorb energy when subjected to radiation of specific wavelength. The absorption of radiation is proportional to the concentration of atoms of the element. The absorption of radiation by the atoms is independent of the wavelength of absorption and temperature of the atoms. These to feature provide AAS a distinct advantage over flame spectroscopy.

Copper (Cu)

Copper in the natural waters also results in higher concentration due to pollution. It is used with sulphate as a pesticides and also separately as an algicide. Although, it passes as such through the body but there is evidence of accumulation of trace quantities in liver. The limits of it in the standards are not due to its toxic effects but are due to its taste producing capacity.

Preparation of 1000 Cu (µg/ml) standard: Dissolve 1 gm of Cu in 50 ml of 6N of HNO_3 and dilute to 1 L to give 1000 µg/ml Cu.

Cadmium (Cd)

Cadmium is fairy dense silver white malleable and most toxic metal, which melts at 320.9°C. Cadmium is used for dye and paint pigment production, batteries, ceramic industries, wooden industries and plastic production. The Cd found in a sample of Ca is likely in a inorganic form and may be relatively less toxic to plants and animals, while same amount of Cd when present in zinc or phosphates may be highly toxic to plants and animals, especially when there is calcium deficiency in soil or livestock rations.

Preparation of 1000 Cd (µg/ml) standard: Dissolve 1 gm of Cd metal in 20 ml of 5N HCl containing 0.5 ml of concentrated HNO_3 and dilute to 1 L to give 1000 µg/ml Cd.

Iron (Fe)

Iron is one of the most abundant elements of the rocks and soil, ranking fourth by weight. All kinds of waters including groundwater have appreciable quantities of iron. Iron has more solubility at acidic pH, therefore large quantities of iron are leached out from the soils by acidic waters. In the alkaline medium iron remain comparatively low in soluble phase.

Preparation of 1000 Fe (µg/ml) standard, dissolve 1.000 g of iron wire in 50 ml of (1+1) HNO_3 and dilute to 1 liter with de-ionized water to give 1000 µg/ml Fe.

Manganese (Mn)

Although not a toxic metal, Mn imparts objectionable and tenacious stains to laundry plumbing fixtures and clothing. It occurs in domestic wastewater, industrial effluents, acid mine drainage and receiving strums and therby enters water bodies. Mn is relatively non-toxic to animals, but toxic to plants at higher levels.

Preparation of 1000 Mn (µg/ml) standard: Dissolve 1 gm of Mn in 50 ml of 6N of HNO_3 and dilute to 1 L to give 1000 µg/ml Mn.

Nickel (Ni)

At high concentrations, nickel has toxic properties. Aquatic organisms have varying sensitivities to nickel salts depending on the water's pH, hardness, alkalinity and type of nickel compound under consideration. Nickel as a metal is a carcinogen.

Preparation of 1000 Ni (µg/ml) standard: Dissolve 1 gm of Ni metal in 50 ml of 6N of HNO_3 and dilute to 1 L to give 1000 µg/ml Ni.

Lead (Pb)

It is a dense soft metal and is quite resistant to corrosion. It has melting point of 327°C. Major source of lead is PbS (Galena). Lead is used for pipes, solders, electrodes, batteries, newsprint, and pigments in paints. Lead is also used in insecticides, beverages, ointments and synthetic dyes. The chemical form of lead determines its solubility in water and biological fluids, the extent of its fixation on soil and type of chemical reactions occurring in atmospheric, aquatic as well as soil environment.

Preparation of 1000 Pb (µg/ml) standard: Dissolve 1 gm of Pb metal in 20 ml of 6N of HNO_3 and dilute to 1 L to give 1000 µg/ml Pb.

Zinc (Zn)

Zinc is present in high concentrations in the wastes from pharmaceuticals, paint, galvanizing, cosmetics, dyes and pigments, etc., and their discharge increases its concentration in appreciable amounts in the waters. Zinc is very essential micronutrient in human beings and only at very high concentration it may cause some toxic effects. Zinc salts produce an undesirable taste to the water and causes water to appear milky and on boiling, a greasy surface scum may also form in the water (Mani *et. al.*, 2005).

Preparation of 1000 Zn (µg/ml) standard: Dissolve 1 gm of Zn metal in 40 ml of 5N HCl and dilute to 1 L to give 1000 µg/ml Zn.

Standards and Calibration

Whenever, possible metals and metals oxides are to be preferred in the preparation of standards. This enables the analyst to dissolve them that samples and standards will contain identical elements and hence minimizes any chemical or physical interference effects.

Only concentrated standards (above 1000 µg/ml) should be held in storage. Working standards should be diluted from standards stock solutions only when needed. At low concentrations (less than 10 µg/ml) solutions have been found to deteriorate quite quickly because of the absorption on the walls of the container. Similar standard addition solutions can be prepared containing all of the required elements. This will avoid multiple splitting of the sample for individual elements and reduce the amount of time spent in preparing standards.

Before weighting, standards materials should be treated to ensure that they are in a standard state.

Metals

Wash with acetone and ether to remove any oil layers. Remove any oxide coating by abrasion with emery cloth or by acid pickling and drying.

Oxides

Dry at 110 °C for two hours. If necessary, heat to evaluated temperatures to remove bound water.

Compounds

Equilibration at constant water content, or drying at 110 °C for two hours to remove any water.

Standards of known concentrations were prepared and evaluated the concentration of metal in water/ soil digested sample. Every metal has some specific properties and atomic weight, so for every metal different condition of standard and calibration was applied for every metal (ECIL methods manual, 2004).

Standard Conditions for Cadmium (Cd)

Stock Standard Solution

Cadmium 1000 µg/ml. Dissolve 1.000 g of cadmium metal in a minimum volume of (1+1) HCl. Dilute to 1 liter with 1 per cent (V/V) HCl.

Operating Parameters

Instruments:	ECIL AAS
Wave length:	228.8 nm
Slit setting:	0.5 nm
Light source:	Hollow cathode lamp
Flame type:	Air-acetylene flame oxidizing (lean, blue)

Sensitivity

For the standard conditions, the sensitivity is about 0.02 µg/ml Cd for 1 per cent absorption. A standard containing 1.5 µg/ml Cd will typically give an absorbance reading of about 0.33 absorbance units (about 53 per cent absorption).

Linear Working Range

For the standard conditions, the working range for Cd is linear up to concentrations of approximately 1.5 µg/ml in aqueous solution.

Other Analytical Lines

Wave Length	*Slit*	*Relative Sensitivity*
22.5 nm	0.5 nm	1
326.1 nm	0.5 nm	410

Standard Conditions for Copper (Cu)

Stock Standard Solution

Copper, 1000 µg/ml. Dissolve 1.000 g of copper metal in a minimum volume of (1+1) HNO_3. Dilute to 1 liter with 1 per cent (V/V) HNO_3.

Operating Parameters

Instruments:	ECIL AAS
Wave length:	324.7 nm
Slit setting:	0.5 nm
Light source:	Hollow cathode lamp
Flame type:	Air-acetylene flame oxidizing (lean, blue)

Sensitivity

The sensitivity is about 0.04 µg/ml Cu for 1 per cent absorption. A standard containing 2 µg/ml Cu will typically give an absorbance reading of about 0.22 absorbance units (39 per cent absorption).

Linear Working Range

For the standard conditions, the working range for Cu is linear upto concentrations of approximately 3 µg/ml in aqueous solution.

Other Analytical Lines

Wave Length	*Slit*	*Relative Sensitivity*
324.7 nm	0.5 nm	1
222.6 nm	0.2 nm	38
244.2 nm	1.0 nm	280

Lamps

With multi element lamps containing nickel or iron, a 0.2 nm spectral slit width should be used with the copper 324.7 nm line.

Standard Conditions for Manganese (Mn)

Stock Standard Solution

Manganese, 1000 µg/ml. Dissolve 1.000 g of manganese metal in a minimum volume of (1+1+ HNO_3). Dilute to 1 liter with 1 per cent (V/V) HCl.

Operating Parameters

Instruments:	ECIL AAS
Wave length:	279.5 nm
Slit setting:	0.2 nm

Light source: Hollow cathode lamp

Flame type: Air-acetylene flame oxidizing (lean, blue)

Sensitivity

For the standard conditions, the sensitivity is about 0.04 µg/ml Mn for 1 per cent absorption. A standard containing 1 µg/ml Mn will typically give an absorbance reading of about 0.11 absorbance units (about 40 per cent absorption).

Linear Working Range

For the standard conditions, the working range for Mn is linear upto concentrations of approximately 2 µg/ml in aqueous solution.

Other Analytical Lines

Wave Length	*Slit*	*Relative Sensitivity*
279.5 nm	0.2 nm	1.0
403.1 nm	0.2 nm	12.5
321.7 nm	0.2 nm	2750.0

Standard Conditions for Nickel (Ni)

Stock Standard Solution

Nickel, 1000 µg/ml. Dissolve 1.000 g of nickel metal in a minimum volume of (1+1) HNO_3. Dilute to 1 liter with 1 per cent (V/V) HNO_3.

Operating Parameters

Instruments: ECIL AAS

Wave length: 232.0 nm

Slit setting: 0.2 nm

Light source: Hollow cathode lamp

Operating current: 3.5 mA

Flame type: Air-acetylene flame oxidizing (lean, blue)

Sensitivity

The sensitivity is about 0.10 µg/ml Ni for 1 per cent absorption. Standard containing 2.5 µg/ml Ni will typically give an absorbance reading of about 0.11 absorbance units (23 per cent absorption).

Linear Working Range

For the standard conditions, the working range for Ni is linear upto concentrations of approximately 3 µg/ml in aqueous solution.

Other analytical lines:

Wave Length	*Slit*	*Relative Sensitivity*
232.0 nm	0.2 nm	1.0
352.5 nm	0.5 nm	5.1
351.5 nm	0.5 nm	11.0
362.5 nm	0.5 nm	540.0

Lamps

With multi-element lamps containing iron, interference may occur when using the 232.0 nm nickel line if nickel is being determined in an iron matrix. A secondary nickel line (352.4 or 341.5 nm) should be used.

Standard Conditions for Lead (Pb)

Stock Standard Solution

Lead, 1000 µg/ml. Dissolve 1.598 g of lead nitrate Pb $(NO_3)_2$ 1 liter of 1 per cent (V/V) HNO_3.

Operating Parameters

Instruments:	ECIL AAS
Wave length:	217.0 nm
Slit setting:	1.0 nm
Light source:	Hollow cathode lamp
Operating current:	5 mA
Flame type:	Air-acetylene flame oxidizing (lean, blue)

Sensitivity

The sensitivity is about 0.25 µg/ml Pb for 1 per cent absorption. Standard containing 6 µg/ml Pb will typically give an absorbance reading of about 0.10 absorbance units (about 34 per cent absorption).

Linear Working Range

For the standard conditions, the working range for Pb is linear upto concentrations of approximately 8 µg/ml in aqueous solution.

Other Analytical Lines

Wave Length	*Slit*	*Relative Sensitivity*
217.0 nm	1.0 nm	1.0
283.3 nm	0.5 nm	2.0
261.4 nm	0.5 nm	36.0
205.3 nm	0.5 nm	340.0

Lamps

With multi-element lamps containing copper, the Cu 216.5 resonance line may interfere with lead determinations at the lead 217.0 nm line. The lead 283.3 nm line should be used instead.

Flame Emission

The most sensitive emission wave length for lead is at 405.8 nm. A nitrous oxide, acetylene flame is recommended. Lead can also be determined at the 368.4, 283.3 and 261.4 nm wave lengths, but with reduced sensitivity. An air acetylene flame may also be used with reduced sensitivity.

Standard Conditions for Zinc (Zn)

Stock Standard Solution

Zinc, 1000 µg/ml. Dissolve 0.500 g of zinc metal in a minimum volume of (1/1) HCl and dilute to 1 liter with 1 per cent (V/V) HCl.

Operating Parameters

Instruments:	ECIL AAS
Wave length:	213.9 nm
Slit setting:	1.0 nm
Light source:	Hollow cathode lamp
Operating current:	5 mA
Flame type:	Air-acetylene flame oxidizing (lean, blue)

Sensitivity

For the standard conditions, the sensitivity is about 0.012 µg/ml Zn for 1 per cent absorption. A standard containing 0.5 µg/ml Zn will typically give an absorbance reading of about 0.18 absorbance units (about 25 per cent absorption).

Linear Working Range

For the standard conditions, the working range for Zn is linear upto concentrations of approximately 1.0 µg/ml in aqueous solution.

Other Analytical Lines

Wave Length	*Slit*	*Relative Sensitivity*
213.9 nm	1.0 nm	1.0
307.6 nm	1.0 nm	8400.0

Standard Conditions for Iron (Fe)

Stock Standard Solution

Iron, 1000 µg/ml. Dissolve 1.000 g of iron wire in 50 ml of (1+1) HNO_3. Dilute to 1 liter with de-ionized water.

Operating Parameters

Instruments:	ECIL AAS
Wave length:	248.3 nm
Slit setting:	0.2 nm
Light source:	Hollow cathode lamp
Operating current:	5 mA
Flame type:	Air-acetylene flame oxidizing (lean, blue)

Sensitivity

For the standard conditions, the sensitivity is about 0.09 μg/ml Fe for 1 per cent absorption. A standard containing 4.0 μg/ml Fe will typically give an absorbance reading of about 0.19 absorbance units (about 35 per cent absorption).

Linear Working Range

For the standard conditions, the working range for Fe is linear up to concentrations of approximately 4.0 μg/ml in aqueous solution.

Other Analytical Lines

Wave Length	*Slit*	*Relative Sensitivity*
248.3 nm	0.2 nm	1.0
392.0 nm	0.2 nm	270.0

Some essential knowledge about heavy metals is given below for basic setup and for understanding the sensitivity of each metal studied.

Table 7.1: Standard Criteria for Analyzing Heavy Metals by Atomic Absorption Spectrophotometer (AAS)

Sl.No.	*Name*	*Symbol*	*Wavelength nm*	*Flame Gases*	*Instrument Detection Limits Mg/l*	*Sensitivity Mg/l*	*Optimum Concentration Range Mg/l*
1.	Cadmium	Cd	228.8	A-Ac	0.002	0.025	0.05-2
2.	Copper	Cu	324.7	A-Ac	0.01	0.1	0.02-10
3.	Lead*	Pb	279.5	A-Ac	0.01	0.05	0.1-10
4.	Manganese	Mn	232.0	A-Ac	0.02	0.15	0.3-10
5.	Nickel	Ni	283.3	A-Ac	0.05	0.5	1-20
6.	Zinc	Zn	213.9	A-Ac	0.005	0.02	0.05-2
7.	Iron	Fe	248.3	A-Ac	0.02	0.12	0.3-10

A-Ac: Air-acetylene.

* The more sensitive 217.0 nm wave length is recommended for instruments with background correction capabilities (APHA, 1998).

Table 7.2: Periodic Position of Selected Heavy Metals

Sl.No.	Name	Symbol	Atomic Number	Atomic Weight
1.	Cadmium	Cd	48	112.411
2.	Copper	Cu	29	63.546
3.	Lead	Pb	82	207.2
4.	Manganese	Mn	25	54.93805
5.	Nickel	Ni	28	58.6934
6.	Zinc	Zn	30	65.39
7.	Iron	Fe	26	55.845

Source: IUPAC, 1994.

About AAS 4129 Model

ECIL's Atomic Absorption Spectrophotometer, AAS 4129 is a PC based instrument for absorption and emission analysis. It is used for quantitative elemental analysis mainly like Copper, Iron, Zinc, Lead etc., by measuring the absorbance of a sample atomized in a flame. Some of the metals like Sodium, Potassium, Calcium, and Lithium are analyzed by emission method. PC does the data processing and partial control of the instrument.

The liquid sample is nebulised (reduce to a spray) by the support gas. The fuel gas is mixed and burnt over the burner head. A modulated light from Hollow Cathode lamp of a pre-selected element travels through the flame and enters the monochromator. The monochromator selects the appropriate resonance line of the selected element and directs it on to the PMT. The DC signal generated by the flame emission is rejected by FE compensation circuit and the original signal of the lamp is amplified, phase sensitively rectified averaged by an integrator to obtain sample signal. A computer processes the sample signal for photometric computations (per cent T, ABS or CONC). Instrument controls like wavelength, lamp current, gain, etc. are achieved and maintained by PC.

The cathode material of Hollow Cathode Lamp corresponds to the element of interest. ECIL's Atomic Absorption Spectrophotometer covers the complete range of Atomic Absorption applications. It incorporates State-of-the-art electronics, safety and operational convenience ensuring reliable and accurate analytical performance.

The data analysis and report making is done by the PC including selection of instrument parameters like wavelength, lamp current selection, and EHT to PMT at various wavelength. The software is user-friendly window based GUI interface using menus, function key, prompts, error messages, enabling ease of operation. Information is presented in easy read menus. In most cases recommended values are filled in.

The instrument was calibrated using regression techniques of linear and quadratic methods against standard solutions and concentration of unknown samples can be quantified. Instrument parameters and calibration parameters can be stored, as applications, which can be retrieved as and when, required. A hard copy of the results can be taken with the aid of a printer connected to one of the parallel ports of the PC.

Operating Steps of AAS

There are some main operating steps used in the operation of AAS (ECIL Instructions manual, 2004):

1. Switch on Stabilizer (Power) and computer.
2. Double click AAS NEW software icon on desktop and click power.
3. Select the required element from list.
4. Adjust selected element lamp on top position and switch on that lamp from instrument.
5. Click HC lamp current, 1, 2, 3, or 4.
6. Click peak, A and Bb options will appear, check the lamp before, if OK then click YES.
7. A peak will appear on screen, it should be sharp and compare element wavelength with present one, and the difference should not be more than ± 1 nm.
8. If peak is not OK or difference is > ±1 then click INDEX, after index again check the peak.
9. If massage "*Signal is too high or too low for alignment*" appears, then check the lamp position, lamp and slit dial.
10. If peak is OK then align the burner with the help of line mark visiting card.
11. Switch on air compressor and open gas cylinder (pressure should be in green zone *i.e.* 5-10).
12. Open air from instant (pressure 15-20 LMP)
13. Open gas (Fuel) and adjust pressure 2-4 LMP and ignite immediately.
14. Switch On exhaust.
15. Click ANALYSIS.
16. Adjust HCl signal to green zone *i.e.* aprox. 3V with the help of EHT increase or decrease.
17. Feed blank in capillary and click zeroing.
18. Click CALIBRATION, and put the number of standard and cover of standard.
19. Feed blank in capillary and click zeroing.
20. Feed standard 1 in capillary and click READ.
21. Feed standard 2 in capillary and click READ.
22. Feed standard 3 in capillary and click READ.

23. Click CALIBRATION GRAPH and close, if graph is not OK then re-calibrate.
24. Click SAMPLE and put the number of samples and name of samples.
25. Feed blank in capillary and click zeroing.
26. Feed No. of samples one by one and click READ.
27. After taking the result of unknown click SAVE.
28. Click QUIT and close the gas cylinder tightly.
29. Switch off Air compressor and close gas from instrument.
30. Switch off lamp 1, 2, 3 or 4.
31. Click POWER, switch of exhaust and click quit.
32. For print out results, double click AAS report, LOAD FILE, RES. OPEN, select the file and print result.
33. For print graph, load graph, print graph, res. open and print graph.
34. Now shut down PC, check the gas cylinder before closing the room and switch off printer and main power (stabilizer).

Precautions

Some precautions and highly attention is required during the handling of AAS 4129 and analyzing the heavy metals:

1. If the current wavelength is differed by ± 0.5 nanometers from the actual wavelength for any element after the peak search program, then the system needs INDEXING for correcting the ZERO error in the wavelength.
2. In case of any emergency while the Air – Acetylene Flame/ N_2O–Acetylene is On, immediately put off the fuel switch and then turn the oxidant knob to off position.
3. While igniting $N_2O - C_2H_2$ flame ensure that the N_2O burner is fixed on the burner assembly.
4. Don't try to light up the flame without switching on the air.
5. Don't try to open or remove the Nebulizer when the flame is ON. Always ensure the Nebulizer is in locked position.
6. Always ensure the drain trap is filled with water to avoid back suction and also water drains out freely from spray chamber through the drain trap to the outlet.
7. Always switch on the exhaust system when the flame is ON.
8. Always keep ready your standards, samples, distilled water and then start working on AAS.
9. Check frequently for the gas leakage at the flow points with the soap solution.
10. After completing the analysis work, always ensure that the gas cylinder, compressor outlet is in closed position. Switch off the power supplies.

7.3 Observations

Experimental Site

Heavy Metals in Textile Effluent

Cadmium (ppm)

The cadmium concentration in textile effluent was recorded maximum 0.052 ± 0.042 in summer 2005 and minimum 0.005 ± 0.004 in winter 2006-07 ranged between 0.001 to 0.1 ppm. The highest value of cadmium in textile effluent was found 0.1 ppm during June 2005 and lowest 0.001 ppm in November 2006. The cadmium variation during the study is given Table 7.3. Monthly variation in cadmium is given in Figure 7.1a and Figure 7.1b.

Copper (ppm)

The concentration of copper in textile effluent was recorded maximum 0.44 ± 0.085 in summer 2005 and minimum 0.165 ± 0.066 in winter 2005-06 ranged between 0.1 to 0.49 ppm. The highest value of copper in textile effluent was found 0.49 ppm during March 2006 and lowest 0.1 ppm in November 2005. The copper variation during the study is given Table 7.3. Monthly variation in copper is given in Figure 7.1a and Figure 7.1b.

Iron (ppm)

The concentration of iron in textile effluent was recorded maximum 0.543 ± 0.17 in summer 2006 and minimum 0.240 ± 0.10 in winter 2006-07 ranged between 0.12 to 0.78 ppm. The highest value of iron in textile effluent was found 0.78 ppm during May 2006 and lowest 0.12 ppm in September 2005. The iron variation during the study is given Table 7.3. Monthly variation in iron is given in Figure 7.1a and Figure 7.1b.

Manganese (ppm)

The concentration of manganese in textile effluent was recorded maximum 0.310 ± 0.014 in summer 2006 and minimum 0.110 ± 0.02 in monsoon 2006 ranged between 0.09 to 0.32 ppm. The highest value of manganese in textile effluent was found 0.32 ppm during February 2007 and lowest 0.09 ppm in September 2005. The manganese variation during the study is given Table 7.3. Monthly variation in manganese is given in Figure 7.1a and Figure 7.1b.

Nickel (ppm)

The concentration of nickel in textile effluent was recorded maximum 0.067 ± 0.037 in summer 2006 and minimum 0.007 ± 0.003 in monsoon 2006 ranged between 0.0012 to 0.061 ppm. The highest value of nickel in textile effluent was found 0.061 ppm during April 2006 and lowest 0.0012 ppm in July 2005. The nickel variation during the study is given Table 7.3. Monthly variation in nickel is given in Figure 7.1a and Figure 7.1b.

Lead (ppm)

The concentration of lead in textile effluent was recorded maximum 0.500 ± 0.21 in summer 2005 and minimum 0.250 ± 0.085 in monsoon 2005 ranged between 0.21

Table 7.3: Seasonal Variations of Heavy Metals in Textile Effluents

Heavy Metals	Unit	Summer 2005		Monsoon 2005		Winter 2005–06		Summer 2006		Monsoon 2006		Winter 2006–07		Summer 2007	
		Mean	*SD*	*Mean*	*SD*	*Mean*	*SD*	*Mean*	*SD*	*Mean*	*SD*	*Mean*	*SD*	*Mean*	*SD*
Cadmium (Cd)	ppm	0.052	0.042	0.025	0.037	0.012	0.008	0.028	0.008	0.009	0.001	0.005	0.004	0.019	0.001
Copper (Cu)	ppm	0.440	0.085	0.343	0.058	0.165	0.066	0.430	0.127	0.367	0.091	0.180	0.030	0.340	0.079
Iron (Fe)	ppm	0.527	0.093	0.253	0.130	0.266	0.079	0.543	0.170	0.287	0.131	0.240	0.100	0.350	0.085
Manganese (Mn)	ppm	0.153	0.050	0.127	0.047	0.224	0.072	0.176	0.060	0.110	0.020	0.215	0.034	0.310	0.014
Nickel (Ni)	ppm	0.037	0.016	0.016	0.015	0.031	0.013	0.067	0.037	0.007	0.003	0.029	0.013	0.045	0.007
Lead (Pb)	ppm	0.500	0.207	0.250	0.085	0.422	0.133	0.475	0.111	0.330	0.028	0.408	0.158	0.485	0.092
Zinc (Zn)	ppm	0.173	0.059	0.110	0.012	0.238	0.132	0.222	0.091	0.165	0.064	0.338	0.146	0.275	0.035

Table 7.3b: Seasonal Variations of Heavy Metals in Drain Sediments

Heavy Metals	Unit	Summer 2005		Monsoon 2005		Winter 2005–06		Summer 2006		Monsoon 2006		Winter 2006–07		Summer 2007	
		Mean	*SD*	*Mean*	*SD*	*Mean*	*SD*	*Mean*	*SD*	*Mean*	*SD*	*Mean*	*SD*	*Mean*	*SD*
Cadmium (Cd)	ppm	5.783	0.231	5.103	0.220	4.250	0.544	5.468	0.553	5.157	0.475	5.110	0.266	5.020	0.184
Copper (Cu)	ppm	49.247	4.963	45.897	6.950	46.454	6.678	54.825	7.333	47.527	10.398	49.435	7.303	45.865	5.819
Iron (Fe)	ppm	8467.667	2448.183	5656.000	1216.719	8090.000	938.127	6469.500	1179.615	6362.000	501.728	7921.250	1751.800	5813.500	185.969
Manganese (Mn)	ppm	1047.333	76.002	791.333	51.733	978.600	52.624	1105.000	107.083	699.333	124.243	1028.250	67.869	1021.000	36.770
Nickel (Ni)	ppm	75.000	3.000	58.000	3.000	60.600	5.941	74.250	4.113	58.667	6.028	58.250	8.139	68.000	4.243
Lead (Pb)	ppm	70.000	4.359	60.667	4.509	47.200	3.834	64.000	11.916	60.667	2.517	52.500	2.646	49.500	10.607
Zinc (Zn)	ppm	260.000	83.540	208.000	87.000	151.400	47.184	251.000	96.620	174.667	67.889	143.500	32.296	196.000	197.799

Table 7.3c: Seasonal Variations of Heavy Metals in Pond Water

Heavy Metals	Unit	Summer 2005		Monsoon 2005		Winter 2005–06		Summer 2006		Monsoon 2006		Winter 2006–07		Summer 2007	
		Mean	SD	Mean	SD	Mean	SD	Mean	SD	Mean	SD	Mean	SD	Mean	SD
Cadmium (Cd)	ppm	0.009	0.001	0.001	0.002	0.004	0.002	0.012	0.006	0.002	0.003	0.004	0.002	0.007	0.001
Copper (Cu)	ppm	0.427	0.045	0.397	0.136	0.252	0.084	0.423	0.053	0.403	0.150	0.245	0.070	0.370	0.014
Iron (Fe)	ppm	9.700	3.035	4.233	1.450	4.820	2.961	9.450	1.529	4.567	1.501	3.850	2.111	8.750	0.636
Manganese (Mn)	ppm	0.530	0.061	0.463	0.070	0.172	0.062	0.490	0.122	0.470	0.056	0.170	0.059	0.295	0.035
Nickel (Ni)	ppm	0.127	0.025	0.073	0.033	0.054	0.045	0.130	0.014	0.082	0.022	0.047	0.038	0.115	0.007
Lead (Pb)	ppm	1.800	0.557	0.610	0.412	0.504	0.364	1.725	0.287	0.727	0.448	0.503	0.346	1.165	0.332
Zinc (Zn)	ppm	0.156	0.031	0.110	0.020	0.158	0.066	0.165	0.017	0.113	0.012	0.183	0.025	0.195	0.007

Table 7.3d: Seasonal Variations of Heavy Metals in Pond Sediments

Heavy Metals	Unit	Summer 2005		Monsoon 2005		Winter 2005–06		Summer 2006		Monsoon 2006		Winter 2006–07		Summer 2007	
		Mean	SD	Mean	SD	Mean	SD	Mean	SD	Mean	SD	Mean	SD	Mean	SD
Cadmium (Cd)	ppm	5.717	0.451	4.780	0.455	4.054	0.595	5.410	0.677	4.993	0.359	5.073	0.153	4.795	0.120
Copper (Cu)	ppm	51.520	5.098	43.880	6.877	43.834	7.332	52.463	6.505	43.477	12.357	44.450	13.422	38.645	9.228
Iron (Fe)	ppm	5640.333	2179.050	4312.667	463.092	6013.400	856.579	4356.500	998.172	4157.667	403.026	5844.000	1878.244	3803.500	113.844
Manganese (Mn)	ppm	942.000	48.135	724.000	56.471	836.400	38.708	928.000	56.291	605.000	161.707	883.500	27.791	927.500	4.950
Nickel (Ni)	ppm	63.667	8.083	52.333	9.074	52.800	3.768	63.750	11.117	54.333	7.024	49.000	5.944	56.500	2.121
Lead (Pb)	ppm	69.000	6.245	55.333	9.292	44.000	5.385	62.000	13.491	57.333	6.028	49.250	3.403	44.500	10.607
Zinc (Zn)	ppm	240.000	92.601	142.333	59.341	123.200	42.032	207.250	86.496	153.667	62.963	116.250	32.755	166.500	12.021

to 0.81 ppm. The highest value of lead in textile effluent was found 0.81 ppm during June 2005 and lowest 0.21 ppm in September 2005. The lead variation during the study is given Table 7.3. Monthly variation in lead is given in Figure 7.1a and Figure 7.1b.

Zinc (ppm)

The concentration of zinc in textile effluent was recorded maximum 0.338 ± 0.146 in winter 2006-07 and minimum 0.110 ± 0.012 in monsoon 2005 ranged between 0.09 to 0.51 ppm. The highest value of zinc in textile effluent was found 0.51 ppm during November 2006 and lowest 0.09 ppm in September 2005. The zinc variation during the study is given Table 7.3. Monthly variation in zinc is given in Figure 7.1a and Figure 7.1b.

Heavy Metals in Pond Water

Cadmium (ppm)

The cadmium concentration in pond water was recorded maximum 0.012 ± 0.006 in summer 2006 and minimum 0.001 ± 0.002 in monsoon 2005 ranged between 0.00 to 0.02 ppm. The highest value of cadmium in pond water was found 0.02 ppm during June 2006 and lowest 0.00 ppm in August 2005. The cadmium variation during the study is given Table 7.3c. Monthly variation in cadmium is given in Figure 7.3a.

Copper (ppm)

The concentration of copper in pond water was recorded maximum 0.427 ± 0.045 in summer 2005 and minimum 0.245 ± 0.07 in winter 2006-07 ranged between 1.9 to 13.2 ppm. The highest value of copper in pond water was found 13.2 ppm during April 2005 and lowest 1.9 ppm in October 2005. The copper variation during the study is given Table 7.3c. Monthly variation in copper is given in Figure 7.3a.

Iron (ppm)

The concentration of iron in pond water was recorded maximum 9.70 ± 3.04 in summer 2005 and minimum 4.23 ± 1.45 in monsoon 2005 ranged between 0.09 to 0.6 ppm. The highest value of iron in pond water was found 0.6 ppm during May 2005 and lowest 0.09 ppm in December 2005. The iron variation during the study is given Table 7.3c. Monthly variation in iron is given in Figure 7.3b.

Manganese (ppm)

The concentration of manganese in pond water was recorded maximum 0.530 ± 0.061 in summer 2005 and minimum 0.170 ± 0.059 in winter 2006-07 ranged between 0.09 to 0.6 ppm. The highest value of manganese in pond water was found 0.6 ppm during May 2005 and lowest 0.09 ppm in December 2005. The manganese variation during the study is given Table 7.3c. Monthly variation in manganese is given in Figure 7.3a.

Nickel (ppm)

The concentration of nickel in pond water was recorded maximum 0.130 ± 0.014 in summer 2006 and minimum 0.047 ± 0.038 in winter 2006-07 ranged between 0.008

to 0.015 ppm. The highest value of nickel in pond water was found 0.15 ppm during May 2006 and lowest 0.008 ppm in November 2005. The nickel variation during the study is given Table 7.3c. Monthly variation in nickel is given in Figure 7.3a.

Lead (ppm)

The concentration of lead in pond water was recorded maximum 1.80 ± 0.56 in summer 2005 and minimum 0.503 ± 0.346 in winter 2006-07 ranged between 0.05 to 2.3 ppm. The highest value of lead in pond water was found 2.3 ppm during May 2005 and lowest 0.05 ppm in October 2005. The lead variation during the study is given Table 7.3c. Monthly variation in lead is given in Figure 7.3b.

Zinc (ppm)

The concentration of zinc in pond water was recorded maximum 195.0 ± 0.007 in summer 2007 and minimum 0.110 ± 0.02 in monsoon 2005 ranged between 0.05 to 0.21 ppm. The highest value of zinc in pond water was found 0.21 ppm during December 2006 and lowest 0.05 ppm in October 2005. The zinc variation during the study is given Table 7.3c. Monthly variation in zinc is given in Figure 7.3a.

Heavy Metals in Groundwater

Cadmium (ppm)

The cadmium concentration in groundwater was recorded maximum 0.01 ± 0.002 in summer 2006 and minimum 0.002 ± 0.002 in monsoon 2006 ranged between 0.00 to 0.01 ppm. The highest value of cadmium in groundwater was found 0.01 ppm during March 2007 and lowest 0.00 ppm in August 2005. The cadmium variation during the study is given Table 7.4. Monthly variation in cadmium is given in Figure 7.5, Figure 7.6 and Figure 7.7 for 40 ft, 80 ft and more than 120 ft depth respectively.

Copper (ppm)

The concentration of copper in groundwater was recorded maximum 0.443 ± 0.067 in summer 2005 and minimum 0.288 ± 0.093 in winter 2006-07 ranged between 0.137 to 0.58 ppm. The highest value of copper in groundwater was found 0.58 ppm during July 2006 and lowest 0.137 ppm in October 2005. The copper variation during the study is given Table 7.4. Monthly variation in copper is given in Figure 7.5a, Figure 7.6a and Figure 7.7a for 40 ft, 80 ft and more than 120 ft depth respectively.

Iron (ppm)

The concentration of iron in groundwater was recorded maximum 3.82 ± 0.59 in summer 2006 and minimum 1.50 ± 0.24 in monsoon 2005 ranged between 0.46 to 4.6 ppm. The highest value of iron in groundwater was found 4.6 ppm during April 2005 and lowest 0.46 ppm in October 2005. The iron variation during the study is given Table 7.4. Monthly variation in iron is given in Figure 7.5b, Figure 7.6b and Figure 7.7b for 40 ft, 80 ft and more than 120 ft depth respectively.

Manganese (ppm)

The concentration of manganese in groundwater was recorded maximum 0.550 ± 0.08 in summer 2005 and minimum 0.284 ± 0.078 in winter 2005-06 ranged between

0.19 to 0.63 ppm. The highest value of manganese in groundwater was found 0.63 ppm during June 2006 and lowest 0.19 ppm in November 2005. The manganese variation during the study is given Table 7.4. Monthly variation in manganese is given in Figure 7.5b, Figure 7.6b and Figure 7.7b for 40 ft, 80 ft and more than 120 ft depth respectively.

Nickel (ppm)

The concentration of nickel in groundwater was recorded maximum 0.153 ± 0.01 in summer 2006 and minimum 0.044 ± 0.035 in winter 2006-07 ranged between 0.009 to 0.16 ppm. The highest value of nickel in groundwater was found 0.16 ppm during May 2006 and lowest 0.009 ppm in November 2005. The nickel variation during the study is given Table 7.4. Monthly variation in nickel is given in Figure 7.5a, Figure 7.6a and Figure 7.7a for 40 ft, 80 ft and more than 120 ft depth respectively.

Lead (ppm)

The concentration of lead in groundwater was recorded maximum 1.84 ± 0.31 in summer 2005 and minimum 0.850 ± 0.24 in winter 2005-06 ranged between 0.55 to 2.12 ppm. The highest value of lead in groundwater was found 2.12 ppm during May 2005 and lowest 0.55 ppm in October 2005. The lead variation during the study is given Table 7.4. Monthly variation in lead is given in Figure 7.5b, Figure 7.6b and Figure 7.7b for 40 ft, 80 ft and more than 120 ft depth respectively.

Zinc (ppm)

The concentration of zinc in groundwater was recorded maximum 0.215 ± 0.03 in summer 2006 and minimum 0.126 ± 0.051 in winter 2006-07 ranged between 0.06 to 0.25 ppm. The highest value of zinc in groundwater was found 0.25 ppm during April 2005 and lowest 0.06 ppm in October 2005. The zinc variation during the study is given Table 7.4. Monthly variation in zinc is given in Figure 7.5a., Figure 7.6a and Figure 7.7a for 40 ft, 80 ft and more than 120 ft depth respectively.

Heavy Metals in Effluent Drain Sediment

Cadmium (ppm)

The cadmium concentration in drain sediment was recorded maximum 5.78 ± 0.231 in summer 2005 and minimum 4.25 ± 0.544 in winter 2005-06 ranged between 3.54 to 6.12 ppm. The highest value of cadmium in drain sediment was found 6.12 ppm during June 2006 and lowest 3.54 ppm in November 2005. The cadmium variation during the study is given Table 7.3b. Monthly variation in cadmium is given in Figure 7.2a.

Copper (ppm)

The concentration of copper in drain sediment was recorded maximum 54.83 ± 7.33 in summer 2006 and minimum 45.87 ± 5.82 in summer 2007 ranged between 37.85 to 55.36 ppm. The highest value of copper in drain sediment was found 55.36 ppm during April 2006 and lowest 37.85 ppm in August 2006. The copper variation during the study is given Table 7.3b. Monthly variation in copper is given in Figure 7.7a.

Iron (ppm)

The concentration of iron in drain sediment was recorded maximum 8467.67 ± 2448.18 in summer 2005 and minimum 5656.0 ± 1216.72 in monsoon 2005 ranged between 4256.0 to 11235.0 ppm. The highest value of iron in drain sediment was found 11235.0 ppm during June 2005 and lowest 4256.0 ppm in August 2005. The iron variation during the study is given Table 7.3b. Monthly variation in iron is given in Figure 7.2b.

Manganese (ppm)

The concentration of manganese in drain sediment was recorded maximum 1105.00 ± 107.08 in summer 2006 and minimum 791.33 ± 51.73 in monsoon 2005 ranged between 568.0 to 1105.0 ppm. The highest value of manganese in drain sediment was found 1105.0 ppm during April 2006 and lowest 568.0 ppm in August 2006. The manganese variation during the study is given Table 7.3b. Monthly variation in manganese is given in Figure 7.2b.

Nickel (ppm)

The concentration of nickel in drain sediment was recorded maximum 75.00 ± 3.0 in summer 2005 and minimum 58.0 ± 3.0 in monsoon 2005 ranged between 49.0 to 79.0 ppm. The highest value of nickel in drain sediment was found 79.0 ppm during June 2006 and lowest 49.0 ppm in October 2006. The nickel variation during the study is given Table 7.3b. Monthly variation in nickel is given in Figure 7.2a.

Lead (ppm)

The concentration of lead in drain sediment was recorded maximum 70.00 ± 4.36 in summer 2005 and minimum 47.20 ± 3.83 in winter 2005-06 ranged between 42.0 to 78.0 ppm. The highest value of lead in drain sediment was found 78.0 ppm during June 2006 and lowest 42.0 ppm in January 2006. The lead variation during the study is given Table 7.3b. Monthly variation in lead is given in Figure 7.2a.

Zinc (ppm)

The concentration of zinc in drain sediment was recorded maximum 260.00 ± 83.54 in summer 2005 and minimum 143.50 ± 32.29 in winter 2006-07 ranged between 95.0 to 385.0 ppm. The highest value of zinc in drain sediment was found 385.0 ppm during June 2006 and lowest 95.0 ppm in October 2005. The zinc variation during the study is given Table 7.3b. Monthly variation in zinc is given in Figure 7.2b.

Heavy Metals in Pond Sediment

Cadmium (ppm)

The cadmium concentration in pond sediment was recorded maximum 5.72 ± 0.45 in summer 2005 and minimum 4.05 ± 0.59 in winter 2005-06 ranged between 3.26 to 6.32 ppm. The highest value of cadmium in pond sediment was found 6.32 ppm during June 2006 and lowest 3.26 ppm in November 2005. The cadmium variation during the study is given Table 7.3d. Monthly variation in cadmium is given in Figure 7.4a.

Copper (ppm)

The concentration of copper in pond sediment was recorded maximum 52.46 ± 6.51 in summer 2006 and minimum 38.65 ± 9.23 in summer 2007 ranged between 24.45 to 57.21 ppm. The highest value of copper in pond sediment was found 57.21 ppm during May 2006 and lowest 24.45 ppm in January 2007. The copper variation during the study is given Table 7.3d. Monthly variation in copper is given in Figure 7.4a.

Iron (ppm)

The concentration of iron in pond sediment was recorded maximum 6013.40 ± 856.58 in winter 2005-06 and minimum 3803.50 ± 113.84 in summer 2007 ranged between 3542.0 to 8514.0 ppm. The highest value of iron in pond sediment was found 8514.0 ppm during December 2006 and lowest 3542.0 ppm in April 2005. The iron variation during the study is given Table 7.3d. Monthly variation in iron is given in Figure 7.4b.

Manganese (ppm)

The concentration of manganese in pond sediment was recorded maximum 942.00 ± 48.14 in summer 2005 and minimum 605.00 ± 161.71 in monsoon 2006 ranged between 465.0 to 998.0 ppm. The highest value of manganese in pond sediment was found 998.0 ppm during June 2006 and lowest 465.0 ppm in August 2006. The manganese variation during the study is given Table 7.3d. Monthly variation in manganese is given in Figure 7.4b.

Nickel (ppm)

The concentration of nickel in pond sediment was recorded maximum 63.75 ± 11.12 in summer 2006 and minimum 49.0 ± 5.94 in winter 2006-07 ranged between 41.0 to 79.0 ppm. The highest value of nickel in pond sediment was found 79.0 ppm during June 2006 and lowest 41.0 ppm in October 2006. The nickel variation during the study is given Table 7.3d. Monthly variation in nickel is given in Figure 7.4a.

Lead (ppm)

The concentration of lead in pond sediment was recorded maximum 69.00 ± 6.24 in summer 2005 and minimum 44.00 ± 5.39 in winter 2005-06 ranged between 37.0 to 75.0 ppm. The highest value of lead in pond sediment was found 75.0 ppm during June 2006 and lowest 37.0 ppm in January 2006. The lead variation during the study is given Table 7.3d. Monthly variation in lead is given in Figure 7.4a.

Zinc (ppm)

The concentration of zinc in pond sediment was recorded maximum 240.00 ± 92.60 in summer 2005 and minimum 116.25 ± 32.76 in winter 2006-07 ranged between 75.0 to 321.0 ppm. The highest value of zinc in pond sediment was found 321.0 ppm during June 2006 and lowest 75.0 ppm in October 2006. The zinc variation during the study is given Table 7.3d. Monthly variation in zinc is given in Figure 7.4b.

Table 7.4: Seasonal Variation of Heavy Metals in Groundwater at 40 ft Depth from Hand Pump in Village Binjhole

Heavy Metals	*Unit*	*Summer 2005*		*Monsoon 2005*		*Winter 2005-06*		*Summer 2006*		*Monsoon 2006*		*Winter 2006-07*		*Summer 2007*	
		Mean	*SD*	*Mean*	*SD*	*Mean*	*SD*	*Mean*	*SD*	*Mean*	*SD*	*Mean*	*SD*	*Mean*	*SD*
Cadmium (Cd)	ppm	0.008	0.002	0.003	0.003	0.007	0.001	0.010	0.002	0.002	0.002	0.007	0.002	0.010	0.000
Copper (Cu)	ppm	0.443	0.067	0.413	0.142	0.296	0.094	0.425	0.068	0.423	0.150	0.288	0.093	0.360	0.028
Iron (Fe)	ppm	3.777	0.891	1.503	0.235	1.628	0.922	3.820	0.589	1.840	0.265	1.513	0.645	3.375	0.389
Manga-nese (Mn)	ppm	0.550	0.079	0.373	0.146	0.284	0.078	0.530	0.083	0.453	0.117	0.313	0.049	0.430	0.028
Nickel (Ni)	ppm	0.147	0.015	0.094	0.020	0.045	0.036	0.153	0.010	0.099	0.012	0.044	0.035	0.125	0.007
Lead (Pb)	ppm	1.840	0.308	1.163	0.214	0.850	0.235	1.820	0.275	1.270	0.401	0.948	0.142	1.295	0.021
Zinc (Zn)	ppm	0.200	0.046	0.133	0.021	0.126	0.051	0.215	0.025	0.153	0.031	0.128	0.043	0.200	0.014

Table 7.5: Seasonal Variation of Heavy Metals in Groundwater at 80 ft Depth from Tube Well and Deep Hand Pump

Heavy Metals	*Unit*	*Summer 2005*		*Monsoon 2005*		*Winter 2005-06*		*Summer 2006*		*Monsoon 2006*		*Winter 2006-07*		*Summer 2007*	
		Mean	*SD*	*Mean*	*SD*	*Mean*	*SD*	*Mean*	*SD*	*Mean*	*SD*	*Mean*	*SD*	*Mean*	*SD*
Cadmium (Cd)	ppm	0.006	0.002	0.001	0.001	0.004	0.002	0.006	0.001	0.000	0.001	0.004	0.001	0.006	0.000
Copper (Cu)	ppm	0.288	0.035	0.219	0.030	0.152	0.024	0.288	0.049	0.218	0.029	0.158	0.025	0.212	0.054
Iron (Fe)	ppm	1.880	0.397	1.657	0.362	0.668	0.224	1.788	0.572	1.783	0.325	0.783	0.311	1.065	0.191
Manga-nese (Mn)	ppm	0.294	0.021	0.215	0.031	0.149	0.047	0.286	0.041	0.228	0.036	0.150	0.036	0.225	0.007
Nickel (Ni)	ppm	0.098	0.017	0.059	0.018	0.016	0.013	0.099	0.028	0.063	0.020	0.013	0.004	0.052	0.013
Lead (Pb)	ppm	0.863	0.076	0.627	0.146	0.530	0.175	0.850	0.130	0.660	0.101	0.505	0.152	0.740	0.014
Zinc (Zn)	ppm	0.163	0.015	0.133	0.025	0.069	0.032	0.170	0.022	0.127	0.021	0.078	0.026	0.135	0.021

Table 7.6: Seasonal Variation of Heavy Metals in Groundwater (more than 120 ft depth) of Tube Well

Heavy Metals	Unit	Summer 2005		Monsoon 2005		Winter 2005-06		Summer 2006		Monsoon 2006		Winter 2006-07		Summer 2007	
		Mean	*SD*	*Mean*	*SD*	*Mean*	*SD*	*Mean*	*SD*	*Mean*	*SD*	*Mean*	*SD*	*Mean*	*SD*
Cadmium (Cd)	ppm	0.000	0.001	0.000	0.000	0.000	0.000	0.002	0.002	0.000	0.000	0.000	0.000	0.000	0.000
Copper (Cu)	ppm	0.200	0.015	0.163	0.008	0.124	0.025	0.206	0.022	0.169	0.017	0.129	0.015	0.168	0.004
Iron (Fe)	ppm	1.017	0.168	0.673	0.117	0.478	0.241	1.048	0.101	0.730	0.092	0.483	0.210	1.030	0.099
Manganese (Mn)	ppm	0.290	0.022	0.187	0.053	0.095	0.066	0.287	0.046	0.209	0.048	0.092	0.051	0.222	0.041
Nickel (Ni)	ppm	0.068	0.014	0.044	0.019	0.009	0.009	0.066	0.020	0.050	0.022	0.009	0.012	0.039	0.009
Lead (Pb)	ppm	0.402	0.069	0.287	0.026	0.289	0.098	0.414	0.029	0.295	0.035	0.281	0.085	0.405	0.030
Zinc (Zn)	ppm	0.113	0.016	0.078	0.007	0.047	0.028	0.117	0.015	0.084	0.008	0.047	0.020	0.103	0.017

Table 7.7: Seasonal Variation of Heavy Metals in Soil Near Textile Industries

Heavy Metals	Unit	Summer 2005		Monsoon 2005		Winter 2005-06		Summer 2006		Monsoon 2006		Winter 2006-07		Summer 2007	
		Mean	*SD*	*Mean*	*SD*	*Mean*	*SD*	*Mean*	*SD*	*Mean*	*SD*	*Mean*	*SD*	*Mean*	*SD*
Cadmium (Cd)	ppm	3.533	0.586	0.773	0.560	1.920	0.396	3.350	0.597	1.133	0.651	2.100	0.432	2.800	0.424
Copper (Cu)	ppm	36.867	7.407	31.300	4.652	15.400	5.439	33.925	9.215	32.233	4.644	16.800	4.574	24.200	2.970
Iron (Fe)	ppm	77.267	14.408	32.700	11.701	28.460	12.233	76.450	10.959	39.000	12.957	25.450	5.298	58.650	12.516
Manga-nese (Mn)	ppm	13.767	0.907	10.967	2.501	6.540	2.221	13.175	2.451	11.833	2.219	7.375	1.417	10.350	1.202
Nickel (Ni)	ppm	10.400	1.179	7.567	1.079	5.940	1.242	10.675	1.670	8.400	1.015	5.975	0.670	8.150	0.919
Lead (Pb)	ppm	62.467	9.804	40.833	11.545	26.820	9.166	60.750	9.793	46.167	13.674	27.075	7.168	47.550	3.182
Zinc (Zn)	ppm	11.267	2.950	16.967	3.037	9.580	2.647	11.675	3.914	17.467	1.966	10.975	2.579	8.000	1.131

Table 7.8: Seasonal Variation of Heavy Metals in Agricultural Irrigated Soil

Heavy Metals	*Unit*	*Summer 2005*		*Monsoon 2005*		*Winter 2005-06*		*Summer 2006*		*Monsoon 2006*		*Winter 2006-07*		*Summer 2007*	
		Mean	*SD*	*Mean*	*SD*	*Mean*	*SD*	*Mean*	*SD*	*Mean*	*SD*	*Mean*	*SD*	*Mean*	*SD*
Cadmium (Cd)	ppm	3.333	0.611	0.667	0.551	1.780	0.444	3.350	0.686	1.200	1.153	1.925	0.350	2.650	0.212
Copper (Cu)	ppm	36.300	7.422	30.200	5.345	13.400	5.968	34.875	9.309	32.333	5.498	14.350	4.916	20.850	3.748
Iron (Fe)	ppm	76.667	16.805	26.767	10.772	28.800	13.221	82.825	18.097	30.400	12.375	27.450	10.051	62.100	13.859
Manga-nese (Mn)	ppm	13.567	1.401	10.233	2.380	5.900	2.799	13.075	1.910	11.033	1.960	5.775	1.676	10.450	1.768
Nickel (Ni)	ppm	10.467	0.757	7.633	1.137	5.780	1.244	10.525	1.812	8.067	1.504	5.325	1.056	9.050	1.061
Lead (Pb)	ppm	62.400	10.922	39.333	11.675	25.340	9.226	61.300	11.385	44.933	15.791	25.100	6.733	46.150	4.738
Zinc (Zn)	ppm	13.167	3.911	18.533	5.150	7.680	2.805	13.025	3.471	19.400	4.747	9.150	3.101	8.000	1.697

Table 7.9: Seasonal Variation of Heavy Metals in Groundwater of 80 ft at Control Site

Heavy Metals	Unit	Summer 2005		Monsoon 2005		Winter 2005-06		Summer 2006		Monsoon 2006		Winter 2006-07		Summer 2007	
		Mean	*SD*	*Mean*	*SD*	*Mean*	*SD*	*Mean*	*SD*	*Mean*	*SD*	*Mean*	*SD*	*Mean*	*SD*
Cadmium (Cd)	ppm	0.000	0.000	0.000	0.000	0.000	0.000	0.000	0.000	0.000	0.000	0.000	0.000	0.000	0.000
Copper (Cu)	ppm	0.157	0.032	0.149	0.012	0.090	0.052	0.152	0.032	0.149	0.007	0.090	0.061	0.684	0.800
Iron (Fe)	ppm	0.857	0.119	0.570	0.122	0.400	0.194	0.868	0.070	0.593	0.074	0.363	0.128	0.800	0.071
Manga nese (Mn)	ppm	0.204	0.030	0.135	0.026	0.079	0.053	0.200	0.041	0.140	0.024	0.075	0.037	0.148	0.004
Nickel (Ni)	ppm	0.044	0.011	0.026	0.017	0.000	0.000	0.043	0.011	0.027	0.010	0.000	0.000	0.018	0.025
Lead (Pb)	ppm	0.265	0.015	0.203	0.028	0.191	0.112	0.266	0.012	0.210	0.034	0.178	0.127	0.282	0.013
Zinc (Zn)	ppm	0.099	0.015	0.065	0.005	0.044	0.021	0.099	0.008	0.072	0.008	0.043	0.013	0.087	0.016

Table 7.10: Seasonal Variation of Heavy Metals in Soil at Control Site

Heavy Metals	Unit	Summer 2005		Monsoon 2005		Winter 2005-06		Summer 2006		Monsoon 2006		Winter 2006-07		Summer 2007	
		Mean	*SD*	*Mean*	*SD*	*Mean*	*SD*	*Mean*	*SD*	*Mean*	*SD*	*Mean*	*SD*	*Mean*	*SD*
Cadmium (Cd)	ppm	2.233	0.306	0.733	0.473	0.460	0.351	2.000	0.770	1.233	0.451	0.225	0.206	1.400	0.141
Copper (Cu)	ppm	11.533	15.396	23.000	4.036	9.980	4.141	20.525	13.495	22.733	4.966	10.250	5.637	15.800	4.667
Iron (Fe)	ppm	68.633	17.539	19.000	11.207	20.940	12.563	66.450	15.315	21.133	11.208	18.400	9.038	48.250	9.263
Manganese (Mn)	ppm	11.500	1.114	9.200	1.833	4.962	2.028	10.775	1.791	9.533	2.159	4.800	1.490	7.950	1.202
Nickel (Ni)	ppm	8.433	1.172	6.533	1.026	3.380	1.846	8.050	1.313	6.767	1.026	2.600	1.867	6.300	0.283
Lead (Pb)	ppm	42.267	3.121	30.633	8.999	19.000	7.316	41.925	3.677	31.400	7.602	18.275	5.934	34.600	4.101
Zinc (Zn)	ppm	12.100	1.500	15.400	3.830	7.460	2.475	11.700	2.655	16.467	3.275	8.950	2.885	7.250	0.778

Table 7.11: Mean Values (±SD) of Heavy Metals in Groundwater at Control and Experiment Sites

Heavy Metals	*Unit*	*2005-06*				*2006-07*			
		Control Site	*SD*	*Experimental Sites*	*SD*	*Control Site*	*SD*	*Experimental Sites*	*SD*
Cadmium (Cd)	ppm	0	0	0.004	0.002	0.0001	0.0002	0.004	0.003
Copper (Cu)	ppm	0.123	0.047	0.244	0.069	0.223	0.327	0.254	0.071
Iron (Fe)	ppm	0.597	0.25	1.414	0.719	0.618	0.231	1.562	0.697
Manganese (Mn)	ppm	0.13	0.063	0.257	0.101	0.139	0.061	0.283	0.093
Nickel (Ni)	ppm	0.022	0.021	0.059	0.038	0.023	0.021	0.066	0.04
Lead (Pb)	ppm	0.218	0.077	0.735	0.248	0.226	0.082	0.785	0.237
Zinc (Zn)	ppm	0.067	0.028	0.115	0.045	0.072	0.026	0.126	0.042

Table 7.12: Mean Values (±SD) of Heavy Metals in Soil at Control and Experiment Sites

Heavy Metals	*Unit*	*2005-06*				*2006-07*			
		Control Site	*SD*	*Experimental Sites*	*SD*	*Control Site*	*SD*	*Experimental Sites*	*SD*
Cadmium (Cd)	ppm	1.033	0.819	1.950	1.098	1.183	0.893	2.292	1.099
Copper (Cu)	ppm	14.317	9.218	24.100	11.451	17.050	9.744	26.075	11.391
Iron (Fe)	ppm	35.267	25.354	43.150	25.433	36.975	24.866	49.417	29.212
Manganese (Mn)	ppm	7.984	3.215	9.283	3.850	8.158	3.086	9.908	3.601
Nickel (Ni)	ppm	5.675	2.527	7.608	2.161	5.767	2.731	8.133	2.611
Lead (Pb)	ppm	29.317	11.729	39.967	17.818	30.508	11.188	43.750	18.264
Zinc (Zn)	ppm	10.667	4.148	11.867	5.640	11.525	4.195	12.833	5.460

Table 7.13a: Correlation between Heavy Metals and Physico-chemical Parameters of Groundwater at 40 ft Depth

Parameters	*Temperature*	*pH*	*TS*	*TSS*	*TDS*	*EC*	*DO*	*Free CO_2*	*BOD*	*COD*	*Total Alkalinity*	*Ca*	*Mg*	*K*	*Cl*	*Na*
Cadmium (Cd)	-0.4720	0.4400	-0.0844	-0.1491	-0.0780	-0.0780	-0.4459	0.4060	-0.3829	-0.3896	-0.2270	-0.5395	-0.4221	-0.8034	-0.7787	-0.5654
Copper (Cu)	0.7176	0.6031	0.6712	0.6589	0.6714	0.6714	-0.4396	0.5115	0.5203	0.4632	0.6306	0.6677	0.6445	0.1376	0.2027	0.5021
Iron (Fe)	0.1340	0.8102	0.4987	0.4064	0.5069	0.5069	-0.6278	0.6729	0.0648	0.0647	0.2490	0.0946	0.0987	-0.5786	-0.4117	0.0366
Manganese (Mn)	0.6892	0.7867	0.8508	0.8001	0.8544	0.8544	-0.6314	0.6306	0.5396	0.5222	0.7074	0.6099	0.6586	0.0178	0.1352	0.5094
Nickel (Ni)	0.4747	0.7167	0.7983	0.7469	0.8021	0.8021	-0.4856	0.5276	0.3021	0.3030	0.6393	0.4971	0.4080	-0.1919	-0.0715	0.4255
Lead (Pb)	0.6106	0.8593	0.8296	0.7709	0.8340	0.8340	-0.6519	0.6839	0.4734	0.4754	0.5818	0.5440	0.6065	-0.0930	0.0378	0.4703
Zinc (Zn)	0.1239	0.7175	0.4302	0.3529	0.4370	0.4370	-0.4412	0.5164	0.0132	-0.0055	0.2999	0.1556	0.0547	-0.5448	-0.4111	0.0297

Table 7.13b: Correlation between Heavy Metals and Physico-chemical Parameters of Groundwater at 80 ft Depth

Para-meters	*Tempe-rature*	*pH*	*TS*	*TSS*	*TDS*	*EC*	*DO*	*Free CO_2*	*BOD*	*COD*	*Total Alkalinity*	*Ca*	*Mg*	*K*	*Cl*	*Na*
Cadmium (Cd)	-0.4249	0.4498	-0.7675	-0.7338	-0.7699	-0.7699	-0.5169	0.5048	-0.3579	-0.3655	0.0237	-0.6322	-0.4514	-0.8700	-0.7527	-0.4974
Copper (Cu)	0.6860	-0.6296	-0.3314	-0.3518	-0.3288	-0.3288	-0.7029	0.7443	0.5230	0.5269	0.8099	0.5265	0.6353	-0.0677	-0.4178	0.6069
Iron (Fe)	0.9188	-0.8515	0.0718	0.0396	0.0750	0.0750	-0.4897	0.5458	0.7191	0.7550	0.8896	0.8398	0.8894	0.3575	-0.0309	0.8503
Manga-nese (Mn)	0.6670	-0.6006	-0.4141	-0.4318	-0.4116	-0.4116	-0.6248	0.6930	0.4701	0.4793	0.7664	0.4874	0.5708	-0.1260	-0.4779	0.5763
Nickel (Ni)	0.7044	-0.7071	-0.2883	-0.3056	-0.2860	-0.2860	-0.7447	0.7536	0.5755	0.5795	0.7918	0.5774	0.6618	-0.0759	-0.3709	0.6491
Lead (Pb)	0.4699	-0.3677	-0.6339	-0.6469	-0.6316	-0.6316	-0.5968	0.6693	0.3066	0.2903	0.5841	0.2398	0.3342	-0.3304	-0.7127	0.3610
Zinc (Zn)	0.6489	-0.7275	-0.2520	-0.2637	-0.2504	-0.2504	-0.6222	0.6518	0.5733	0.5755	0.7069	0.5727	0.5782	-0.1022	-0.2697	0.6108

Table 7.13c: Correlation between Heavy Metals and Physico-chemical Parameters of Groundwater (more than 120 ft)

Parameters	*Temperature*	*pH*	*TS*	*TSS*	*TDS*	*EC*	*DO*	*Free CO_2*	*BOD*	*COD*	*Total Alkalinity*	*Ca*	*Mg*	*K*	*Cl*	*Na*
Cadmium (Cd)	0.2586	-0.1922	-0.1886	-0.1927	-0.1882	-0.1882	-0.1208	0.1808	0.1577	0.1501	0.1851	0.2236	0.2761	0.0105	-0.1648	0.2274
Copper (Cu)	0.5657	-0.7221	-0.7972	-0.7977	-0.7970	-0.7970	-0.5948	0.6392	0.5026	0.5084	0.8666	0.4787	0.4949	-0.3674	-0.3668	0.5099
Iron (Fe)	0.3286	-0.3743	-0.9582	-0.9592	-0.9579	-0.9579	-0.6330	0.6815	0.1580	0.1551	0.8783	0.1801	0.1629	-0.6196	-0.6129	0.2244
Manganese (Mn)	0.6144	-0.6217	-0.9099	-0.9097	-0.9098	-0.9098	-0.6125	0.6554	0.4566	0.4558	0.8834	0.4620	0.4791	-0.4055	-0.4809	0.5253
Nickel (Ni)	0.7675	-0.7133	-0.7984	-0.7932	-0.7988	-0.7988	-0.5782	0.5974	0.5868	0.5877	0.8302	0.6059	0.6627	-0.2177	-0.3703	0.6663
Lead (Pb)	0.0368	-0.0109	-0.8292	-0.8222	-0.8297	-0.8297	-0.6002	0.6347	-0.1034	-0.0908	0.7136	-0.1830	-0.1350	-0.7297	-0.7721	-0.0983
Zinc (Zn)	0.4445	-0.6682	-0.8084	-0.8107	-0.8080	-0.8080	-0.5886	0.6340	0.3849	0.3854	0.8653	0.3815	0.3368	-0.4709	-0.3812	0.3780

Table 7.14: Correlation between Heavy Metals and Physico-chemical Parameters of Agricultural Soil

Parameters	*Temperature*	*WHC*	*Bulk Density*	*Soil Moisture*	*pH*	*EC*	*Cl*	*Na*	*K*
Cadmium (Cd)	0.2211	0.8421	-0.5639	-0.8066	-0.2848	0.3965	0.3654	0.7499	0.6852
Copper (Cu)	0.9134	0.5300	-0.0815	0.0914	0.3754	0.9454	0.8479	0.0853	-0.1689
Iron (Fe)	0.3455	0.8736	-0.6445	-0.6449	-0.4051	0.4198	0.3249	0.8436	0.7116
Manganese (Mn)	0.8162	0.6657	-0.2403	-0.0712	0.2034	0.9076	0.7254	0.3169	0.0611
Nickel (Ni)	0.7218	0.7798	-0.3816	-0.2585	0.0187	0.8286	0.6603	0.4813	0.2623
Lead (Pb)	0.7091	0.8483	-0.4542	-0.2607	0.0462	0.8596	0.6972	0.4876	0.2369
Zinc (Zn)	0.8513	0.0750	0.3246	0.5031	0.6535	0.7586	0.7744	-0.4581	-0.6269

Table 7.15: Correlation between Heavy Metals and Physico-chemical Parameters of Groundwater at Control Site

Parameters	*Temperature*	*pH*	*TS*	*TSS*	*TDS*	*EC*	*DO*	*Free CO_2*	*BOD*	*COD*	*Total Alkalinity*	*Ca*	*Mg*	*K*	*Cl*	*Na*
Cadmium (Cd)	0.3314	-0.3195	-0.1977	-0.1917	-0.1982	-0.1982	-0.1820	0.1921	0.2838	0.2944	0.3888	0.3900	0.3991	0.0728	-0.2197	0.3398
Copper (Cu)	0.0236	0.1308	-0.1525	-0.1720	-0.1505	-0.1505	-0.0006	0.0907	0.0053	-0.0221	0.2028	-0.1738	-0.1293	-0.1285	-0.1667	-0.0372
Iron (Fe)	0.5000	-0.1850	-0.9750	-0.9773	-0.9745	-0.9745	-0.7472	0.8049	0.2986	0.3181	0.8597	0.2190	0.1777	-0.4871	-0.7837	0.2632
Manganese (Mn)	0.7006	-0.3744	-0.8475	-0.8573	-0.8464	-0.8464	-0.5798	0.6515	0.4829	0.5029	0.8739	0.4496	0.4510	-0.2603	-0.6374	0.5588
Nickel (Ni)	0.8099	-0.5217	-0.7634	-0.7685	-0.7627	-0.7627	-0.5530	0.6664	0.6405	0.6720	0.8175	0.5739	0.6263	-0.0985	-0.5878	0.6268
Lead (Pb)	0.2533	0.3462	-0.6613	-0.6535	-0.6619	-0.6619	-0.5997	0.6927	-0.1630	-0.1323	0.8009	-0.1149	-0.1161	-0.4930	-0.8937	-0.0558
Zinc (Zn)	0.5352	-0.4305	-0.8369	-0.8382	-0.8365	-0.8365	-0.6323	0.6614	0.5439	0.5513	0.7411	0.3509	0.3381	-0.3646	-0.5368	0.4302

Table 7.16: Correlation between Heavy Metals and Physico-chemical Parameters of Soil at Control Site

Parameters	*Temperature*	*WHC*	*Bulk Density*	*Soil Moisture*	*pH*	*EC*	*Cl*	*Na*	*K*
Cadmium (Cd)	0.6763	0.8199	-0.3977	-0.3107	-0.3093	0.7773	0.6980	0.5320	0.0836
Copper (Cu)	0.6307	0.2015	0.2982	0.4074	0.4782	0.0989	0.6089	-0.2909	-0.4336
Iron (Fe)	0.3246	0.8397	-0.6903	-0.6830	-0.6779	0.8251	0.3079	0.8795	0.5500
Manganese (Mn)	0.8814	0.6726	-0.1244	0.0151	0.0432	0.6419	0.7925	0.2525	-0.1647
Nickel (Ni)	0.7578	0.6737	-0.2465	-0.0108	-0.0862	0.6505	0.6530	0.3863	-0.1443
Lead (Pb)	0.6596	0.8608	-0.4316	-0.2208	-0.2642	0.8055	0.6111	0.5749	0.1158
Zinc (Zn)	0.8704	0.1744	0.3951	0.5580	0.4955	0.2569	0.7613	-0.3999	-0.5542

Table 7.17: Correlation between Heavy Metals and Physico-chemical Parameters of Textile Effluents

Parameters	*Temperature*	*pH*	*TS*	*TSS*	*TDS*	*EC*	*DO*	*Free CO_2*	*BOD*	*COD*	*Total Alkalinity*	*Ca*	*Mg*	*K*	*Cl*	*Na*
Cadmium (Cd)	0.5954	0.2054	0.4065	0.1194	0.4967	0.4949	-0.6152	0.6627	0.4508	0.4810	0.5408	0.4782	0.4821	0.2125	0.0739	0.0502
Copper (Cu)	0.5386	-0.0945	0.6047	0.2598	0.6126	0.5849	-0.5741	0.6112	0.2584	0.2011	0.5988	0.4728	0.3539	0.0231	-0.0995	0.1084
Iron (Fe)	0.4191	-0.2336	0.3555	-0.0584	0.6848	0.6947	-0.8010	0.7573	0.3639	0.1431	0.3816	0.3389	0.3894	-0.1857	-0.2576	-0.1650
Manganese (Mn)	-0.7252	-0.5784	-0.5025	-0.7485	0.3101	0.2820	0.1536	-0.0984	-0.3822	-0.7932	-0.6307	-0.6879	-0.8048	-0.7144	-0.8178	-0.7276
Nickel (Ni)	-0.1097	-0.4854	-0.1497	-0.3875	0.3455	0.3433	-0.4807	0.4410	0.0597	-0.2737	-0.0827	-0.1454	-0.0609	-0.5091	-0.5042	-0.4635
Lead (Pb)	0.1387	-0.1111	0.1279	-0.3402	0.7373	0.7303	-0.2132	0.3252	-0.0972	-0.1625	0.0515	-0.0132	-0.0200	-0.0134	-0.2652	-0.2585
Zinc (Zn)	-0.4590	-0.1986	-0.4412	-0.5178	0.0579	0.0003	0.0685	-0.0395	-0.0593	-0.2533	-0.4103	-0.4126	-0.3176	-0.3053	-0.2827	-0.3781

Table 7.18: Correlation between Heavy Metals of Textile Effluents and Heavy Metals of Pond Water

	Cadmium (Cd)	*Copper (Cu)*	*Iron (Fe)*	*Manganese (Mn)*	*Nickel (Ni)*	*Lead (Pb)*	*Zinc (Zn)*
Cadmium (Cd)	0.3867	0.6257	0.3811	0.5244	0.4933	0.3803	-0.1229
Copper (Cu)	0.4993	0.7851	0.6926	0.7488	0.7762	0.7975	0.0633
Iron (Fe)	0.7229	0.6959	0.7251	0.6304	0.8110	0.8506	0.2362
Manganese (Mn)	0.1607	-0.1164	0.4070	-0.5011	0.2191	0.1712	0.8434
Nickel (Ni)	0.5772	0.2261	0.5584	0.2070	0.5204	0.5855	0.4325
Lead (Pb)	0.6835	0.4987	0.5301	0.0962	0.5946	0.4128	0.4019
Zinc (Zn)	0.0683	-0.2977	-0.1108	-0.4107	-0.2036	-0.1970	0.4195

Table 7.19a: Correlation between Heavy Metals of Pond Water and Heavy Metals of Groundwater

	Cadmium (Cd)	*Copper (Cu)*	*Iron (Fe)*	*Manganese (Mn)*	*Nickel (Ni)*	*Lead (Pb)*	*Zinc (Zn)*
Cadmium (Cd)	0.5955	0.4925	0.7118	0.7155	0.6233	0.7618	0.5954
Copper (Cu)	-0.0073	0.9116	0.5612	0.8907	0.7902	0.8271	0.6287
Iron (Fe)	0.6647	0.4978	0.9580	0.6654	0.8211	0.7114	0.9422
Manganese (Mn)	-0.1682	0.6457	0.5135	0.7864	0.8076	0.7998	0.4612
Nickel (Ni)	0.4472	0.6508	0.8689	0.7913	0.9223	0.8042	0.9085
Lead (Pb)	0.5029	0.5822	0.9387	0.7889	0.8411	0.8863	0.8754
Zinc (Zn)	0.6681	0.1114	0.4908	0.0796	0.0789	0.1328	0.5405

Table 7.19b: Correlation between Heavy Metals of Groundwater at 40 ft Depth

	Cadmium (Cd)	*Copper (Cu)*	*Iron (Fe)*	*Manganese (Mn)*	*Nickel (Ni)*	*Lead (Pb)*	*Zinc (Zn)*
Cadmium (Cd)	1						
Copper (Cu)	-0.0297	1					
Iron (Fe)	0.6622	0.4766	1				
Manganese (Mn)	0.1973	0.8465	0.7032	1			
Nickel (Ni)	0.3374	0.5842	0.8490	0.8207	1		
Lead (Pb)	0.2640	0.7652	0.7915	0.9365	0.8564	1	
Zinc (Zn)	0.5500	0.5634	0.9363	0.6657	0.8227	0.73721449	1

Table 7.19c: Correlation between Heavy Metals of Groundwater at 80 ft Depth

	Cadmium (Cd)	*Copper (Cu)*	*Iron (Fe)*	*Manganese (Mn)*	*Nickel (Ni)*	*Lead (Pb)*	*Zinc (Zn)*
Cadmium (Cd)	1						
Copper (Cu)	0.2862	1					
Iron (Fe)	-0.1759	0.8387	1				
Manganese (Mn)	0.2938	0.9623	0.8108	1			
Nickel (Ni)	0.2378	0.9548	0.8665	0.9446	1		
Lead (Pb)	0.4973	0.8559	0.5784	0.9195	0.8326	1	
Zinc (Zn)	0.1943	0.9000	0.8367	0.9207	0.9394	0.784857601	1

Table 7.19d: Correlation between Heavy Metals of Groundwater (more than 120 ft depth)

	Cadmium (Cd)	*Copper (Cu)*	*Iron (Fe)*	*Manganese (Mn)*	*Nickel (Ni)*	*Lead (Pb)*	*Zinc (Zn)*
Cadmium (Cd)	1						
Copper (Cu)	0.3286	1					
Iron (Fe)	0.2481	0.8289	1				
Manganese (Mn)	0.3517	0.9427	0.9104	1			
Nickel (Ni)	0.3877	0.9122	0.7884	0.9536	1		
Lead (Pb)	0.1820	0.5557	0.8446	0.6508	0.5299	1	
Zinc (Zn)	0.3245	0.9509	0.8660	0.9143	0.8346	0.578293375	1

Table 7.20a: Correlation between Heavy Metals of Groundwater and Heavy Metals of Irrigated Agriculture Soil

	Cadmium (Cd)	*Copper (Cu)*	*Iron (Fe)*	*Manganese (Mn)*	*Nickel (Ni)*	*Lead (Pb)*	*Zinc (Zn)*
Cadmium (Cd)	0.7611	0.6056	0.3012	0.6082	0.5303	0.6466	0.4632
Copper (Cu)	-0.0836	0.8941	0.9733	0.8702	0.9121	0.6607	0.8910
Iron (Fe)	0.7266	0.7917	0.4410	0.8074	0.7946	0.8628	0.7463
Manganese (Mn)	0.0953	0.9025	0.9132	0.9145	0.9279	0.7404	0.9548
Nickel (Ni)	0.2968	0.9387	0.8307	0.9458	0.9475	0.8383	0.9350
Lead (Pb)	0.2862	0.9482	0.8397	0.9664	0.9199	0.8732	0.8856
Zinc (Zn)	-0.5899	0.4847	0.8065	0.4190	0.5301	0.2027	0.4890

Table 7.20b: Correlation between Heavy Metals of Agricultural Soil

	Cadmium (Cd)	*Copper (Cu)*	*Iron (Fe)*	*Manganese (Mn)*	*Nickel (Ni)*	*Lead (Pb)*	*Zinc (Zn)*
Cadmium (Cd)	1						
Copper (Cu)	0.3719	1					
Iron (Fe)	0.7981	0.5348	1				
Manganese (Mn)	0.4834	0.9502	0.6730	1			
Nickel (Ni)	0.6368	0.8904	0.8093	0.9581	1		
Lead (Pb)	0.6911	0.8811	0.7809	0.9274	0.9566	1	
Zinc (Zn)	-0.1615	0.7484	-0.0190	0.5876	0.4374	0.466065186	1

Table 7.21: Correlation between Heavy Metals of Irrigated Agriculture Soil and Heavy Metals of Drain Sediments

	Cadmium (Cd)	*Copper (Cu)*	*Iron (Fe)*	*Manganese (Mn)*	*Nickel (Ni)*	*Lead (Pb)*	*Zinc (Zn)*
Cadmium (Cd)	0.5246	0.7650	0.5272	0.7646	0.7149	0.7497	0.4546
Copper (Cu)	0.2703	0.2117	0.2542	0.1501	0.1793	0.0689	0.0597
Iron (Fe)	0.2989	-0.1898	-0.1413	-0.2625	-0.2321	-0.1173	-0.1968
Manganese (Mn)	0.7224	-0.1082	0.5619	-0.0121	0.1935	0.1732	-0.6052
Nickel (Ni)	0.8617	0.4783	0.8799	0.5761	0.7417	0.7788	-0.0353
Lead (Pb)	0.3275	0.8432	0.4109	0.7390	0.7063	0.7123	0.6133
Zinc (Zn)	0.6465	0.7171	0.5620	0.7216	0.7597	0.8553	0.4458

Table 7.22: Correlation between Heavy Metals of Pond Sediment and Heavy Metals of Drain Sediment

	Cadmium (Cd)	*Copper (Cu)*	*Iron (Fe)*	*Manganese (Mn)*	*Nickel (Ni)*	*Lead (Pb)*	*Zinc (Zn)*
Cadmium (Cd)	0.9493	0.2589	0.0530	0.2225	0.4803	0.7144	0.5922
Copper (Cu)	0.3647	0.9273	0.2261	0.2643	0.1140	0.2962	0.0239
Iron (Fe)	-0.1269	0.0944	0.9466	0.1694	-0.0400	-0.0980	0.0671
Manganese (Mn)	0.1856	0.3550	0.2388	0.9223	0.5129	0.0492	0.1831
Nickel (Ni)	0.5374	0.0246	0.2511	0.2658	0.8060	0.6183	0.9359
Lead (Pb)	0.6434	0.3742	0.0489	0.1409	0.3669	0.9388	0.4618
Zinc (Zn)	0.6764	-0.0189	0.2305	0.2456	0.7559	0.6630	0.9563

Table 7.23a: Correlation between Physico-chemical Parameters and Heavy Metals of Effluent Drain Sediment

Parameters	*Temperature*	*WHC*	*Bulk Density*	*Soil Moisture*	*pH*	*EC*	*Cl*	*Na*	*K*
Cadmium (Cd)	0.6827	0.4937	0.7226	-0.0967	-0.7117	0.7258	0.6894	0.2038	-0.0080
Copper (Cu)	0.2637	-0.0428	0.3623	-0.2021	-0.3459	0.2671	0.2181	-0.0577	0.0405
Iron (Fe)	-0.0671	0.0108	-0.0119	-0.3035	-0.0500	-0.1298	0.1399	-0.1357	0.1114
Manganese (Mn)	-0.2886	0.4659	-0.2402	-0.8078	-0.1016	-0.1484	-0.2057	0.6220	0.7253
Nickel (Ni)	0.3379	0.8752	0.2655	-0.6915	-0.5829	0.4826	0.3521	0.7610	0.5300
Lead (Pb)	0.8233	0.3379	0.8137	0.0752	-0.8377	0.8331	0.8293	-0.0199	-0.3465
Zinc (Zn)	0.6377	0.7598	0.5957	-0.2807	-0.7645	0.6900	0.7570	0.3600	0.0915

Table 7.23b: Correlation between Heavy Metals of Effluent Drain Sediment

	Cadmium (Cd)	Copper (Cu)	Iron (Fe)	Manganese (Mn)	Nickel (Ni)	Lead (Pb)	Zinc (Zn)
Cadmium (Cd)	1						
Copper (Cu)	0.3195	1					
Iron (Fe)	-0.0188	0.1213	1				
Manganese (Mn)	0.1399	0.3559	0.2154	1			
Nickel (Ni)	0.4724	0.1088	0.1213	0.5474	1		
Lead (Pb)	0.7101	0.2241	-0.0060	0.0275	0.4277	1	
Zinc (Zn)	0.6378	-0.0334	0.1473	0.1706	0.7211	0.6439	1

Table 7.23c: Correlation between Physico-chemical Parameters and Heavy Metals of Pond Sediment

Parameters	Temperature	WHC	Bulk Density	Soil Moisture	pH	EC	Cl	Na	K
Cadmium (Cd)	0.5929	0.4568	0.5963	-0.0794	-0.6108	0.6199	0.6032	0.1886	0.1151
Copper (Cu)	0.3937	-0.0270	0.2926	-0.1381	-0.3878	0.3652	0.2819	-0.0854	0.0318
Iron (Fe)	-0.1524	-0.1202	-0.0516	-0.1906	0.0855	-0.2443	0.0477	-0.2204	-0.0107
Manganese (Mn)	-0.2233	0.3758	-0.2722	-0.7337	0.0100	-0.1751	-0.2210	0.6170	0.7442
Nickel (Ni)	0.5642	0.8307	0.6211	-0.4386	-0.6914	0.6017	0.6717	0.4295	0.3376
Lead (Pb)	0.7460	0.2088	0.6861	0.0758	-0.7389	0.7095	0.7158	-0.0899	-0.2138
Zinc (Zn)	0.6040	0.8126	0.6614	-0.3505	-0.7260	0.6359	0.7137	0.3865	0.2800

Table 7.23d: Correlation between Heavy Metals of Pond Sediment

	Cadmium (Cd)	*Copper (Cu)*	*Iron (Fe)*	*Manganese (Mn)*	*Nickel (Ni)*	*Lead (Pb)*	*Zinc (Zn)*
Cadmium (Cd)	1						
Copper (Cu)	0.3044	1					
Iron (Fe)	-0.0569	0.2003	1				
Manganese (Mn)	0.2751	0.3340	0.1939	1			
Nickel (Ni)	0.5113	0.0523	0.1490	0.2330	1		
Lead (Pb)	0.7033	0.4419	-0.0253	0.1549	0.4747	1	
Zinc (Zn)	0.6544	0.0345	0.1314	0.2260	0.9485	0.524939198	1

Table 7.24: Transfer Factor for Heavy Metals from Textile Effluent to Groundwater in Experimental Sites

	Summer 2005	*Monsoon 2005*	*Winter 2005-06*	*Summer 2006*	*Monsoon 2006*	*Winter 2006-07*	*Summer 2007*
Cadmium (Cd)	0.16	0.12	0.62	0.36	0.19	1.50	0.53
Copper (Cu)	1.01	1.20	1.79	0.99	1.15	1.60	1.06
Iron (Fe)	7.17	5.93	6.12	7.04	6.42	6.30	9.64
Manganese (Mn)	3.59	2.95	1.27	3.01	4.12	1.45	1.39
Nickel (Ni)	3.93	5.83	1.45	2.28	14.14	1.52	2.78
Lead (Pb)	3.68	4.65	2.01	3.83	3.85	2.33	2.67
Zinc (Zn)	1.16	1.21	0.53	0.97	0.93	0.38	0.73

Table 7.25: Transfer Factor for Heavy Metals from Textile Effluent to Pond Water in Experimental Sites

	Summer 2005	*Monsoon 2005*	*Winter 2005-06*	*Summer 2006*	*Monsoon 2006*	*Winter 2006-07*	*Summer 2007*
Cadmium (Cd)	0.17	0.05	0.32	0.41	0.19	0.83	0.37
Copper (Cu)	0.97	1.16	1.53	0.98	1.10	1.36	1.09
Iron (Fe)	18.42	16.71	18.12	17.42	15.93	16.04	25.00
Manganese (Mn)	3.46	3.66	0.77	2.78	4.27	0.79	0.95
Nickel (Ni)	3.39	4.52	1.74	1.95	11.76	1.64	2.56
Lead (Pb)	3.60	2.44	1.19	3.63	2.20	1.23	2.40
Zinc (Zn)	0.90	1.00	0.66	0.74	0.69	0.54	0.71

Table 7.26: Transfer Factor for Heavy Metals from Pond Water to Groundwater (40ft) in Experimental Sites

	Summer 2005	*Monsoon 2005*	*Winter 2005-06*	*Summer 2006*	*Monsoon 2006*	*Winter 2006-07*	*Summer 2007*
Cadmium (Cd)	0.93	2.20	1.93	0.87	1.00	1.80	1.43
Copper (Cu)	1.04	1.04	1.17	1.01	1.05	1.17	0.97
Iron (Fe)	0.39	0.36	0.34	0.40	0.40	0.39	0.39
Manganese (Mn)	1.04	0.81	1.65	1.08	0.96	1.84	1.46
Nickel (Ni)	1.16	1.29	0.83	1.17	1.20	0.93	1.09
Lead (Pb)	1.02	1.91	1.69	1.06	1.75	1.89	1.11
Zinc (Zn)	1.28	1.21	0.80	1.30	1.35	0.70	1.03

Table 7.27: Transfer Factor for Heavy Metals from Pond Water to Groundwater (80 ft) in Experimental Sites

	Summer 2005	*Monsoon 2005*	*Winter 2005-06*	*Summer 2006*	*Monsoon 2006*	*Winter 2006-07*	*Summer 2007*
Cadmium (Cd)	0.64	0.50	1.05	0.53	0.18	1.11	0.82
Copper (Cu)	0.68	0.55	0.60	0.68	0.54	0.64	0.57
Iron (Fe)	0.19	0.39	0.14	0.19	0.39	0.20	0.12
Manganese (Mn)	0.55	0.46	0.87	0.58	0.48	0.88	0.76
Nickel (Ni)	0.77	0.82	0.29	0.76	0.77	0.27	0.45
Lead (Pb)	0.48	1.03	1.05	0.49	0.91	1.00	0.64
Zinc (Zn)	1.05	1.21	0.44	1.03	1.12	0.43	0.69

Table 7.28: Transfer Factor for Heavy Metals from Pond Water to Groundwater (deep than 120 ft) in Experimental Sites

	Summer 2005	*Monsoon 2005*	*Winter 2005-06*	*Summer 2006*	*Monsoon 2006*	*Winter 2006-07*	*Summer 2007*
Cadmium (Cd)	0.04	0.00	0.00	0.13	0.00	0.00	0.00
Copper (Cu)	0.47	0.41	0.49	0.49	0.42	0.52	0.45
Iron (Fe)	0.10	0.16	0.10	0.11	0.16	0.13	0.12
Manganese (Mn)	0.55	0.40	0.55	0.58	0.44	0.54	0.75
Nickel (Ni)	0.54	0.61	0.17	0.51	0.60	0.19	0.33
Lead (Pb)	0.22	0.47	0.57	0.24	0.41	0.56	0.35
Zinc (Zn)	0.72	0.71	0.29	0.71	0.74	0.25	0.53

Table 7.29: Transfer Factor for Heavy Metals from Groundwater (40 ft) to Agricultural Soil in Experimental Sites

	Summer 2005	*Monsoon 2005*	*Winter 2005-06*	*Summer 2006*	*Monsoon 2006*	*Winter 2006-07*	*Summer 2007*
Cadmium (Cd)	400.00	227.27	243.17	335.00	720.00	285.19	265.00
Copper (Cu)	81.94	73.18	45.33	82.11	76.38	49.91	57.92
Iron (Fe)	20.30	17.80	17.69	21.68	16.52	18.15	18.40
Manganese (Mn)	24.67	27.41	20.77	24.67	24.34	18.48	24.30
Nickel (Ni)	71.36	81.49	127.88	69.02	81.48	121.71	72.40
Lead (Pb)	33.91	33.81	29.81	33.68	35.38	26.49	35.64
Zinc (Zn)	65.83	139.00	60.95	60.58	126.52	71.76	40.00

Table 7.30: Transfer Factor for Heavy Metals from Groundwater (80 ft) to Agricultural Soil in Experimental Sites

	Summer 2005	*Monsoon 2005*	*Winter 2005-06*	*Summer 2006*	*Monsoon 2006*	*Winter 2006-07*	*Summer 2007*
Cadmium (Cd)	574.71	1000.00	445.00	546.94	4000.00	461.08	460.87
Copper (Cu)	125.90	138.11	88.39	120.99	148.09	90.97	98.58
Iron (Fe)	40.78	16.16	43.11	46.34	17.05	35.08	58.31
Manganese (Mn)	46.20	47.52	39.65	45.76	48.46	38.44	46.44
Nickel (Ni)	106.80	128.65	361.25	106.58	127.37	417.65	175.73
Lead (Pb)	72.28	62.77	47.81	72.12	68.08	49.70	62.36
Zinc (Zn)	80.61	139.00	110.98	76.62	153.16	117.31	59.26

Table 7.31: Transfer Factor for Heavy Metals from Groundwater (deep than 120 ft depth) to Agricultural Soil in Experimental Sites

	Summer 2005	*Monsoon 2005*	*Winter 2005-06*	*Summer 2006*	*Monsoon 2006*	*Winter 2006-07*	*Summer 2007*
Cadmium (Cd)	10000.00	NP	NP	2233.33	NP	NP	NP
Copper (Cu)	181.80	184.90	107.89	169.30	190.94	111.67	124.11
Iron (Fe)	75.41	39.75	60.25	79.07	41.64	56.89	60.29
Manganese (Mn)	46.84	54.63	62.11	45.64	52.79	62.77	47.07
Nickel (Ni)	153.92	173.48	624.19	159.47	162.42	608.57	235.06
Lead (Pb)	155.22	136.89	87.74	147.98	152.32	89.24	114.09
Zinc (Zn)	116.52	238.63	164.81	111.09	231.87	196.77	77.67

NP: Non-point source.

Table 7.32: Transfer Factor for Heavy Metals from Groundwater to Soil in Control Site

	Summer 2005	*Monsoon 2005*	*Winter 2005-06*	*Summer 2006*	*Monsoon 2006*	*Winter 2006-07*	*Summer 2007*
Cadmium (Cd)	67000.00	NP	NP	11428.57	NP	NP	NP
Copper (Cu)	73.62	154.02	110.89	135.26	152.57	114.21	23.10
Iron (Fe)	80.12	33.33	52.35	76.60	35.62	50.76	60.31
Manganese (Mn)	56.37	67.98	62.81	53.94	68.10	64.00	53.72
Nickel (Ni)	193.13	251.28	NP	189.41	250.62	NP	360.00
Lead (Pb)	159.30	150.90	99.58	157.46	149.76	102.52	122.70
Zinc (Zn)	122.22	236.92	169.55	117.88	227.65	210.59	83.82

Table 7.33: Enrichment Factor for Various Metals in Drain Sediment

	Summer 2005	Monsoon 2005	Winter 2005-06	Summer 2006	Monsoon 2006	Winter 2006-07	Summer 2007
Cadmium (Cd)	0.68	0.90	0.53	0.85	0.81	0.65	0.86
Copper (Cu)	5.82	8.11	5.74	8.47	7.47	6.24	7.89
Iron (Fe)	1000.00	1000.00	1000.00	1000.00	1000.00	1000.00	1000.00
Manganese (Mn)	123.69	139.91	120.96	170.80	109.92	129.81	175.63
Nickel (Ni)	8.86	10.25	7.49	11.48	9.22	7.35	11.70
Lead (Pb)	8.27	10.73	5.83	9.89	9.54	6.63	8.51
Zinc (Zn)	30.71	36.78	18.71	38.80	27.45	18.12	33.71

Table 7.34: Enrichment Factor for Various Metals in Pond Sediment

	Summer 2005	Monsoon 2005	Winter 2005-06	Summer 2006	Monsoon 2006	Winter 2006-07	Summer 2007
Cadmium (Cd)	1.01	1.11	0.67	1.24	1.20	0.87	1.26
Copper (Cu)	9.13	10.17	7.29	12.04	10.46	7.61	10.16
Iron (Fe)	1000.00	1000.00	1000.00	1000.00	1000.00	1000.00	1000.00
Manganese (Mn)	167.01	167.88	139.09	213.02	145.51	151.18	243.85
Nickel (Ni)	11.29	12.13	8.78	14.63	13.07	8.38	14.85
Lead (Pb)	12.23	12.83	7.32	14.23	13.79	8.43	11.70
Zinc (Zn)	42.55	33.00	20.49	47.57	36.96	19.89	43.78

Table 7.35a: Geo-accumulation Index for Various Metals in Drain Sediment Samples

	Summer 2005		*Monsoon 2005*		*Winter 2005-06*		*Summer 2006*		*Monsoon 2006*		*Winter 2006-07*		*Summer 2007*	
	Value	I_{geo}	*Value*	I_{geo}	*Value*	I_{geo}	*Value*	I_{geo}	*Value*	I_{geo}	*Value*	I_{geo}	*Value*	I_{geo}
Cadmium (Cd)	12.85	3.68	11.34	3.50	9.44	3.24	12.15	3.60	11.46	3.52	11.36	3.51	11.16	3.48
Copper (Cu)	0.73	-0.45	0.68	-0.56	0.69	-0.54	0.81	-0.30	0.70	-0.51	0.73	-0.45	0.68	-0.56
Iron (Fe)	0.12	-3.06	0.08	-3.65	0.11	-3.13	0.09	-3.45	0.09	-3.48	0.11	-3.16	0.08	-3.61
Manganese (Mn)	0.82	-0.28	0.62	-0.69	0.77	-0.38	0.87	-0.21	0.55	-0.87	0.81	-0.31	0.80	-0.32
Nickel (Ni)	0.74	-0.44	0.57	-0.81	0.59	-0.75	0.73	-0.46	0.58	-0.80	0.57	-0.81	0.67	-0.58
Lead (Pb)	2.33	1.22	2.02	1.02	1.57	0.65	2.13	1.09	2.02	1.02	1.75	0.81	1.65	0.72
Zinc (Zn)	1.82	0.87	1.46	0.55	1.06	0.09	1.76	0.82	1.23	0.29	1.01	0.01	1.38	0.46

Table 7.35b: Geo-accumulation Index for Highest Value of Metals in Drain Sediment Samples

	2005-06		*2006-07*	
	Value	I_{geo}	*Value*	I_{geo}
Cadmium (Cd)	13.22	3.72	13.60	3.77
Copper (Cu)	0.79	-0.33	0.92	-0.12
Iron (Fe)	0.16	-2.66	0.15	-2.76
Manganese (Mn)	0.89	-0.17	0.98	-0.03
Nickel (Ni)	0.76	-0.39	0.77	-0.37
Lead (Pb)	2.43	1.28	2.60	1.38
Zinc (Zn)	2.40	1.26	2.70	1.43

Table 7.36a: Geo-accumulation Index for Various Metals in Pond Sediment Samples

	Summer 2005		Monsoon 2005		Winter 2005-06		Summer 2006		Monsoon 2006		Winter 2006-07		Summer 2007	
	Value	I_{geo}	*Value*	I_{geo}	*Value*	I_{geo}	*Value*	I_{geo}	*Value*	I_{geo}	*Value*	I_{geo}	*Value*	I_{geo}
Cadmium (Cd)	12.70	3.67	10.62	3.41	9.01	3.17	12.02	3.59	11.10	3.47	11.27	3.49	10.66	3.41
Copper (Cu)	0.76	-0.39	0.65	-0.62	0.65	-0.62	0.78	-0.36	0.64	-0.63	0.66	-0.60	0.57	-0.80
Iron (Fe)	0.08	-3.65	0.06	-4.04	0.08	-3.56	0.06	-4.02	0.06	-4.09	0.08	-3.60	0.05	-4.22
Manganese (Mn)	0.74	-0.44	0.57	-0.82	0.66	-0.61	0.73	-0.46	0.47	-1.08	0.69	-0.53	0.73	-0.46
Nickel (Ni)	0.62	-0.68	0.51	-0.96	0.52	-0.95	0.63	-0.68	0.53	-0.91	0.48	-1.06	0.55	-0.85
Lead (Pb)	2.30	1.20	1.84	0.88	1.47	0.55	2.07	1.05	1.91	0.93	1.64	0.72	1.48	0.57
Zinc (Zn)	1.68	0.75	1.00	0.00	0.86	-0.21	1.45	0.54	1.08	0.11	0.82	-0.29	1.17	0.22

Table 7.36b: Geo-accumulation Index for Highest Value Metals in Pond Sediment Samples

	2005-06		2006-07	
	Value	I_{geo}	*Value*	I_{geo}
Cadmium (Cd)	13.67	3.77	14.04	3.81
Copper (Cu)	0.82	-0.28	0.85	-0.24
Iron (Fe)	0.11	-3.17	0.00	-9.70
Manganese (Mn)	0.77	-0.37	0.78	-0.35
Nickel (Ni)	0.70	-0.52	0.77	-0.37
Lead (Pb)	2.47	1.30	2.50	1.32
Zinc (Zn)	2.18	1.12	2.25	1.17

Table 7.37: Metal Accumulation Factor for the Groundwater of Panipat Region

	Summer 2005	*Monsoon 2005*	*Winter 2005-06*	*Summer 2006*	*Monsoon 2006*	*Winter 2006-07*	*Summer 2007*
Cadmium (Cd)	250.00	α	α	57.14	α	α	α
Copper (Cu)	2.83	2.76	3.28	2.80	2.84	3.20	0.53
Iron (Fe)	4.41	2.64	4.07	4.40	3.10	4.17	4.22
Manganese (Mn)	2.70	2.76	3.59	2.65	3.24	4.17	2.91
Nickel (Ni)	3.36	3.60	α	3.59	3.67	α	7.14
Lead (Pb)	6.93	5.73	4.45	6.84	6.06	5.32	4.59
Zinc (Zn)	2.02	2.05	2.86	2.17	2.12	3.00	2.31

α: infinite

Table 7.38: Metal Accumulation Factor for the Agricultural Soil of Panipat Region

	Summer 2005	*Monsoon 2005*	*Winter 2005-06*	*Summer 2006*	*Monsoon 2006*	*Winter 2006-07*	*Summer 2007*
Cadmium (Cd)	1.49	0.91	3.87	1.68	0.97	8.56	1.89
Copper (Cu)	3.15	1.31	1.34	1.70	1.42	1.40	1.32
Iron (Fe)	1.12	1.41	1.38	1.25	1.44	1.49	1.29
Manganese (Mn)	1.18	1.11	1.19	1.21	1.16	1.20	1.31
Nickel (Ni)	1.24	1.17	1.71	1.31	1.19	2.05	1.44
Lead (Pb)	1.48	1.28	1.33	1.46	1.43	1.37	1.33
Zinc (Zn)	1.09	1.20	1.03	1.11	1.18	1.02	1.10

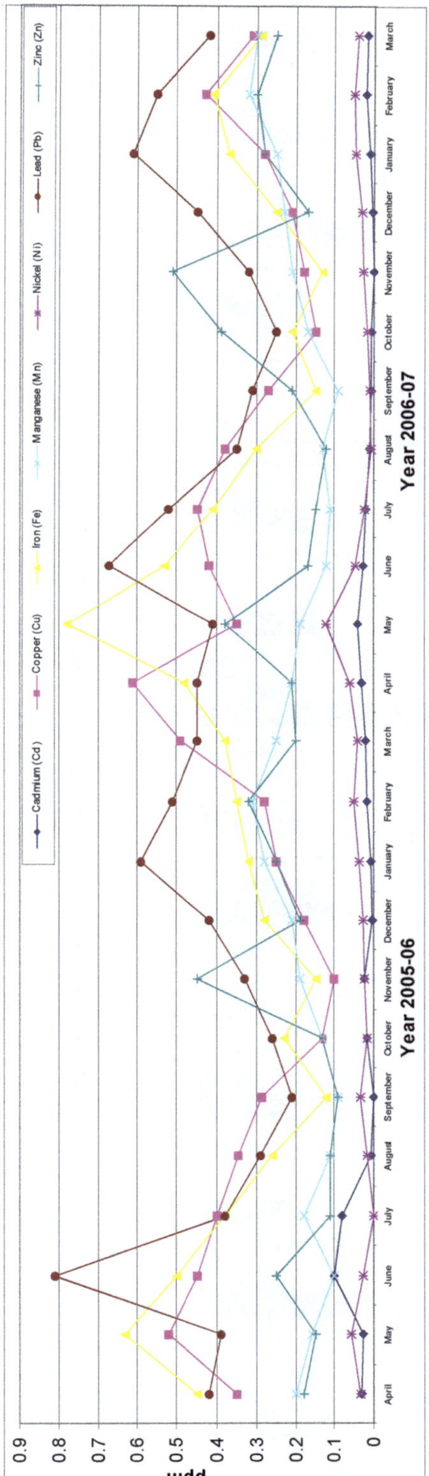

Figure 7.1a: Monthly Variation of Heavy Metals in Textile Industrial Effluent

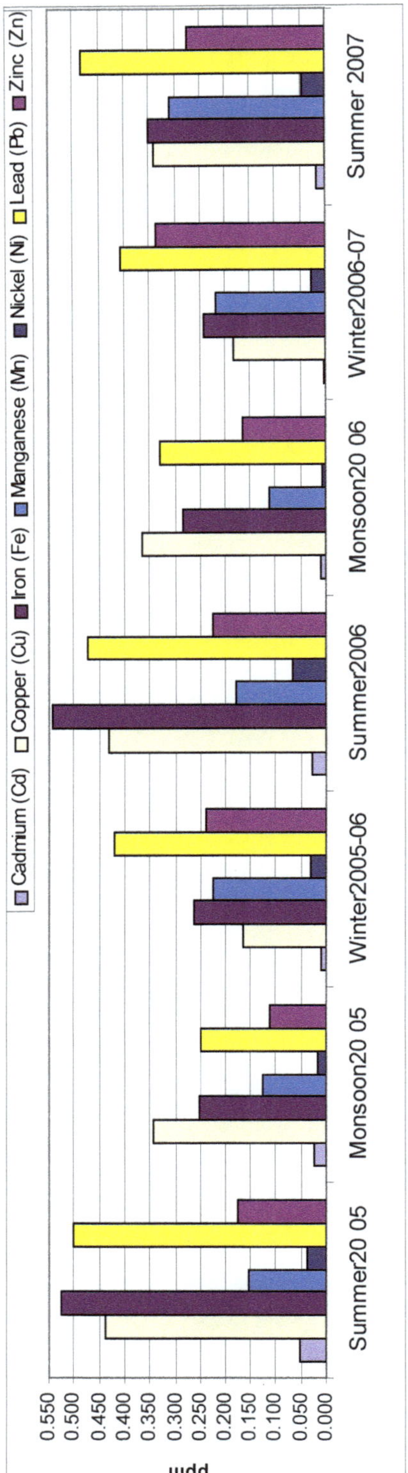

Figure 7.1b: Seasonal Fluctuation of Heavy Metals in Textile Industrial Effluent

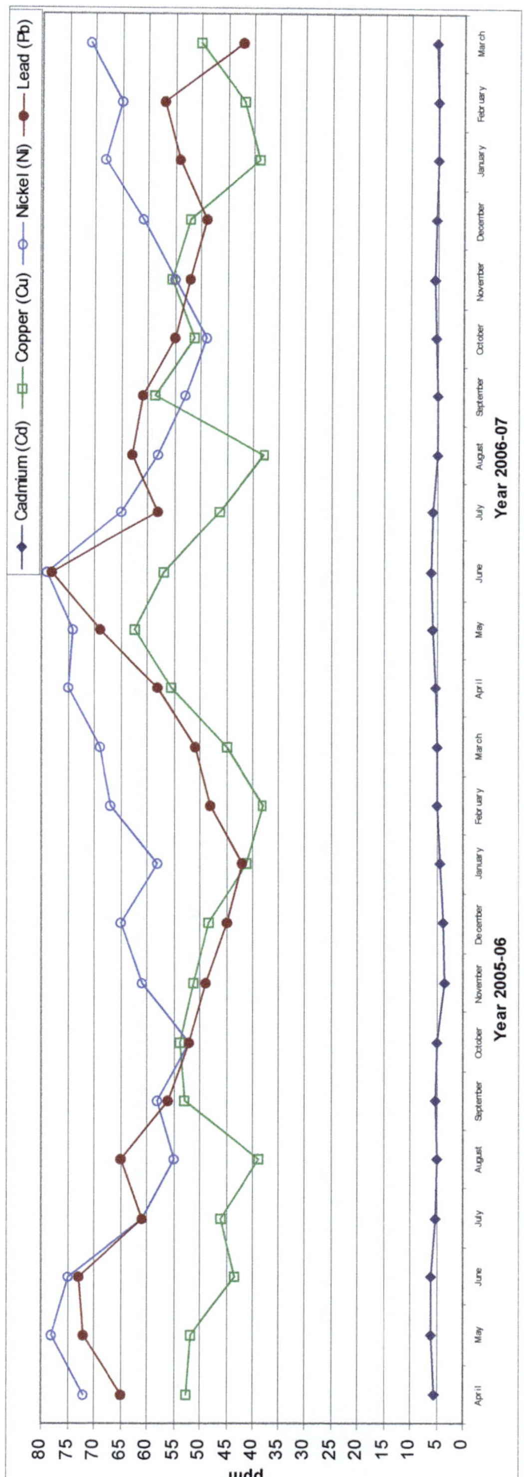

Figure 7.2a: Monthly Fluctuation of Heavy Metals in Effluent Drain Sediment

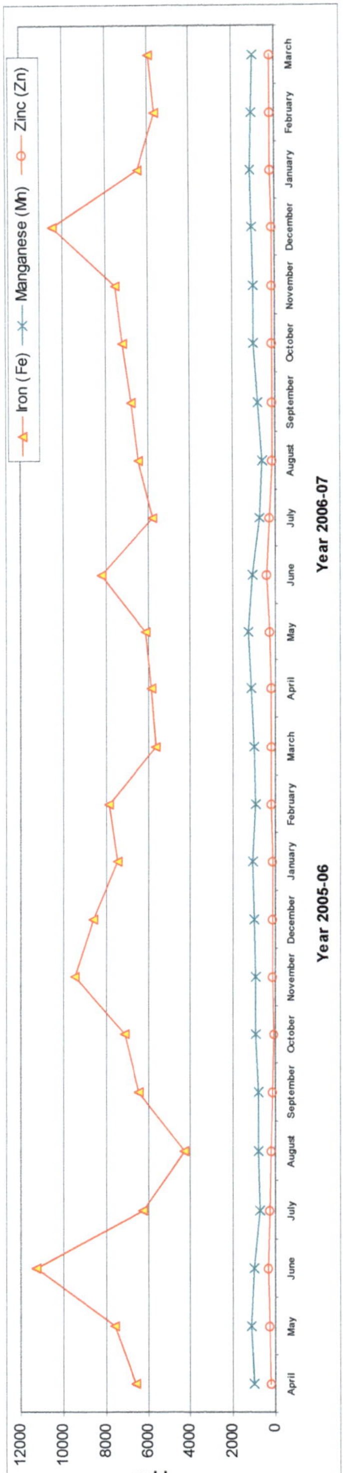

Figure 7.2b: Monthly Fluctuation of Heavy Metals in Effluent Drain Sediment

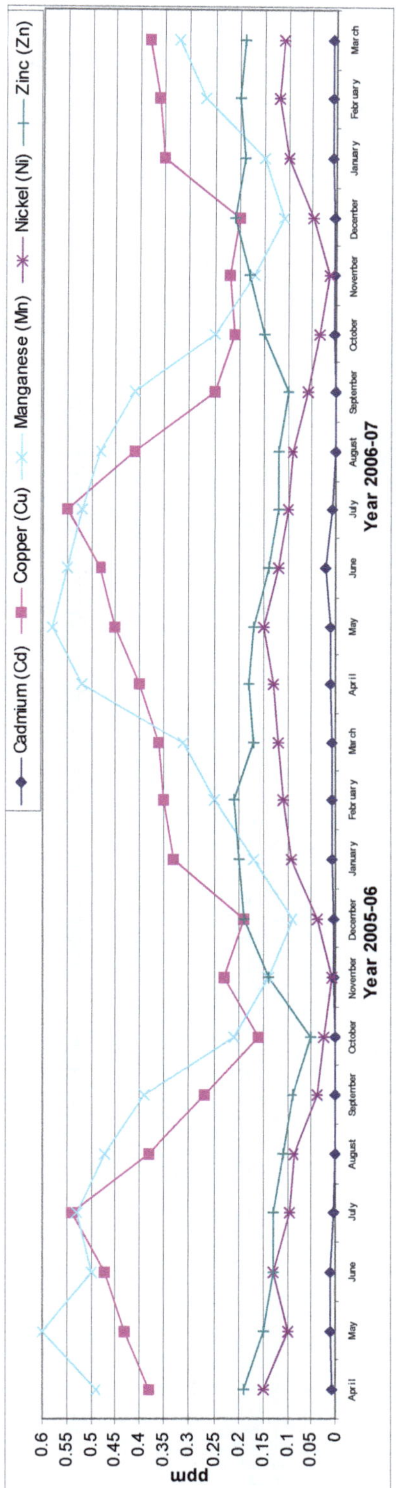

Figure 7.3a: Monthly Fluctuation of Heavy Metals in Pond Water

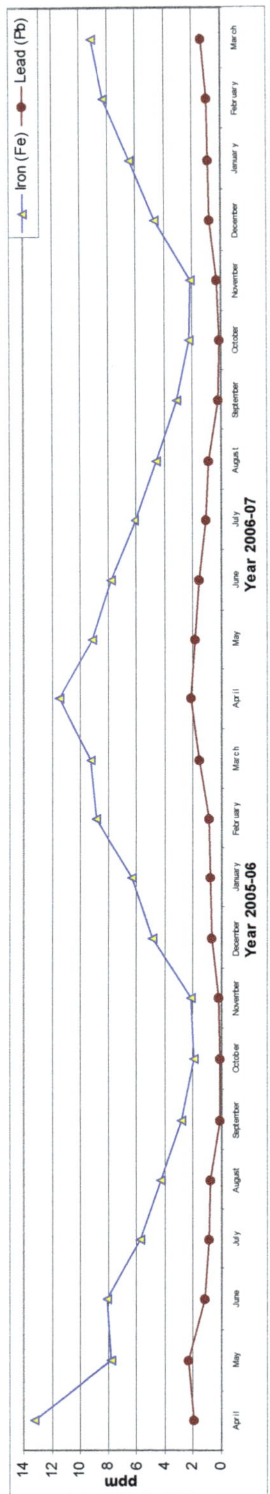

Figure 7.3b: Monthly Fluctuation of Heavy Metals in Pond Water

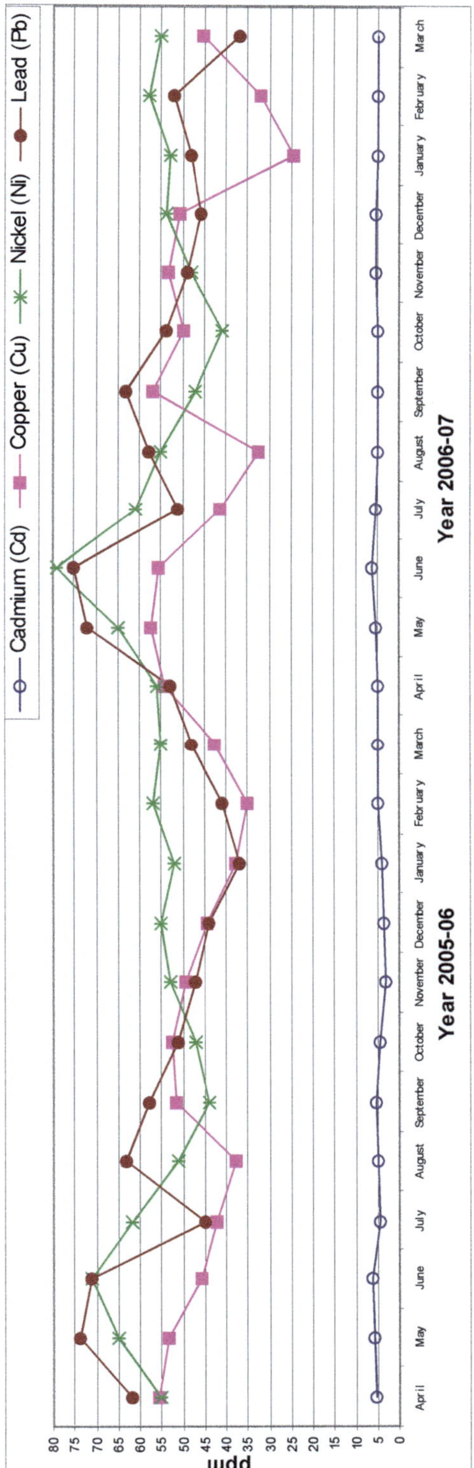

Figure 7.4a: Monthly Variation of Heavy Metals in Pond Sediment

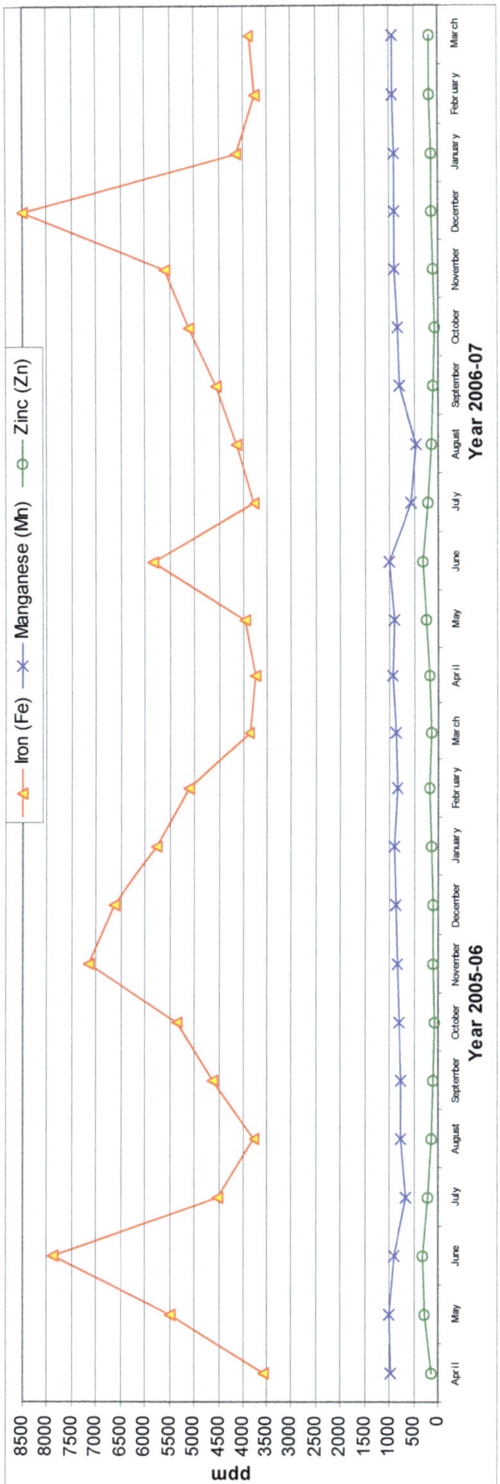

Figure 7.4b: Monthly Fluctuation of Heavy Metals in Pond Sediment

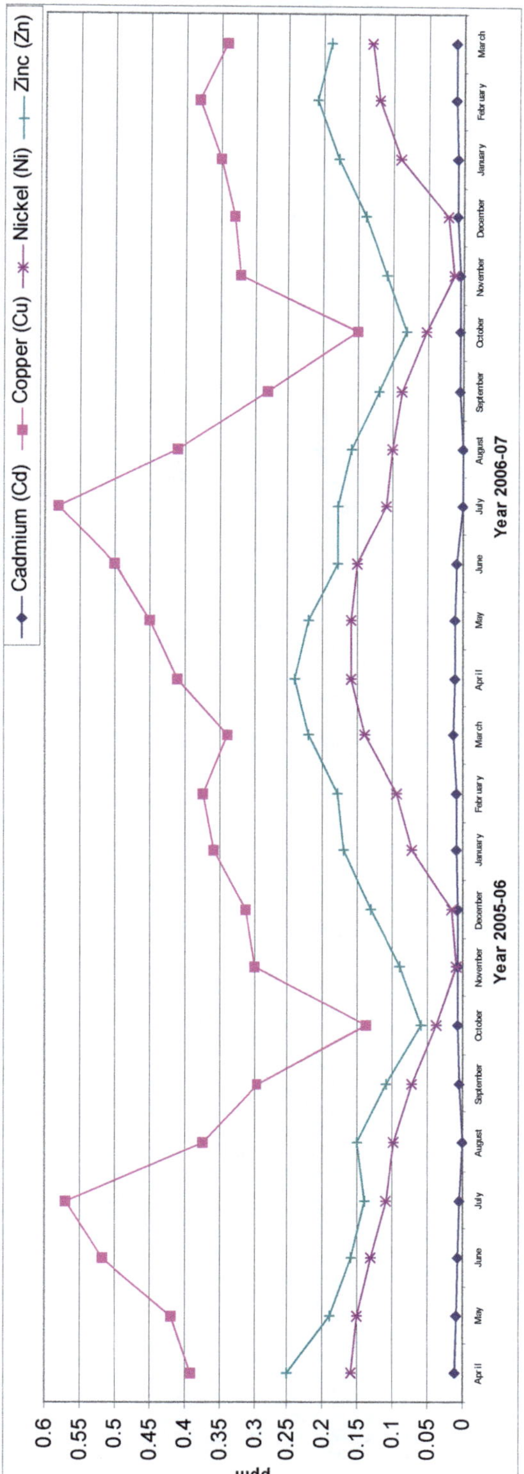

Figure 7.5a: Monthly Variation of Heavy Metals in Groundwater at 40 ft Depth from Hand Pump in Village Binjhole

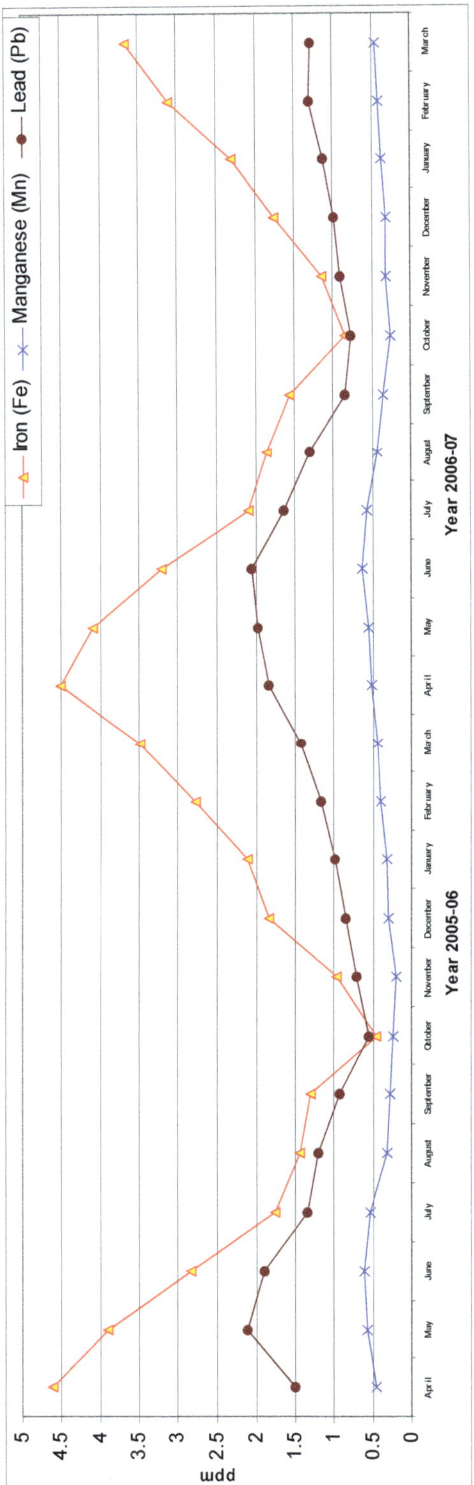

Figure 7.5b: Monthly Fluctuation of Heavy Metals in Groundwater at 40 ft Depth from Hand Pump in Village Binjhole

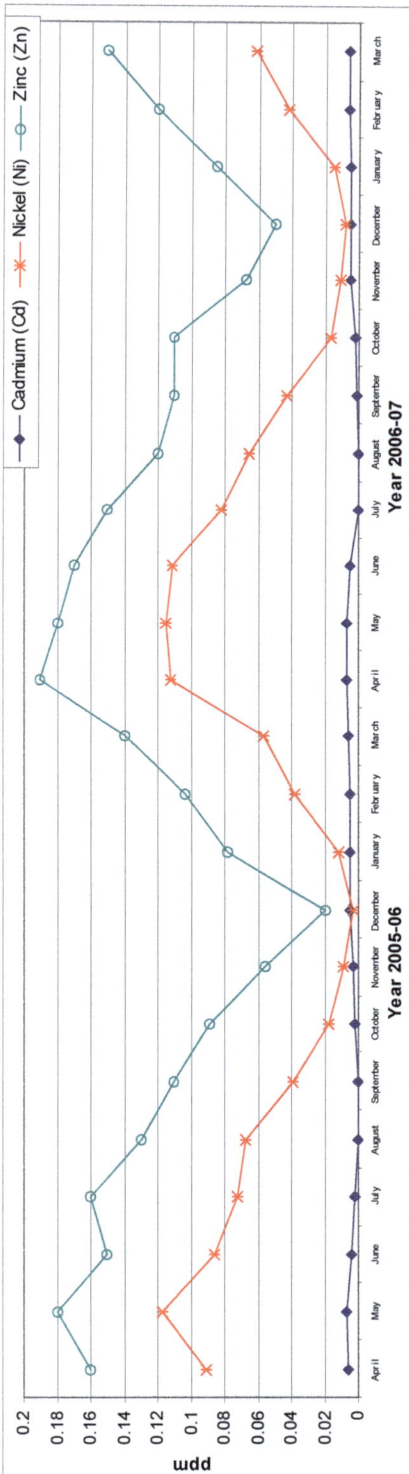

Figure 7.6a: Monthly Variation of Heavy Metals in Groundwater at 80 ft Depth from Tube Well and Deep Hand Pump

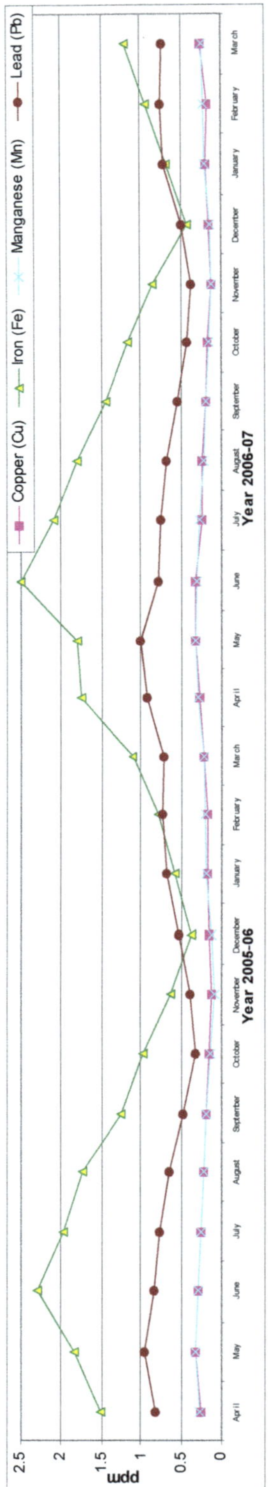

Figure 7.6b: Monthly Fluctuation of Heavy Metals in Groundwater at 80 ft Depth from Tube Well and Deep Hand Pump

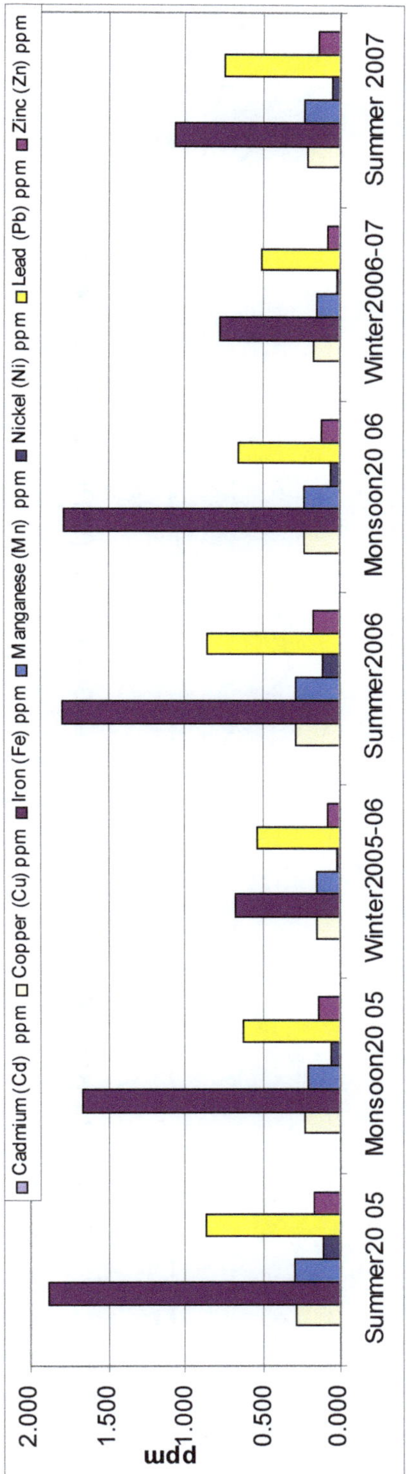

Figure 7.6c: Seasonal Fluctuation of Heavy Metals in Groundwater at 80 ft Depth from Tube Well and Deep Hand Pump

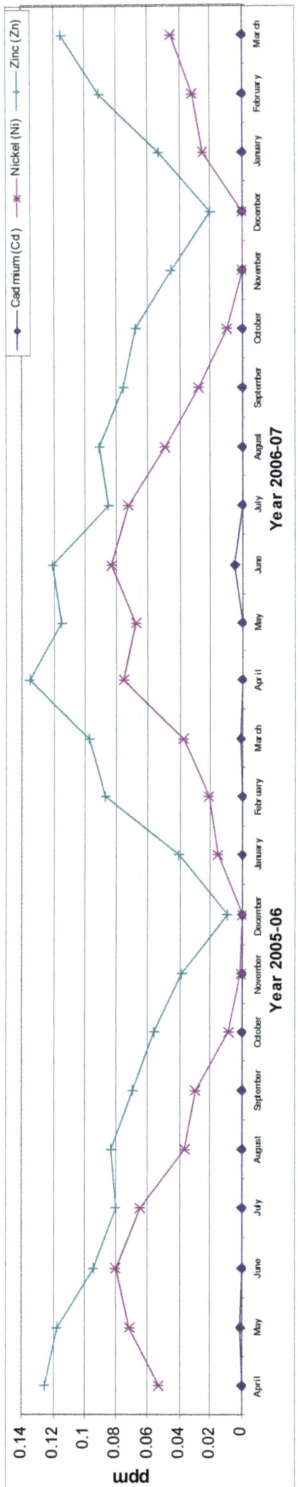

Figure 7.7a: Heavy Metals Fluctuation in Deep Groundwater (120 ft depth) of Tube Well

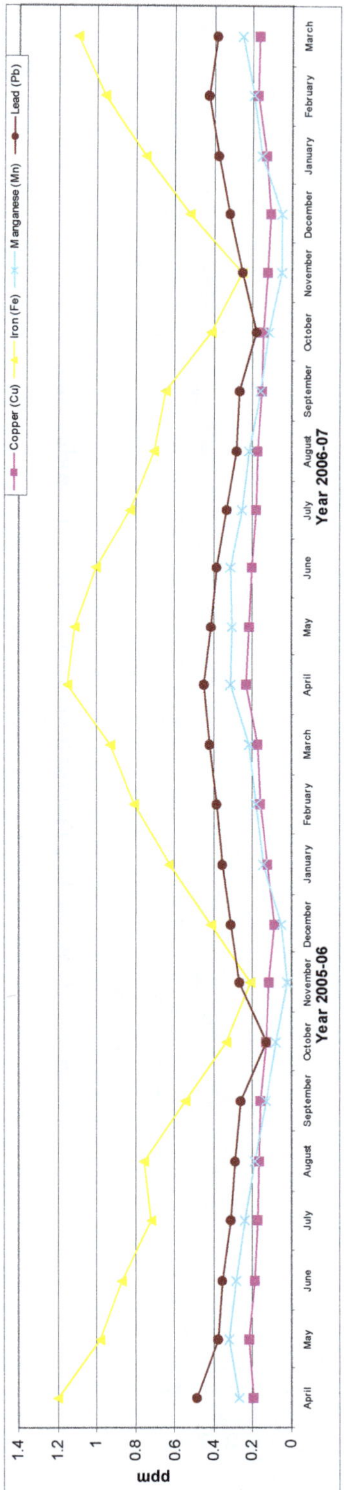

Figure 7.7b: Fluctuation of Heavy Metals in Deep Groundwater (120 ft depth) of Tube Well

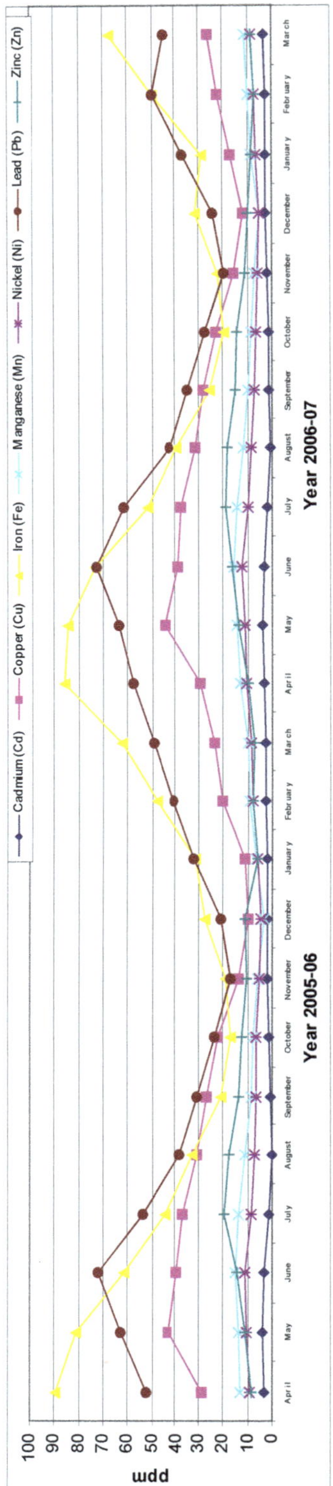

Figure 7.8: Heavy Metals Fluctuation in Soil Near Textile Industries

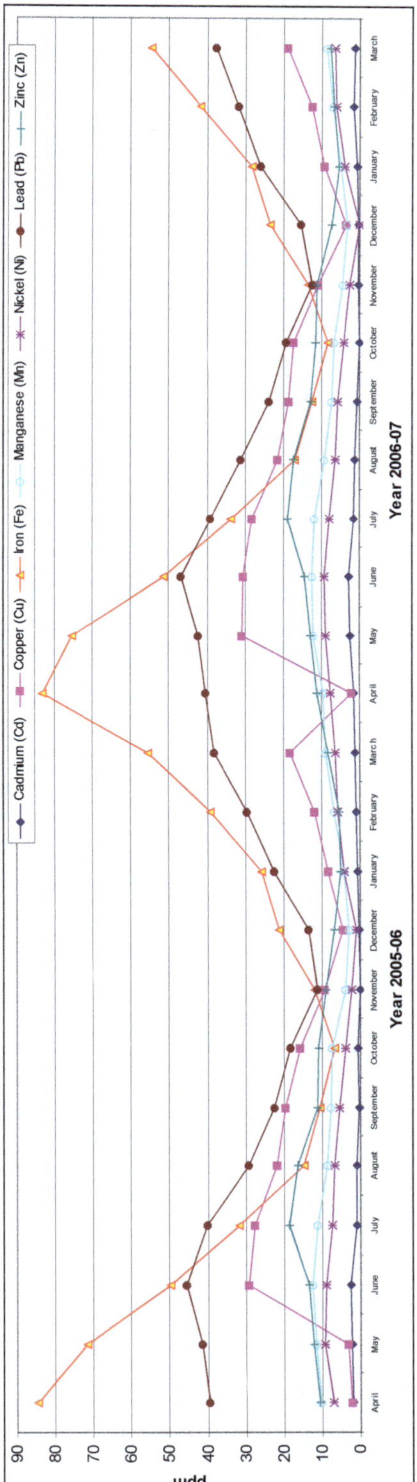

Figure 7.9: Heavy Metals Fluctuation in Agricultural Soil

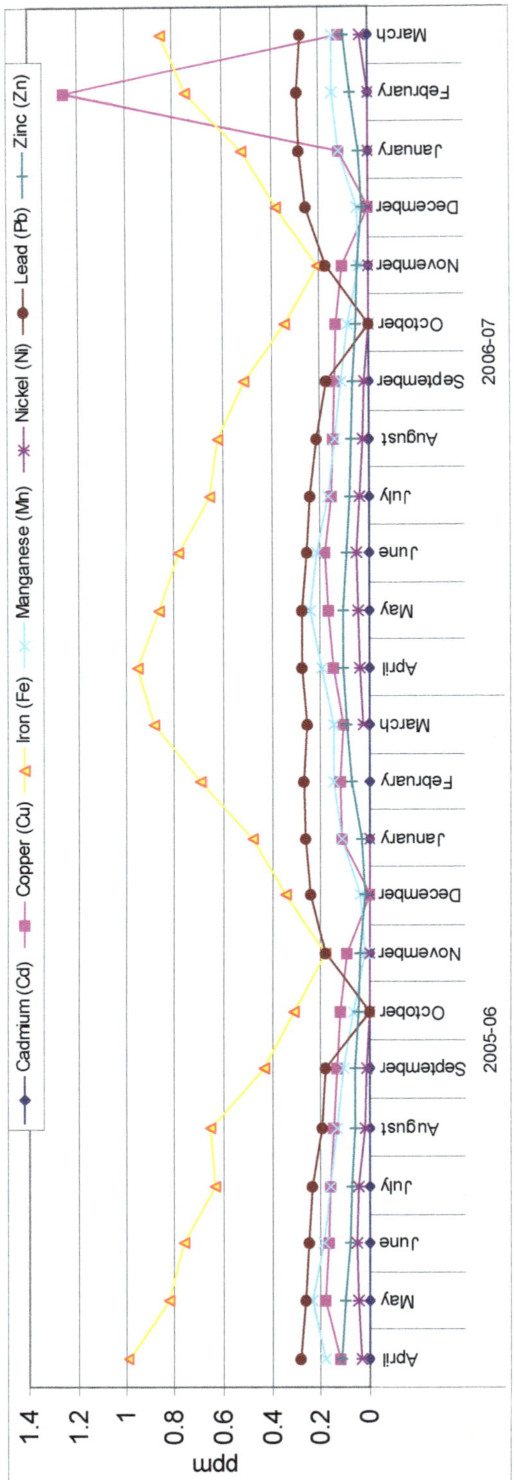

Figure 7.10a: Heavy Metals Fluctuation in Groundwater at Control Site

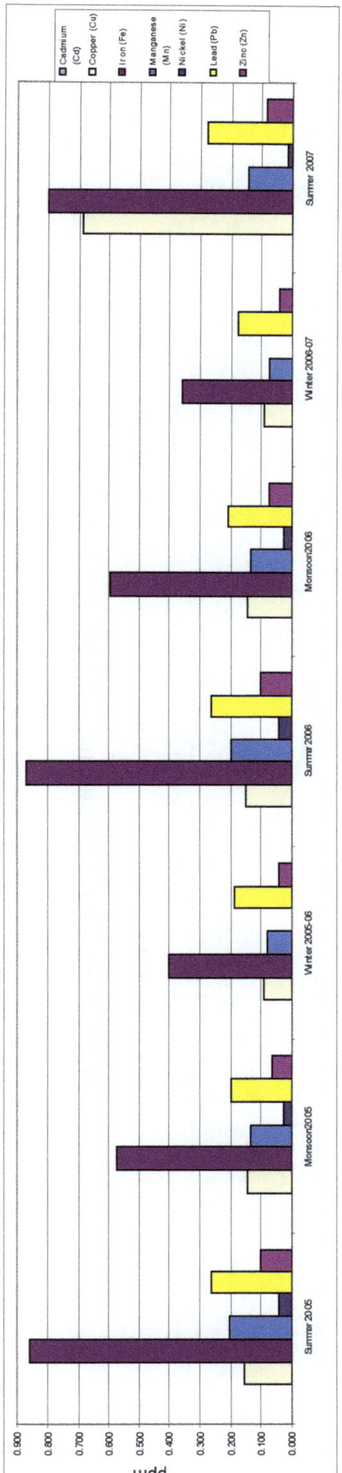

Figure 7.10b: Seasonal Fluctuation of Heavy Metals in Groundwater at Control Site

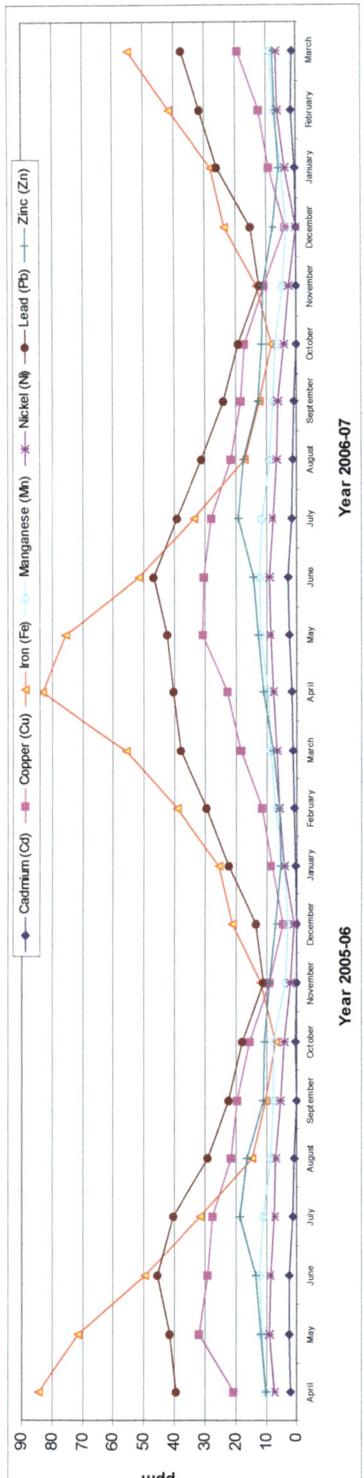

Figure 7.11a: Heavy Metals Fluctuation in Soil at Control Site

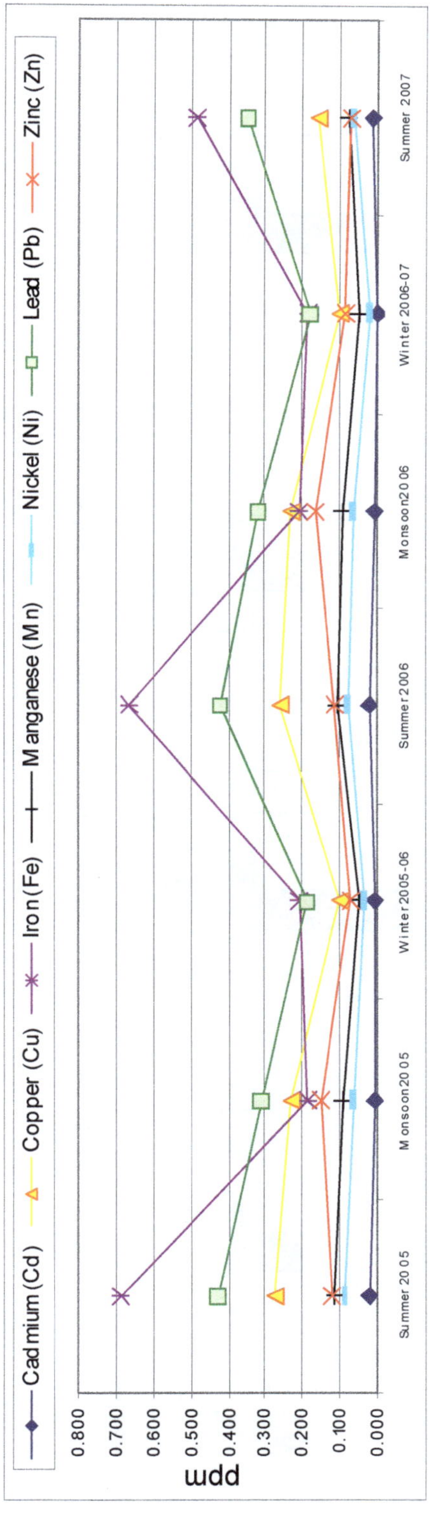

Figure 7.11b: Seasonal Fluctuation of Heavy Metals in Soil at Control Site

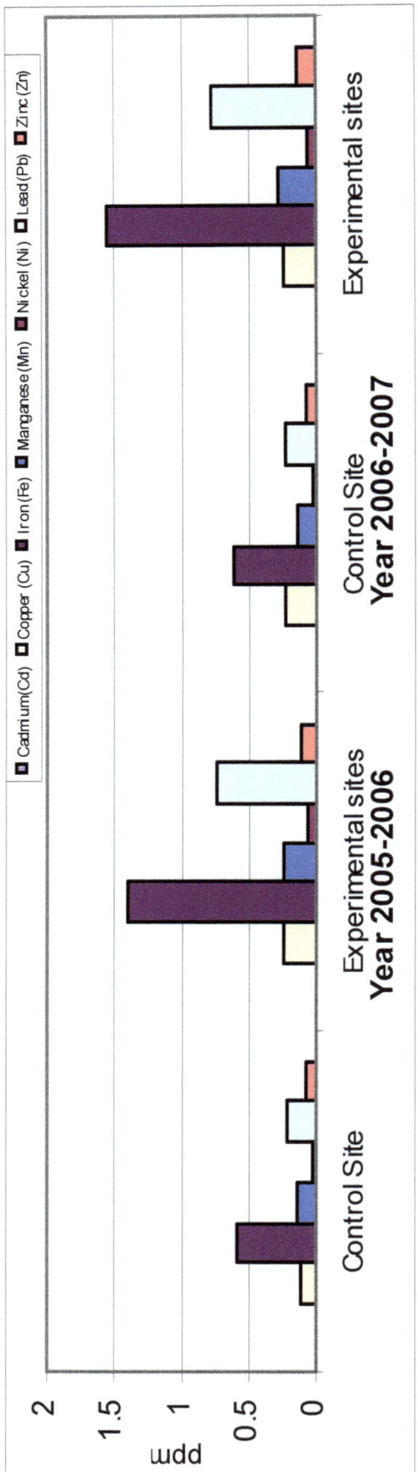

Figure 7.12: Mean Values of Heavy Metals in Groundwater at Control and Experimental Sites

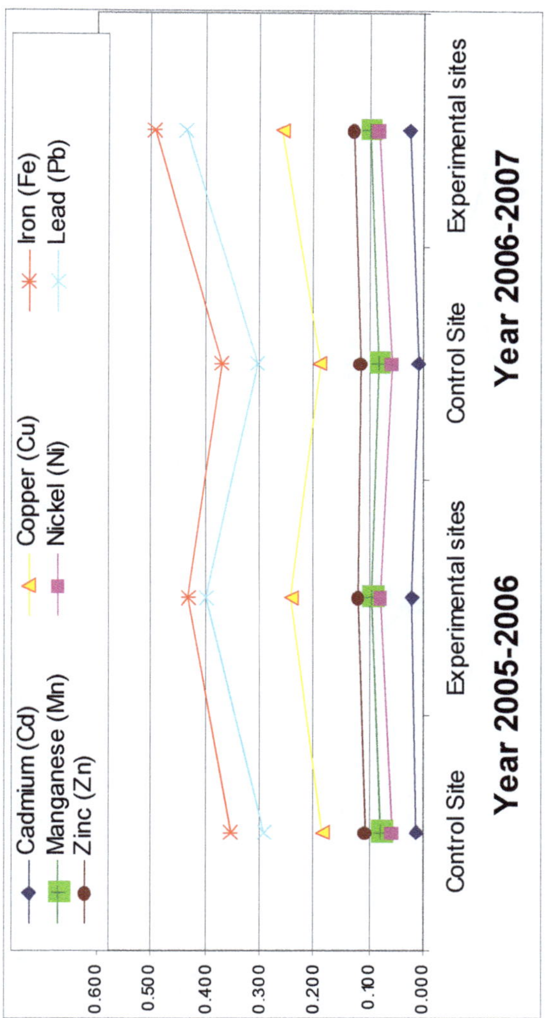

Figure 7.13: Mean Values of Heavy Metals in Soil at Control and Experimental Sites

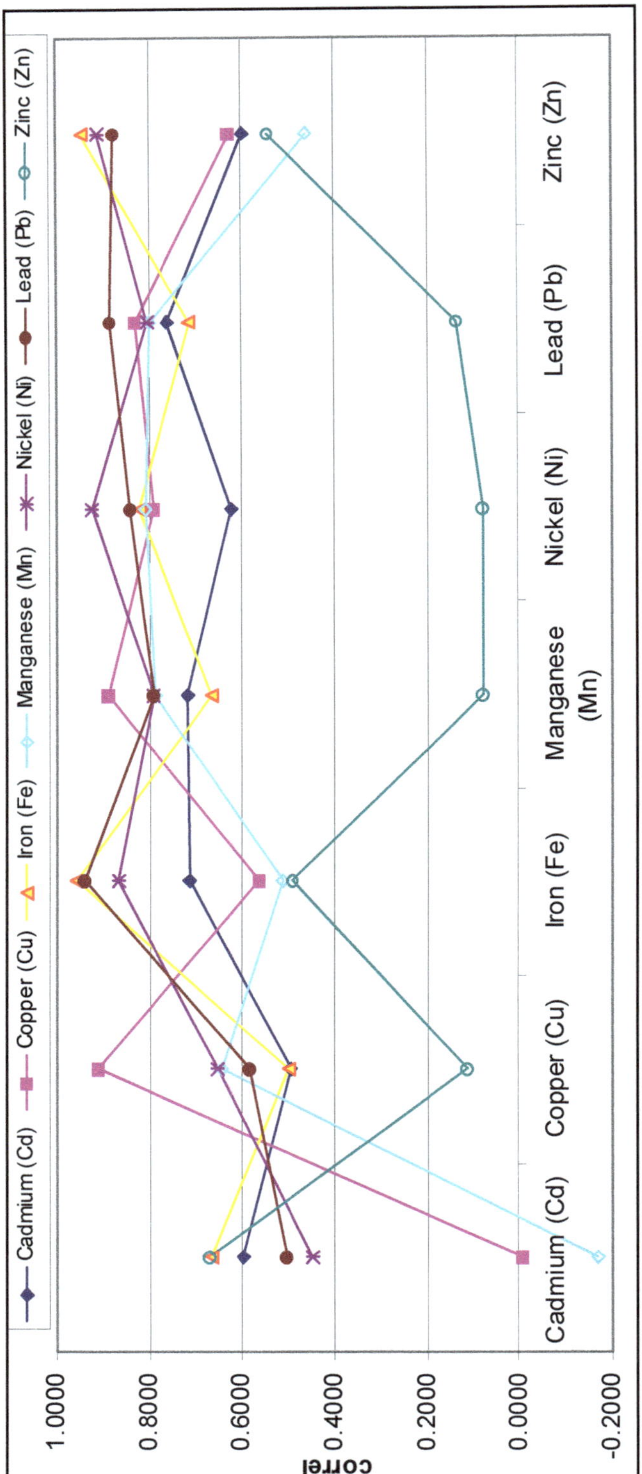

Figure 7.14: Correlation between Heavy Metals of Pond Water and Heavy Metals of Groundwater

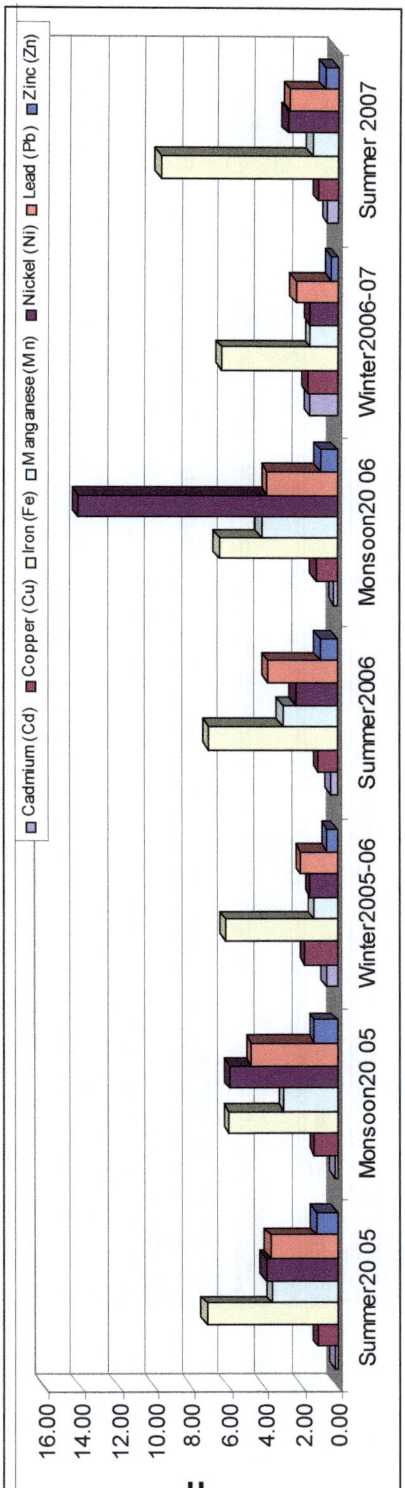

Figure 7.15: Transfer Factor of Heavy Metals from Textile Industrial Effluent to Groundwater

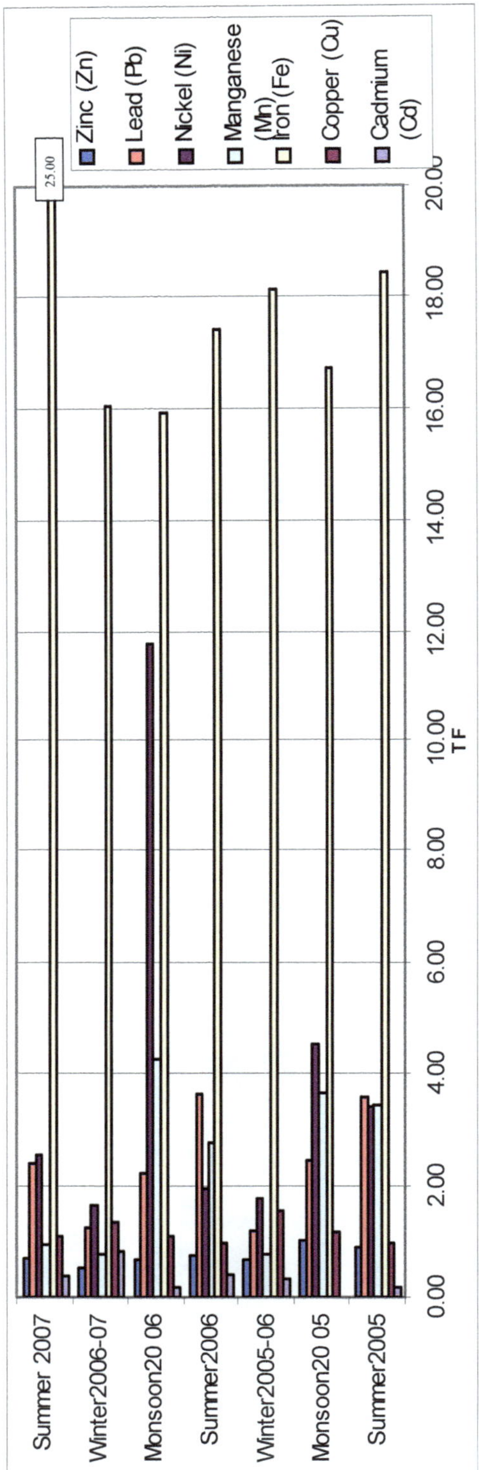

Figure 7.16: Transfer Factor of Heavy Metals from Textile Industrial Effluent to Pond Water

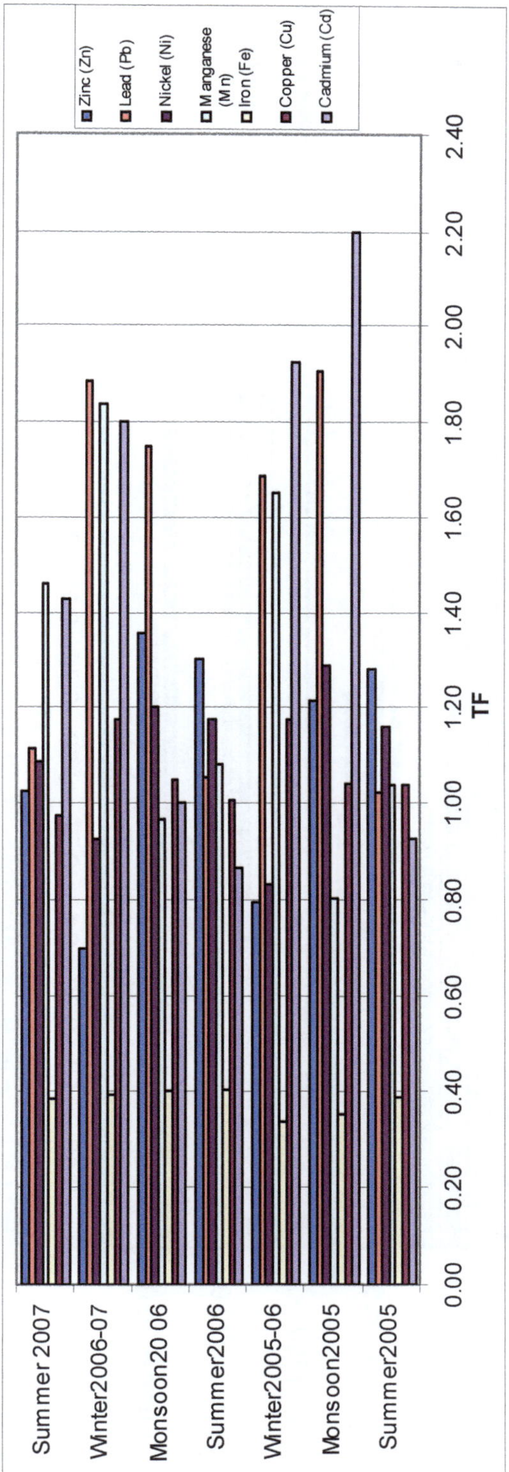

Figure 7.17: Transfer Factor of Heavy Metals from Pond Water to Groundwater of 40 ft Depth

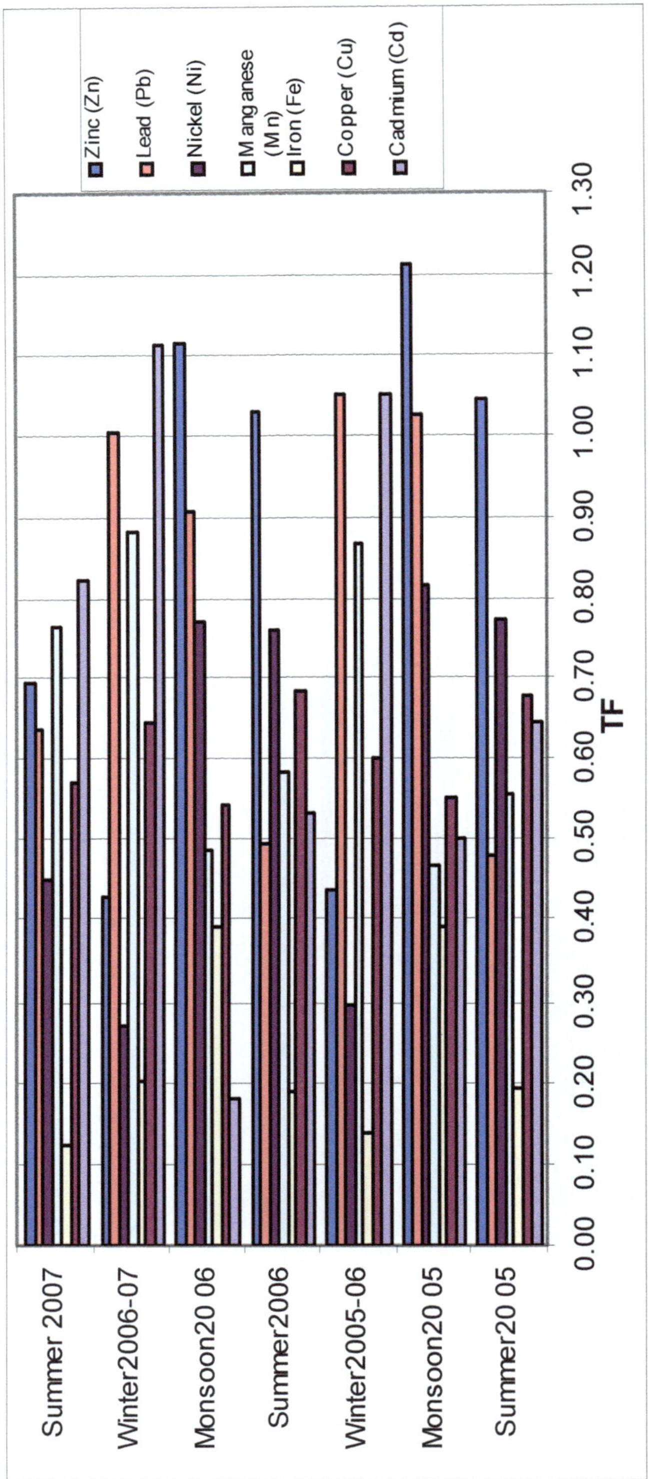

Figure 7.18: Transfer Factor of Heavy Metals from Pond Water to Groundwater of 80 ft Depth

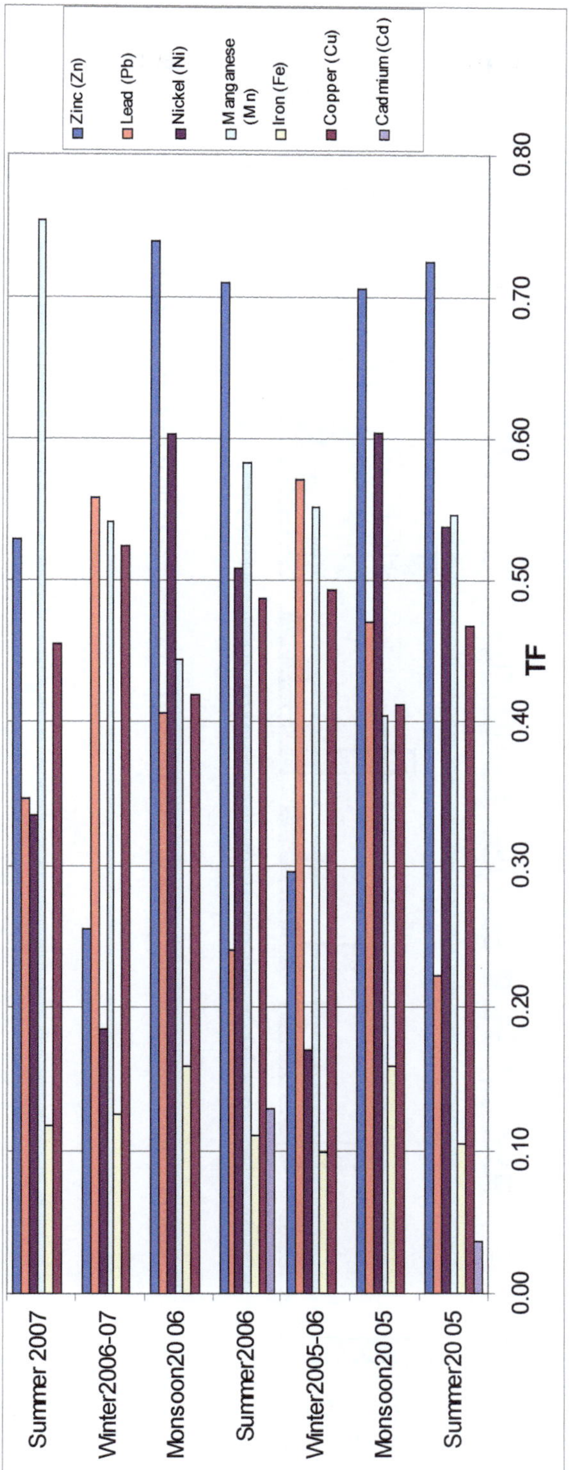

Figure 7.19: Transfer Factor of Heavy Metals from Pond Water to Deep Groundwater (120 ft depth)

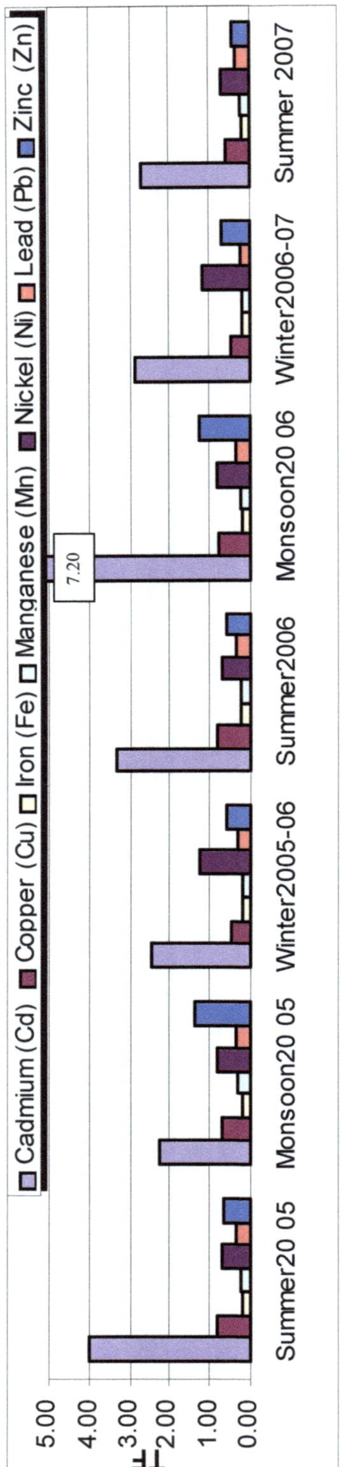

Figure 7.20: Transfer Factor of Heavy Metals from Groundwater of 40 ft to Agricultural Soil

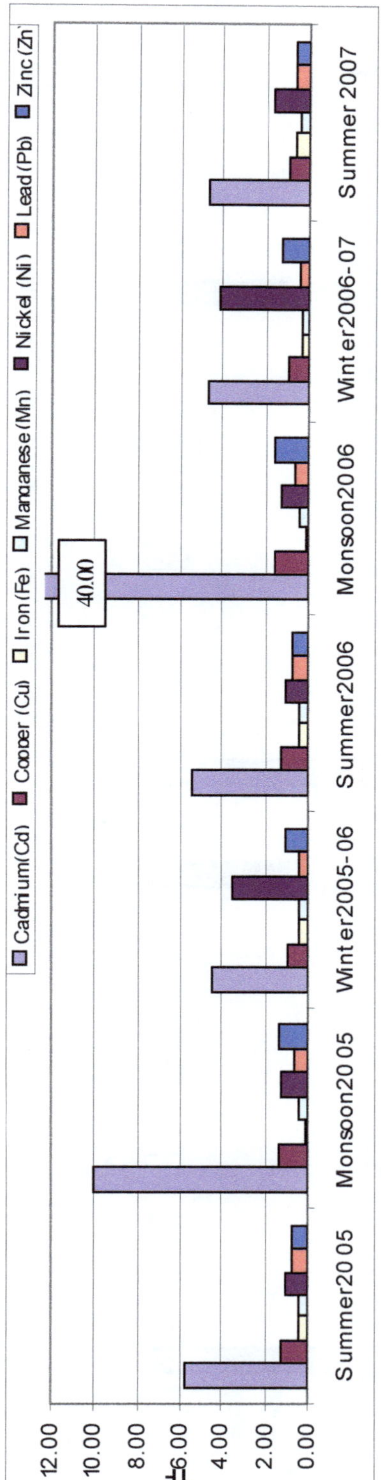

Figure 7.21: Transfer Factor of Heavy Metals from Groundwater of 80 ft to Agricultural Soil

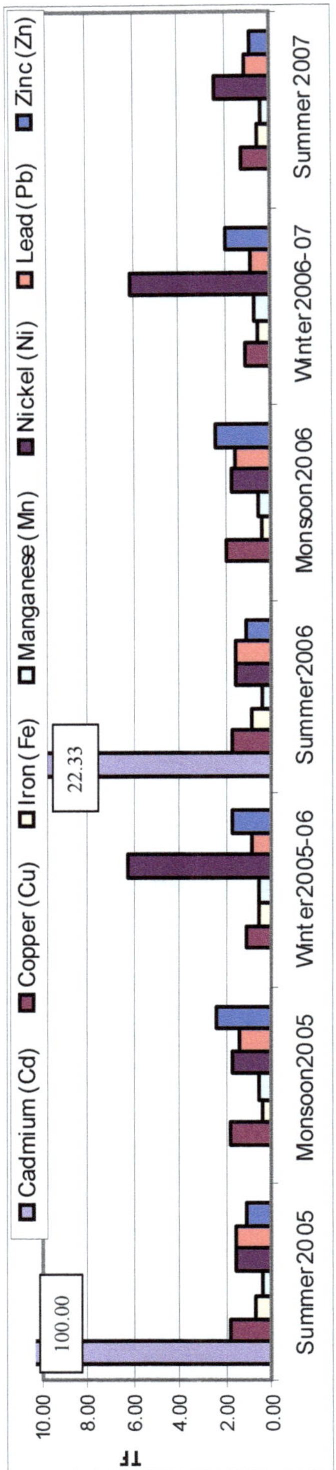

Figure 7.22: Transfer Factor of Heavy Metals from Deep Groundwater of 120 ft to Agricultural Soil

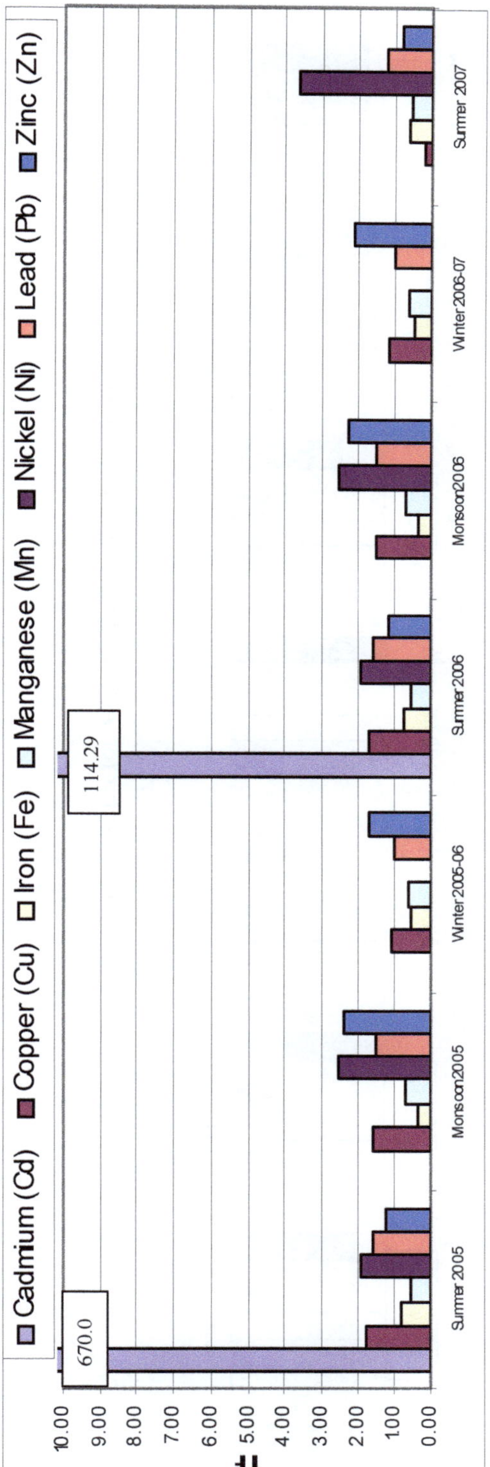

Figure 7.23: Transfer Factor of Heavy Metals from Groundwater to Soil at Control Site

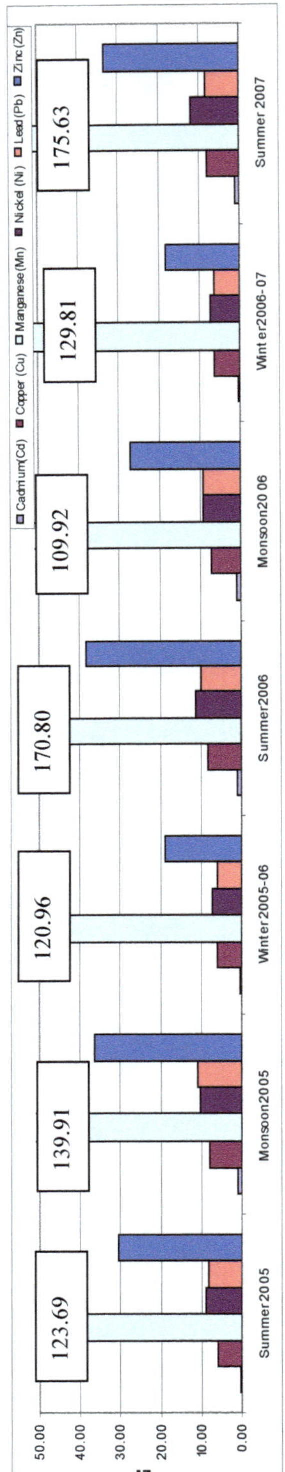

Figure 7.24: Enrichment Factor for Various Metals in Drain Sediment

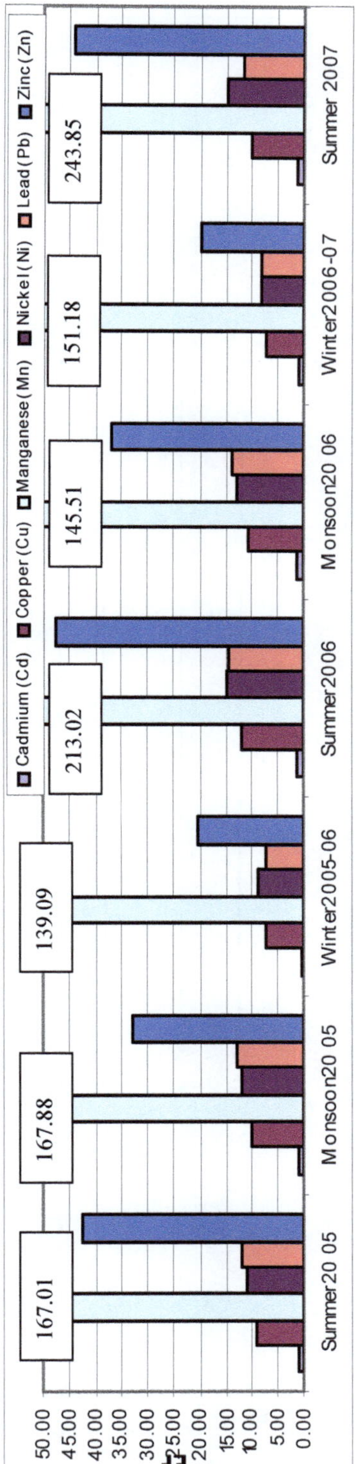

Figure 7.25: Enrichment Factor for Various Metals in Pond Sediment

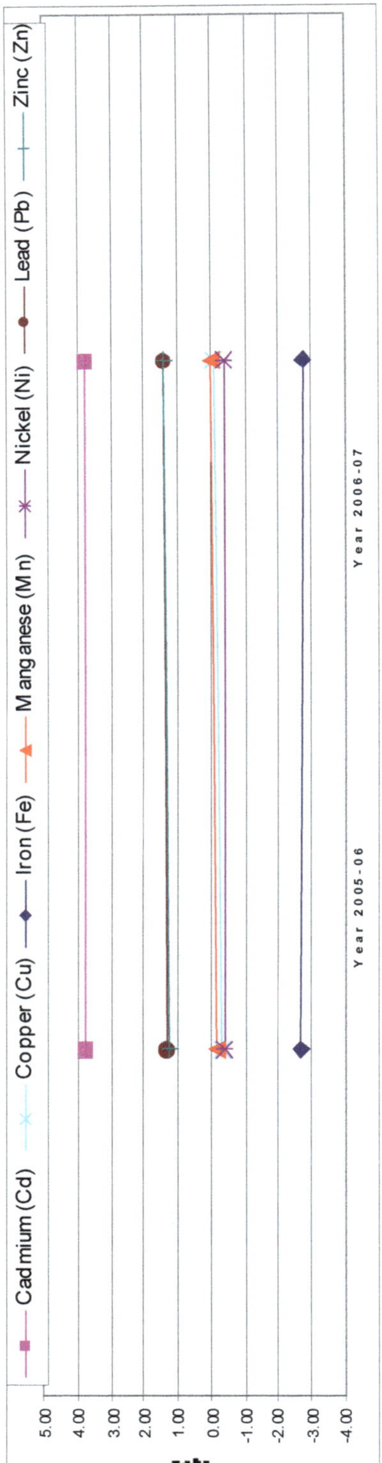

Figure 7.26: Geo-accumulation Index for Various Metals in Effluent Drain Sediment

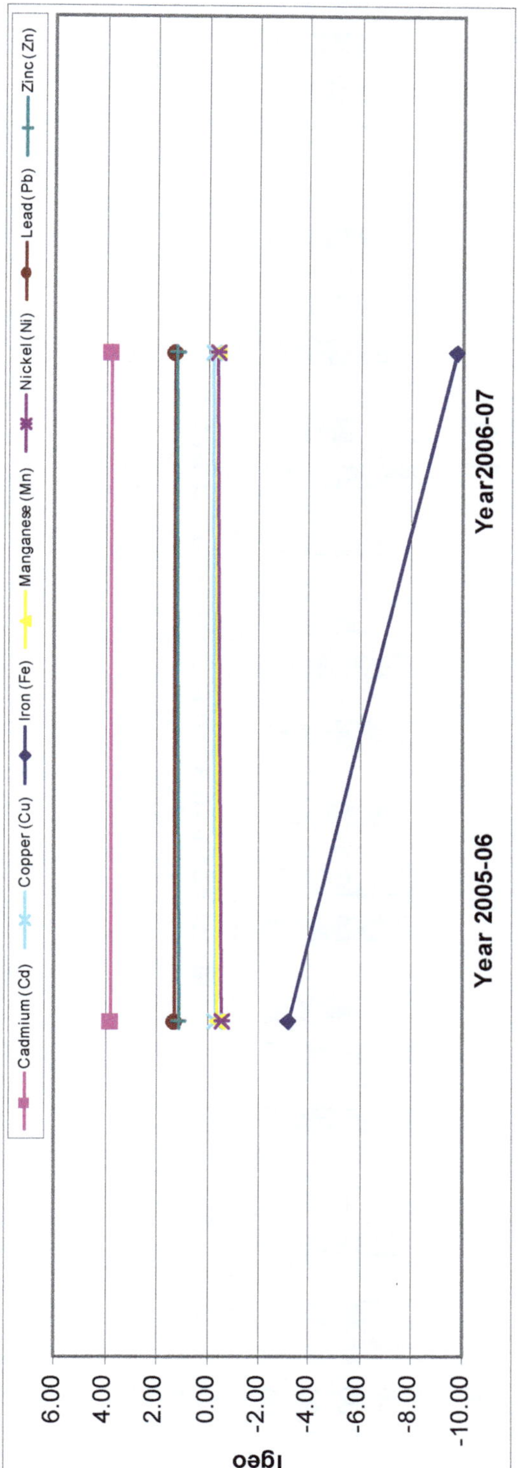

Figure 7.27: Geo-accumulation Index for Various Metals in Pond Sediment

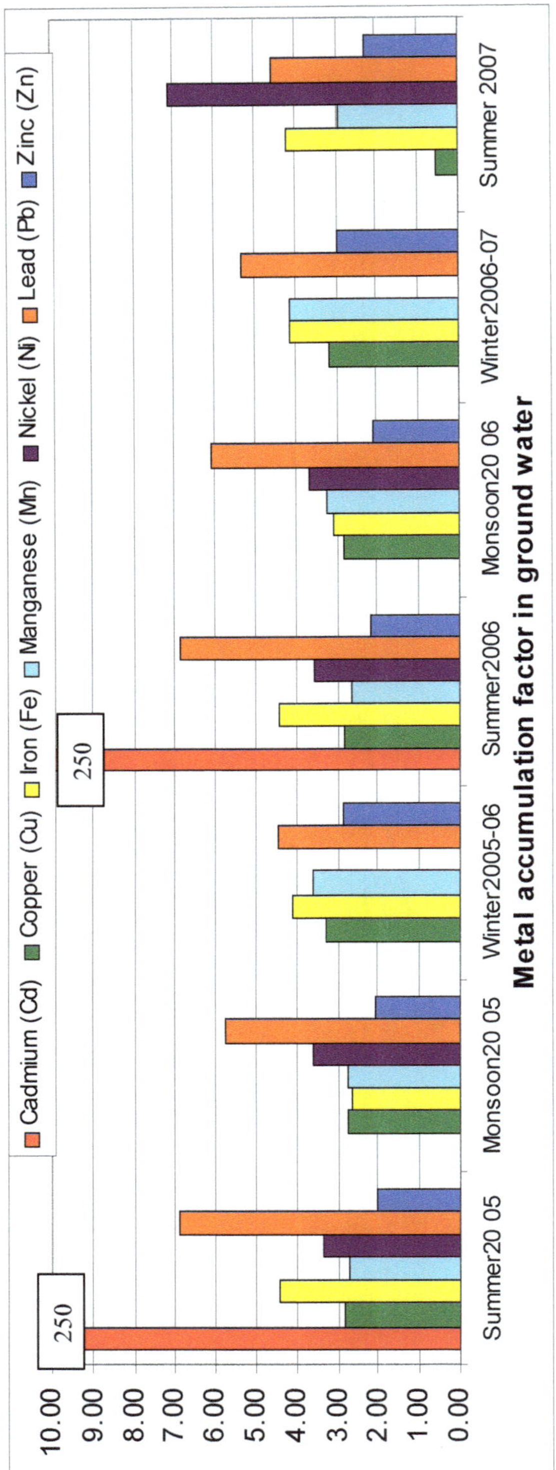

Figure 7.28: Accumulation Factor for Various Metals in Groundwater of Panipat Region

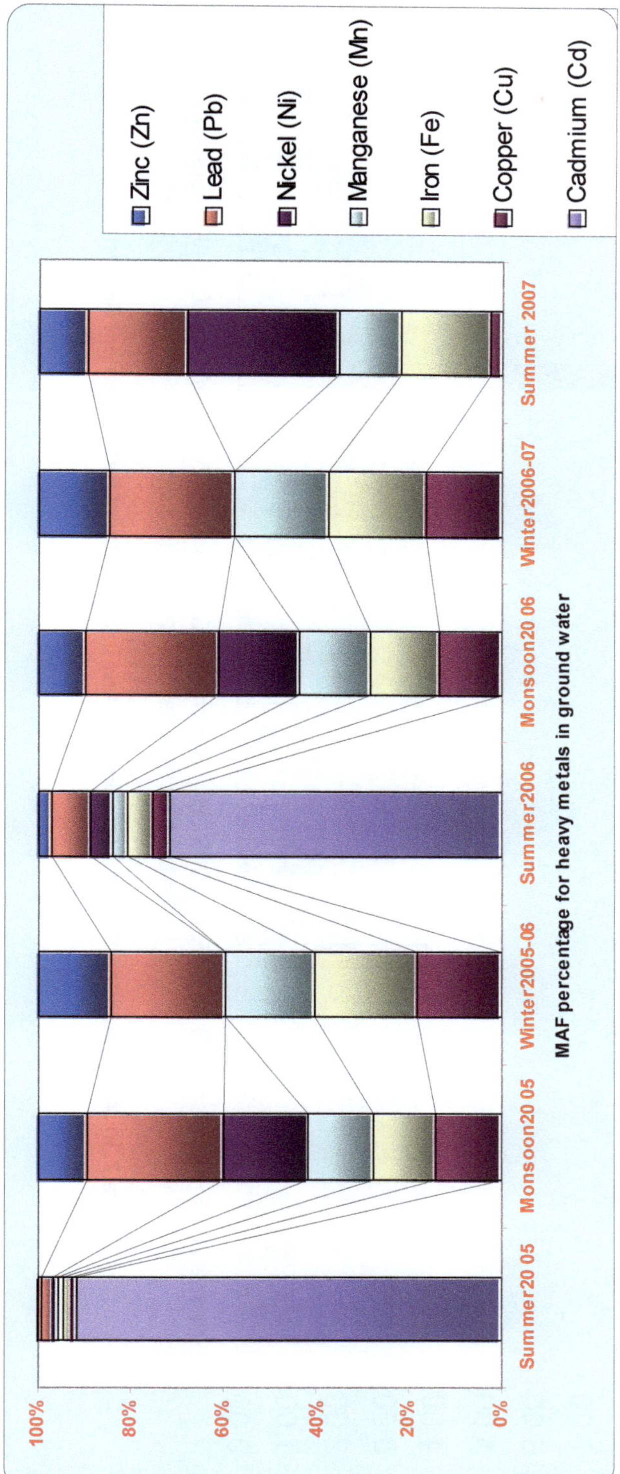

Figure 7.29: Percentage MAF for Various Metals in Groundwater of Panipat Region

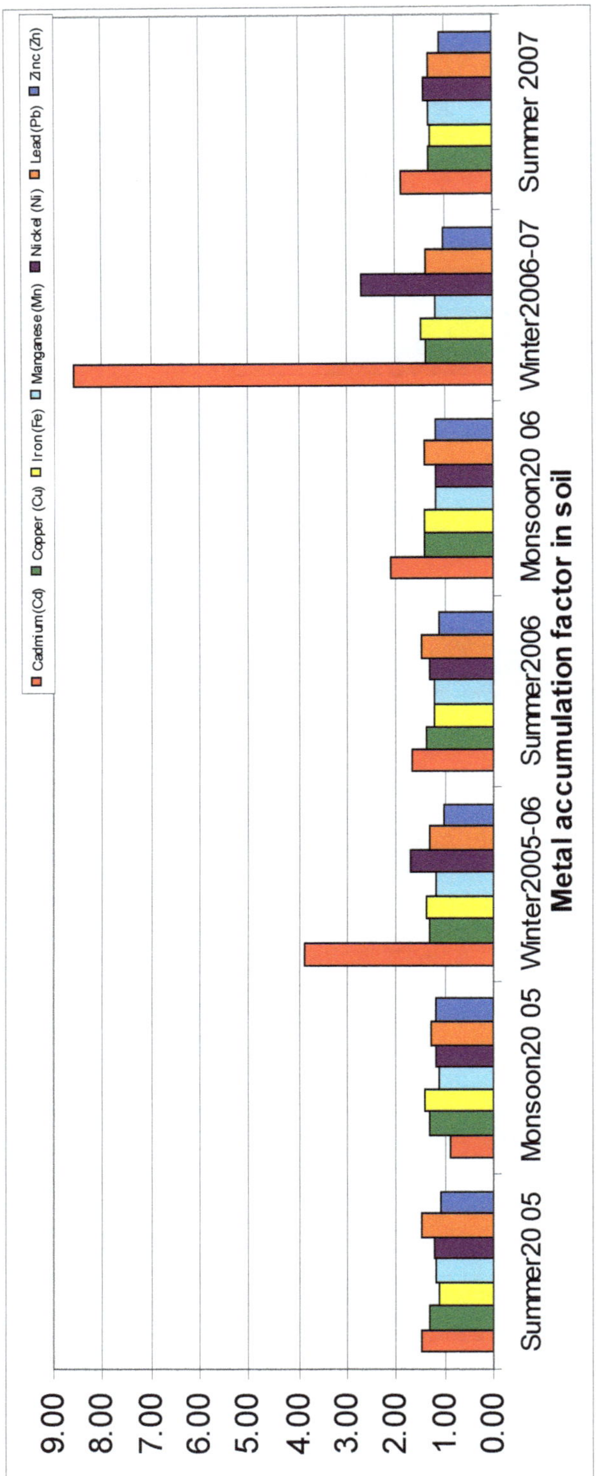

Figure 7.30: Accumulation Factor for Various Metals in Agricultural Soil of Panipat Region

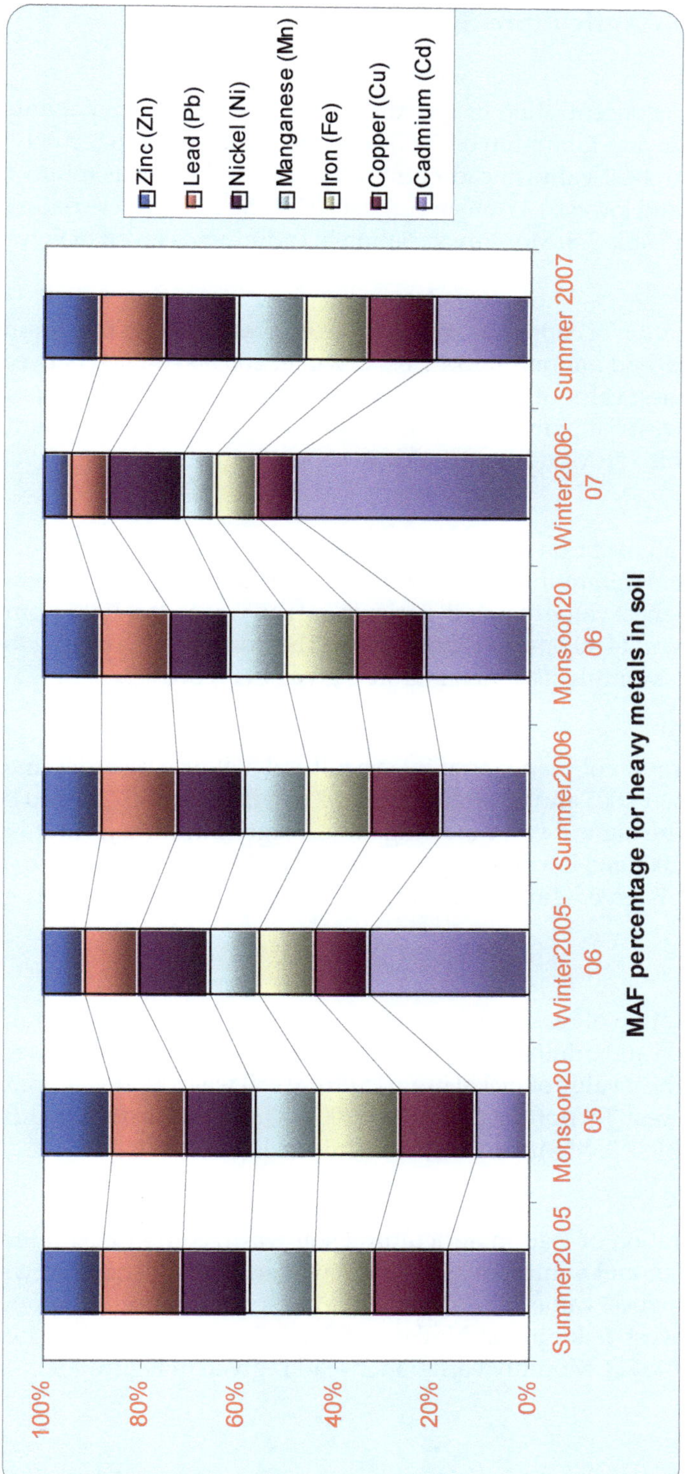

Figure 7.31: Percentage MAF for Various Metals in Agricultural Soil of Panipat Region

Heavy Metals in Agriculture Soil

Cadmium (ppm)

The cadmium concentration in agricultural soil was recorded maximum 3.4 ± 0.7 in summer 2006 and minimum 0.70 ± 0.60 in monsoon 2005 ranged between 0.1 to 4.0 ppm. The highest value of cadmium in agricultural soil was found 4.0 ppm during June 2006 and lowest 0.1 ppm in August 2005. The cadmium variation during the study is given Table 7.8. Monthly variation in cadmium is given in Figure 7.9.

Copper (ppm)

The concentration of copper in agricultural soil was recorded maximum 36.3 ± 7.4 in summer 2005 and minimum 13.4 ± 6.0 in winter 2005-06 ranged between 5.0 to 42.5 ppm. The highest value of copper in agricultural soil was found 42.5 ppm during June 2006 and lowest 5.0 ppm in December 2005. The copper variation during the study is given Table 7.8. Monthly variation in copper is given in Figure 7.9.

Iron (ppm)

The concentration of iron in agricultural soil was recorded maximum 82.8 ± 18.1 in summer 2006 and minimum 26.8 ± 10.8 in monsoon 2005 ranged between 14.40 to 100.5 ppm. The highest value of iron in agricultural soil was found 100.5 ppm during May 2006 and lowest 14.40 ppm in October 2005. The iron variation during the study is given Table 7.8. Monthly variation in iron is given in Figure 7.9.

Manganese (ppm)

The concentration of manganese in agricultural soil was recorded maximum 13.6 ± 1.4 in summer 2005 and minimum 5.8 ± 1.7 in winter 2006-07 ranged between 2.0 to 15.0 ppm. The highest value of manganese in agricultural soil was found 15.0 ppm during June 2005 and lowest 2.0 ppm in December 2005. The manganese variation during the study is given Table 7.8. Monthly variation in manganese is given in Figure 7.9.

Nickel (ppm)

The concentration of nickel in agricultural soil was recorded maximum 10.5 ± 0.8 in summer 2005 and minimum 5.3 ± 1.1 in winter 2006-07 ranged between 3.8 to 12.1 ppm. The highest value of nickel in agricultural soil was found 12.1 ppm during May 2006 and lowest 3.8 ppm in December 2006. The nickel variation during the study is given Table 7.8. Monthly variation in nickel is given in Figure 7.9.

Lead (ppm)

The concentration of lead in agricultural soil was recorded maximum 62.4 ± 10.9 in summer 2005 and minimum 25.1 ± 6.7 in winter 2006-07 ranged between 16.0 to 73.5 ppm. The highest value of lead in agricultural soil was found 73.5 ppm during June 2006 and lowest 16.0 ppm in November 2005. The lead variation during the study is given Table 7.8. Monthly variation in lead is given in Figure 7.9.

Zinc (ppm)

The concentration of zinc in agricultural soil was recorded maximum 19.4 ± 4.7 in monsoon 2006 and minimum 7.7 ± 2.8 in winter 2005-06 ranged between 4.3 to 23.5 ppm. The highest value of zinc in agricultural soil was found 23.5 ppm during July 2006 and lowest 4.3 ppm in January 2006. The zinc variation during the study is given Table 7.8. Monthly variation in zinc is given in Figure 7.9.

Control Site

Heavy Metals in Groundwater

Cadmium (ppm)

The cadmium concentration in groundwater on control site was recorded maximum 0.00 ± 0.00 and minimum 0.00 ± 0.00 ranged between 0.00 to 0.0005 ppm. The highest value of cadmium in groundwater was found 0.0005 ppm during June 2006 and lowest 0.00 ppm in April 2005 (mostly nil). The cadmium variation during the study is given Table 7.9. Monthly variation in cadmium is given in Figure 7.10a and seasonal fluctuation in Figure 7.10b.

Copper (ppm)

The concentration of copper in groundwater on control site was recorded maximum 0.684 ± 0.80 in summer 2007 and minimum 0.90 ± 0.061 in winter 2005-06 ranged between 0.00 to 0.18 ppm. The highest value of copper in groundwater was found 0.18 ppm during May 2005 and lowest 0.00 ppm in December 2005. The copper variation during the study is given Table 7.9. Monthly variation in copper is given in Figure 7.10a and seasonal fluctuation in Figure 7.10b.

Iron (ppm)

The concentration of iron in groundwater on control site was recorded maximum 0.868 ± 0.07 in summer 2006 and minimum 0.363 ± 0.128 in winter 2006-07 ranged between 0.18 to 0.99 ppm. The highest value of iron in groundwater was found 0.99 ppm during April 2005 and lowest 0.18 ppm in November 2005. The iron variation during the study is given Table 7.9. Monthly variation in iron is given in Figure 7.10a and seasonal fluctuation in Figure 7.10b.

Manganese (ppm)

The concentration of manganese in groundwater on control site was recorded maximum 0.204 ± 0.03 in summer 2005 and minimum 0.075 ± 0.037 in winter 2006-07 ranged between 0.02 to 0.245 ppm. The highest value of manganese in groundwater was found 0.245 ppm during May 2006 and lowest 0.02 ppm in November 2005. The manganese variation during the study is given Table 7.9. Monthly variation in manganese is given in Figure 7.10a and seasonal fluctuation in Figure 7.10b.

Nickel (ppm)

The concentration of nickel in groundwater on control site was recorded maximum 0.044 ± 0.011 in summer 2005 and minimum 0.00 ± 0.00 in winter 2005-06 ranged between 0.00 to 0.055 ppm. The highest value of nickel in groundwater was found 0.055 ppm during June 2006 and lowest 0.00 ppm in October 2005. The nickel variation during the study is given Table 7.9. Monthly variation in nickel is given in Figure 7.10a and seasonal fluctuation in Figure 7.10b.

Lead (ppm)

The concentration of lead in groundwater on control site was recorded maximum 0.282 ± 0.013 in summer 2007 and minimum 0.178 ± 0.127 in winter 2006-07 ranged between 0.00 to 0.291 ppm. The highest value of lead in groundwater was found 0.291 ppm during February 2007 and lowest 0.00 ppm in October 2006. The lead variation during the study is given Table 7.9. Monthly variation in lead is given in Figure 7.10a and seasonal fluctuation in Figure 7.10b.

Zinc (ppm)

The concentration of zinc in groundwater on control site was recorded maximum 0.099 ± 0.015 in summer 2005 and minimum 0.043 ± 0.013 in winter 2006-07 ranged between 0.02 to 0.112 ppm. The highest value of zinc in groundwater was found 0.112 ppm during April 2005 and lowest 0.02 ppm in December 2005. The zinc variation during the study is given Table 7.9. Monthly variation in zinc is given in Figure 7.10a and seasonal fluctuation in Figure 7.10b.

Heavy Metals in Agriculture Soil

Cadmium (ppm)

The cadmium concentration in soil on control site was recorded maximum 2.23 ± 0.31 in summer 2005 and minimum 0.225 ± 0.21 in winter 2006-07 ranged between 0.00 to 2.80 ppm. The highest value of cadmium in agricultural soil was found 2.80 ppm during June 2006 and lowest 0.00 ppm in September 2006. The cadmium variation during the study is given Table 7.10. Monthly variation in cadmium is given in Figure 7.11a and seasonal fluctuation in Figure 7.11b.

Copper (ppm)

The concentration of copper in soil on control site was recorded maximum 27.4 ± 5.7 in summer 2005 and minimum 10.0 ± 4.1 in winter 2005-06 ranged between 3.5 to 32.0 ppm. The highest value of copper in agricultural soil was found 32.0 ppm during May 2005 and lowest 3.5 ppm in December 2006. The copper variation during the study is given Table 7.10. Monthly variation in copper is given in Figure 7.11a and seasonal fluctuation in Figure 7.11b.

Iron (ppm)

The concentration of iron in soil on control site was recorded maximum 68.6 ± 17.5 in summer 2005 and minimum 18.4 ± 9.0 in winter 2006-07 ranged between 6.90 to 84.50 ppm. The highest value of iron in agricultural soil was found 84.50 ppm during April 2005 and lowest 6.90 ppm in October 2005. The iron variation during the study is given Table 7.10. Monthly variation in iron is given in Figure 7.11a and seasonal fluctuation in Figure 7.11b.

Manganese (ppm)

The concentration of manganese in soil on control site was recorded maximum 11.5 ± 1.1 in summer 2005 and minimum 4.8 ± 1.5 in winter 2006-07 ranged between

3.2 to 12.5 ppm. The highest value of manganese in agricultural soil was found 12.5 ppm during June 2005 and lowest 3.2 ppm in December 2006. The manganese variation during the study is given Table 7.10. Monthly variation in manganese is given in Figure 7.11a and seasonal fluctuation in Figure 7.11b.

Nickel (ppm)

The concentration of nickel in soil on control site was recorded maximum 8.43 ± 1.2 in summer 2005 and minimum 2.6 ± 1.87 in winter 2006-07 ranged between 0.00 to 9.20 ppm. The highest value of nickel in agricultural soil was found 9.20 ppm during June 2006 and lowest 0.00 ppm in November 2006. The nickel variation during the study is given Table 7.10. Monthly variation in nickel is given in Figure 7.11a and seasonal fluctuation in Figure 7.11b.

Lead (ppm)

The concentration of lead in soil on control site was recorded maximum 42.3 ± 3.1 in summer 2005 and minimum 18.25 ± 5.9 in winter 2006-07 ranged between 11.3 to 46.8 ppm. The highest value of lead in agricultural soil was found 46.8 ppm during June 2006 and lowest 11.3 ppm in November 2005. The lead variation during the study is given Table 7.10. Monthly variation in lead is given in Figure 7.11a and seasonal fluctuation in Figure 7.11b.

Zinc (ppm)

The concentration of zinc in soil on control site was recorded maximum 16.5 ± 3.3 in monsoon 2006 and minimum 7.3 ± 0.8 in summer 2007 ranged between 5.1 to 19.1 ppm. The highest value of zinc in agricultural soil was found 19.1 ppm during July 2006 and lowest 5.1 ppm in January 2006. The zinc variation during the study is given Table 7.10. Monthly variation in zinc is given in Figure 7.11a and seasonal fluctuation in Figure 7.12b.

Correlation Coefficient

Correlation between physico-chemical parameters and heavy metals of groundwater of 40 ft depth are given in Table 7.13a for each metal. For cadmium correlation was found maximum 0.4400 with pH only. Correlation for copper was found maximum 0.7176 and 0.6712 with temperature and total solids respectively. Correlation for iron was found maximum 0.8102 and 0.6729 with pH and free carbon dioxide respectively. Correlation for manganese was found maximum 0.8508, 0.8001 and 0.8544 with total solids, total suspended solids and total dissolved solids respectively. Correlation for nickel was found maximum 0.8021 and 8021 with total dissolved solids and Electric conductivity respectively. Correlation for lead was found maximum 0.8593, 0.8340 and 0.8340 with pH, total dissolved solids and Electric conductivity respectively. Correlation for zinc was found maximum 0.7175 and 0.5164 with pH and free carbon dioxide respectively.

Correlation between physico-chemical parameters and heavy metals of groundwater of 80 ft depth are given in Table 7.13b for each metal. For cadmium correlation was found maximum 0.4448 and 0.5048 with pH and free carbon dioxide respectively. Correlation for copper was found maximum 0.7443 and 0.8099 with free carbon dioxide and total alkalinity respectively. Correlation for iron was found maximum 0.9188 and 0.8896 with temperature and total alkalinity. Correlation for manganese was found maximum 0.6930 and 0.7664 with free carbon dioxide and total alkalinity respectively. Correlation for nickel was found maximum 0.7536 and 0.7918 with free carbon dioxide and total alkalinity respectively. Correlation for lead was found maximum 0.6693 and 0.5841 with free carbon dioxide and total alkalinity respectively. Correlation for zinc was found maximum 0.6518 and 0.7069 with free carbon dioxide and total alkalinity respectively.

Correlation between physico-chemical parameters and heavy metals of deep groundwater of more than 120 ft depth are given in Table 7.13c for each metal. For cadmium correlation was found maximum 0.2586 and 2761 with temperature and magnesium respectively. Correlation for copper was found maximum 0.6392 and 0.8666 with free carbon dioxide and total alkalinity respectively. Correlation for iron was found maximum 0.6815 and 0.8783 with free carbon dioxide and total alkalinity respectively. Correlation for manganese was found maximum 0.6144 and 0.8834 with temperature and total alkalinity respectively. Correlation for nickel was found maximum 0.7675 and 0.8302 with temperature and total alkalinity respectively. Correlation for lead was found maximum 0.6347 and 0.7136 with free carbon dioxide and total alkalinity respectively. Correlation for zinc was found maximum 0.6340 and 0.8653 with free carbon dioxide and total alkalinity respectively.

Correlation between physico-chemical parameters and heavy metals of groundwater at control site are given in Table 7.15 for each metal. For cadmium correlation was found maximum 0.3900 and 0.3991 with calcium and magnesium respectively. Correlation for copper was found maximum 0.1308 and 0.2028 with pH and total alkalinity respectively. Correlation for iron was found maximum 0.8049 and 0.8597 with free carbon dioxide and total alkalinity respectively. Correlation for manganese was found maximum 0.7006 and 0.8739 with temperature and total alkalinity respectively. Correlation for nickel was found maximum 0.8099 and 0.8175 with temperature and total alkalinity respectively. Correlation for lead was found maximum 0.8009 and 0.6927 with total alkalinity and free carbon dioxide respectively. Correlation for zinc was found maximum 0.6614 and 0.7411 with free carbon dioxide and total alkalinity respectively.

Transfer Factor (Coefficient) for Heavy Metals

From textile industrial effluent to groundwater, transfer factors for heavy metals are given in Table 7.24 for experimental sites. Transfer factor for cadmium was found maximum 1.50 in winter 2006-07 and minimum 0.12 in monsoon 2005, while for copper transfer factor was found maximum 1.79 in winter 2005-06 and minimum 0.99 in summer 2006. Transfer factor for iron was found maximum 9.64 in summer 2007 and minimum 5.93 in monsoon 2005. Transfer factor for manganese was found maximum 4.12 in monsoon 2006 and minimum 1.27 in winter 2005-06. Transfer factor for nickel was found maximum 14.14 in monsoon 2006 and minimum 1.45 in winter 2005-06. Transfer factor for lead was found maximum 4.65 in monsoon 2005 and minimum 2.01 in winter 2005-06. Transfer factor for zinc was found maximum 1.21 in monsoon 2005 and minimum 0.38 in winter 2006-07.

From textile industrial effluent to pond water, transfer factors for heavy metals are given in Table 7.25. Transfer factor for cadmium was found maximum 0.83 in winter 2006-07 and minimum 0.05 in monsoon 2005, while for copper transfer factor was found maximum 1.53 in winter 2005-06 and minimum 0.97 in summer 2005. Transfer factor for iron was found maximum 25.0 in summer 2007 and minimum 15.93 in monsoon 2006. Transfer factor for manganese was found maximum 4.27 in monsoon 2006 and minimum 0.77 in winter 2005-06. Transfer factor for nickel was found maximum 11.76 in monsoon 2006 and minimum 1.64 in winter 2006-07. Transfer factor for lead was found maximum 3.63 in summer 2006 and minimum 1.19 in winter 2005-06. Transfer factor for zinc was found maximum 1.00 in monsoon 2005 and minimum 0.54 in winter 2006-07.

From pond water to groundwater of 40 ft depth, transfer factors for heavy metals are given in Table 7.26. Transfer factor for cadmium was found maximum 2.20 in monsoon 2005 and minimum 0.87 in summer 2006, while for copper transfer factor was found maximum 1.17 in winter 2005-06 and minimum 0.97 in summer 2007. Transfer factor for iron was found maximum 0.40 in monsoon 2006 and minimum 0.34 in winter 2005-06. Transfer factor for manganese was found maximum 1.84 in winter 2006-07 and minimum 0.96 in monsoon 2006. Transfer factor for nickel was found maximum 1.29 in monsoon 2005 and minimum 0.83 in winter 2005-06. Transfer factor for lead was found maximum 1.91 in monsoon 2005 and minimum 1.02 in summer 2005. Transfer factor for zinc was found maximum 1.35 in monsoon 2006 and minimum 0.70 in winter 2006-07.

From pond water to deep groundwater of 120 ft depth, transfer factors for heavy metals are given in Table 7.28 for various metals. Transfer factor for cadmium was found maximum 0.13 in summer 2006 and minimum 0.00 in monsoon 2005, while for copper transfer factor was found maximum 0.52 in winter 2006-07 and minimum 0.41 in monsoon 2005. Transfer factor for iron was found maximum 0.16 in monsoon 2006 and minimum 0.10 in winter 2005-06. Transfer factor for manganese was found maximum 0.75 in summer 2007 and minimum 0.40 in monsoon 2005. Transfer factor for nickel was found maximum 0.61 in monsoon 2005 and minimum 0.19 in winter 2006-07. Transfer factor for lead was found maximum 0.57 in winter 2005-06 and minimum 0.22 in summer 2005. Transfer factor for zinc was found maximum 0.74 in monsoon 2006 and minimum 0.25 in winter 2006-07.

From groundwater of 40 ft depth to agricultural soil, transfer factors for heavy metals are given in Table 7.29. Transfer factor for cadmium was found maximum 7.20 in monsoon 2006 and minimum 2.27 in monsoon 2005, while for copper transfer factor was found maximum 0.82 in summer 2005 and minimum 0.45 in winter 2005-06. Transfer factor for iron was found maximum 0.22 in summer 2006 and minimum 0.17 in monsoon 2006. Transfer factor for manganese was found maximum 0.27 in monsoon 2005 and minimum 0.18 in winter 2006-07. Transfer factor for nickel was found maximum 1.28 in winter 2005-06 and minimum 0.69 in summer 2006. Transfer factor for lead was found maximum 0.36 in summer 2007 and minimum 0.26 in winter 2006-07. Transfer factor for zinc was found maximum 1.39 in monsoon 2005 and minimum 0.40 in summer 2007.

From groundwater of 80 ft depth to agricultural soil, transfer factors for heavy metals are given in Table 7.30. Transfer factor for cadmium was found maximum 40.00 in monsoon 2006 and minimum 4.45 in winter 2005-06, while for copper transfer factor was found maximum 1.48 in monsoon 2006 and minimum 0.88 in winter 2005-06. Transfer factor for iron was found maximum 0.58 in summer 2007 and minimum 0.16 in monsoon 2005. Transfer factor for manganese was found maximum 0.48 in monsoon 2005 and minimum 0.38 in winter 2006-07. Transfer factor for nickel was found maximum 4.18 in winter 2006-07 and minimum 1.07 in summer 2006. Transfer factor for lead was found maximum 0.72 in summer 2006 and minimum 0.48 in winter 2005-06. Transfer factor for zinc was found maximum 1.53 in monsoon 2006 and minimum 0.59 in summer 2007. From deep groundwater of more than 120 ft depth to agricultural soil, transfer factors for heavy metals are given in Table 7.31. Transfer factor for cadmium was found maximum 100.00 in summer 2005 (almost NP) and minimum 22.33 in summer 2006, while for copper transfer factor was found maximum 1.91 in monsoon 2006 and minimum 1.08 in winter 2005-06. Transfer factor for iron was found maximum 0.79 in summer 2006 and minimum 0.40 in monsoon 2005. Transfer factor for manganese was found maximum 0.63 in winter 2006-07 and minimum 0.46 in summer 2006. Transfer factor for nickel was found maximum 6.24 in winter 2005-06 and minimum 1.54 in summer 2005. Transfer factor for lead was found maximum 1.55 in summer 2005 and minimum 0.88 in winter 2005-06. Transfer factor for zinc was found maximum 2.39 in monsoon 2005 and minimum 0.78 in summer 2007.

From groundwater of control site to soil of control site, transfer factors for heavy metals are given in Table 7.32. Transfer factor for cadmium was found maximum 670.00 in summer 2005 (mostly NP) and minimum 114.29 in summer 2006, while for copper transfer factor was found maximum 1.75 in summer 2005 and minimum 0.23 in summer 2007. Transfer factor for iron was found maximum 0.80 in summer 2005 and minimum 0.33 in monsoon 2005. Transfer factor for manganese was found maximum 0.68 in monsoon 2005 and minimum 0.54 in summer 2006. Transfer factor for nickel was found maximum 2.51 in monsoon 2005 (abundantly NP) and minimum 1.89 in summer 2006. Transfer factor for lead was found maximum 1.59 in summer 2005 and minimum 1.00 in winter 2005-06. Transfer factor for zinc was found maximum 2.37 in monsoon 2005 and minimum 0.84 in summer 2007.

Enrichment Factor for Heavy Metals

In the drain sediment, metal enrichment factor for heavy metals are given in Table 7.33. Enrichment factor for cadmium was found maximum 0.90 in monsoon 2005 and minimum 0.53 in winter 2005-06. For copper enrichment factor was found maximum 8.47 in summer 2006 and minimum 5.74 in winter 2005-06. Enrichment factor for iron was maximum 1000 due to the comparison base of study in the evaluation of enrichment factor of metals in drain sediment. For manganese enrichment factor was found maximum 175.63 in summer 2007 and minimum 109.92 in monsoon 2006. Enrichment factor for nickel was maximum 11.70 in summer 2007 and minimum 7.35 in winter 2006-07. For lead enrichment factor was found maximum 10.73 in monsoon 2005 and minimum 5.83 in winter 2005-06. Enrichment factor for zinc was found maximum 38.80 in summer 2006 and minimum 18.12 in winter 2006-07.

In the pond sediment, metal enrichment factor for heavy metals are given in Table 7.34. Metal enrichment factor for cadmium was found maximum 1.26 in summer 2007 and minimum 0.67 in winter 2005-06. For copper enrichment factor was found maximum 12.04 in summer 2006 and minimum 7.61 in winter 2006-07. Enrichment factor for iron was maximum 1000 due to the comparison base of study in the evaluation of enrichment factor of metals in pond sediment. For manganese enrichment factor was found maximum 243.85 in summer 2007 and minimum 139.09 in winter 2005-06. Enrichment factor for nickel was maximum 14.85 in summer 2007 and minimum 8.38 in winter 2006-07. For lead enrichment factor was found maximum 14.23 in summer 2006 and minimum 7.32 in winter 2005-06. Enrichment factor for zinc was found maximum 47.57 in summer 2006 and minimum 19.89 in winter 2006-07.

Geo-accumulation Index for Heavy Metals

Geo-accumulation index and values for heavy metals in effluent drain sediment are given in Table 7.35a and in Table 7.35b for highest values. Value of cadmium in geo-accumulation index was found maximum 12.85 in summer 2005 and minimum 3.24 in winter 2005-06. For copper geo-accumulation index value was found maximum 0.81 in summer 2006 and minimum 0.68 in monsoon 2006. Geo-accumulation index value for iron was found maximum 0.12 in summer 2005 and minimum 0.08 in summer 2007. For manganese geo-accumulation index value was found maximum 0.87 in summer 2006 and minimum 0.62 in monsoon 2005. Geo-accumulation index value for nickel was found maximum 0.74 in summer 2005 and minimum 0.57 in monsoon 2005. For lead geo-accumulation index value was found maximum 2.33 in summer 2005 and minimum 1.57 in winter 2005-06. Geo-accumulation index value for zinc was maximum 1.82 in summer 2005 and minimum 1.01 in winter 2006-07.

Rank of cadmium in geo-accumulation index was found maximum 3.68 in summer 2005 and minimum 3.24 in winter 2005-06. For copper, iron, manganese and nickel geo-accumulation index rank was found in negative values below 0.00 For lead geo-accumulation index rank was found maximum 1.22 in summer 2005 and minimum 0.65 in winter 2005-06. Geo-accumulation index rank for zinc was maximum 0.87 in summer 2005 and minimum 0.01 in winter 2006-07. While for highest concentrations of heavy metals, highest values for cadmium, copper, iron, manganese, nickel, lead and zinc was found 13.60, 0.92, 0.16, 0.98, 0.77, 2.60 and 2.70 respectively. Geo-accumulation index rank was found 3.77, 1.38 and 1.43 for cadmium, lead and zinc respectively.

In pond sediment, geo-accumulation index and values of heavy metals are given in Table 7.36a and in Table 7.36b for highest values. Value of cadmium in geo-accumulation index was found maximum 12.70 in summer 2005 and minimum 9.01 in winter 2005-06. For copper geo-accumulation index value was found maximum 0.78 in summer 2006 and minimum 0.57 in summer 2007. Geo-accumulation index value for iron was found maximum 0.08 in winter 2005-06 and minimum 0.05 in summer 2007. For manganese geo-accumulation index value was found maximum

0.74 in summer 2005 and minimum 0.47 in monsoon 2006. Geo-accumulation index value for nickel was found maximum 0.63 in summer 2006 and minimum 0.48 in winter 2006-07. For lead geo-accumulation index value was found maximum 2.30 in summer 2005 and minimum 1.47 in winter 2005-06. Geo-accumulation index value for zinc was maximum 1.68 in summer 2005 and minimum 0.82 in winter 2006-07.

Rank of cadmium in geo-accumulation index was found maximum 3.67 in summer 2005 and minimum 3.17 in winter 2005-06. For copper, iron, manganese and nickel geo-accumulation index rank was found in negative values below 0.00. For lead geo-accumulation index rank was found maximum 1.20 in summer 2005 and minimum 0.55 in winter 2005-06. Geo-accumulation index rank for zinc was maximum 0.75 in summer 2005 and minimum 0.00 in monsoon 2005. While for highest concentrations of heavy metals, highest values for cadmium, copper, iron, manganese, nickel, lead and zinc was found 14.04, 0.85, 0.11, 0.78, 0.77, 2.50 and 2.25 respectively. Geo-accumulation index rank was found 3.81, 1.32 and 1.17 for cadmium, lead and zinc respectively.

Metal Accumulation Factor

In the present study of groundwater, metal accumulation factor for heavy metals are given in Table 7.37. Metal accumulation factor for cadmium was found maximum 250.00 in summer 2005 (almost infinite) and minimum 57.14 in summer 2006. For copper metal accumulation factor was found maximum 3.28 in winter 2005-06 and minimum 0.53 in summer 2007. Metal accumulation factor for iron was maximum 4.41 in summer 2005 and minimum 2.64 in monsoon 2005. For manganese metal accumulation factor was maximum 4.17 in winter 2006-07 and minimum 2.65 in summer 2006. Metal accumulation factor for nickel was found maximum 7.14 in summer 2007 and minimum 3.36 in summer 2005. For lead metal accumulation factor was found maximum 6.84 in summer 2006 and minimum 4.45 in winter 2005-06. Metal accumulation factor for zinc was maximum 3.00 in winter 2006-07 and minimum 2.02 in summer 2005. The value of metal accumulation factor for groundwater is reflecting with Figure 7.28 and Figure 7.29.

For soil, metal accumulation factor of heavy metals are depicted in Table 7.38 and Figure 7.30 and Figure 7.31. Metal accumulation factor for cadmium was found maximum 8.56 in winter 2006-07 and minimum 0.91 in monsoon 2005. For copper metal accumulation factor was found maximum 3.15 in summer 2005 and minimum 1.31 in monsoon 2005. Metal accumulation factor for iron was maximum 1.49 in winter 2006-07 and minimum 1.12 in summer 2005. For manganese metal accumulation factor was maximum 1.31 in summer 2007 and minimum 1.11 in monsoon 2005. Metal accumulation factor for nickel was found maximum 2.05 in winter 2006-07 and minimum 1.17 in monsoon 2005. For lead metal accumulation factor was found maximum 1.48 in summer 2005 and minimum 1.28 in monsoon 2005. Metal accumulation factor for zinc was maximum 1.20 in monsoon 2005 and minimum 1.02 in winter 2006-07.

Discussion

Heavy metals in textile effluent, pond water, groundwater, control groundwater, effluent drain sediment, pond sediment, agriculture soil and control soil-

An increasing number of groundwater wells were found to be contaminated by chemical waste from point sources and also widespread known point source pollution from fertilizers was threatening the water resources in extensive areas (Lagas *et al.*, 1989 and Meier and Mull, 1989).

Wastewaters from textile industries have the high amount of the heavy metals entering them most hazardous for soil plant and other organisms including human beings (Sial *et al.*, 2006). Many effects of Cd action results from interactions of various micro and macro elements as it antagonizes to Cu, Zn and Fe and also interferes with Ca absorption. Cadmium found in a sample of Ca is likely in an inorganic form and may be relatively less toxic to flora and fauna, while same amount of Cd when present in zinc or phosphates may be highly toxic to plants and animals, especially in calcium deficiency in soil. The use of contaminated water into irrigation may be increase the cadmium status in agricultural soil (Mani *et al.*, 2005).

Cadmium concentrations in textile industrial effluent (0.005-0.052 ppm), pond water (0.001-0.012 ppm) and agricultural soil (0.7-3.4 ppm) turn down in rainy season due to the dilution factor, while in the groundwater system its impact seems after monsoon and cadmium concentration reduces. Cadmium slightly fluctuated and remains mostly constant in pond sediment (4.0-5.7 ppm) and drain sediment (4.2-5.8 ppm). Transfer factor for cadmium (0.12-1.5) from textile industrial effluent to groundwater was recorded high in winter months (after monsoon). In geo-accumulation index cadmium (3.77-3.81) laid in between 3 to 4 mean strongly polluted by cadmium. Metal accumulation factor for groundwater (57.1-250.0) was found mostly very high or infinite for cadmium, because in most samples of control sites cadmium was absent.

Transfer factor of cadmium from groundwater to agricultural soil on control site was found infinite, because cadmium was almost absent in groundwater and agricultural soil at control site. So, any correlation between groundwater and agricultural soil was not noticed. Cadmium transfer is due to some other non-point sources of cadmium pollution. Figure 7.21 showing the transfer factor of cadmium from groundwater to agricultural soil clearly indicated the transfer level of metal. During the last decade, the industrial use of Cd has increased and can create both acute and chronic cases of clinically identifiable toxicity in humans. In human, absorption of dietary Cd is limited up to 15 per cent and that of inhaled Cd is high 40 per cent and not absorbed from skin (Mani *et al.*, 2005). When the animal kept on contaminated feed along with the industrial polluted environment for long time, they suffer from its toxicity symptoms which may be sub clinical or clinical. The retention time of Cd is quite high, half life being 40 years. Cadmium is produced as inevitable by-product of zinc (or occasionally lead) refining (lenntech.com). Cd was found slightly high in groundwater for drinking purposes, because it should not be more than 0.003 mg/l for health views, while it should be less than 1 µg/l generally in drinking water (WHO, 2006). Helena (1999), Umar and Ahmad (2000), Kumar *et al.* (2001) and Yadav *et al.* (2002) observed the high Cd concentrations in industrial area like the present study of groundwater, while Krebs *et al.* (1999), Singh *et al.* (1994) and Fakayode and Oniyawa (2002) in soil and Bordas and Bourg (1998), Vazquez *et al.* (2007) and Korfali and Davies (2004) in sediment.

Concentrations of copper in textile industrial effluent are due to the residues of some dyes especially used for blue or green colours. Copper concentrations in textile industrial effluent (0.16-0.44 ppm) were found parallel or sometimes higher than iron. Most of the part of copper content reach in pond water (0.25-0.43 ppm) and rest settle down on drain sediment (45.8-54.8 ppm). Copper in pond sediment (38.6-52.4 ppm) also settled down after some time. Transfer factor of copper from pond water to groundwater (0.97-1.17) and groundwater to agricultural soil (0.45-0.82) was found higher than other microelement except cadmium. In geo-accumulation index copper was not recorded in polluted condition. Metal accumulation factor for copper was recorded high in winter season (3.28) for groundwater; while in agricultural soil high metal accumulation factor was found in summer months (1.20).

Copper is sometimes found in insufficient quantity in soil (Miller and Turk, 2002) but it may be added in to soil system by some anthropogenic activities. Copper may accumulate in living organisms and their various body parts (Kudesia, 1992). High amount if heavy metals like Cu and Zn may harm to living organism of existing ecosystem (Aslam *et al.*, 2004).

However, Cu concentration was not detected high in groundwater samples as described by WHO (2006) in drinking water *i.e.* 2.0 mg/l. The Cu concentrations level in study area have compared with those observed by Yusus and Sonibare (2004), Aslam *et al.* (2004), Helena (1999), Umar and Ahmad (2000), Kumar *et al.* (2001) and Yadav *et al.* (2002) in groundwater; Krebs *et al.* (1999), Singh *et al.* (1994) and Fakayode and Oniyawa (2002) in soil and Bordas and Bourg (1998), Nasr *et al.* (2006), Vazquez *et al.* (2007) and Korfali and Davies (2004) in sediments. Ather and Vohra (1985) reported that heavy metal enters in water system by various ways in natural resources like erosion, weathering, volcanic activities and anthropogenic activities that include direct emissions of heavy metal by different industries.

Iron is one of the most abundant elements of the rocks and soil, ranking fourth by weight. All kinds of waters including groundwater have appreciable quantities of iron. Iron has more solubility at acidic pH, therefore large quantities of iron are leached out from the soils by acidic waters. In the alkaline medium iron remain comparatively low in soluble phase. Excess calcium can reduced the activity of iron in soil and soil aeration may influence the availability of iron to plants (Miller and Turk, 2002).

In groundwaters most of the iron remains in ferrous state due to general lack of oxygen. In alkaline conditions in groundwaters, the iron is mostly ferrous bicarbonate, which is a colourless substance. When the groundwater with higher concentrations of iron is tapped, it quickly oxidizes to ferric state in the form of insoluble ferric hydroxide, a brown substance. If appreciable quantities of ferrous bicarbonates are present in groundwater, and they come in contact with oxygen at surface, the hydroxides are formed, CO_2 is released and hence it increases the pH, facilitating the oxidation processes (Minhas and Gupta, 1992). Generally, movement of iron compounds take place downward through the soil profile and reach in to groundwater table (Miller and Turk, 2002). However, iron was not found in high concentration in textile industrial effluents (0.24-0.54 ppm), but the high concentrations of iron were observed in effluent drain sediment (5656.0-8467.7 ppm) due to the continuous settlement and abundantly natural occurrence in earth crust materials. Iron in shallow groundwater samples (1.5-3.8 ppm) was found higher to textile effluents, while iron in deep groundwater (0.48-1.05 ppm) and control groundwater (0.363-0.868 mg/l) was not detected in high concentrations. Iron was present in high quantity in pond sediment (3803.5-6013.4 ppm) and effluent drain sediment (5656.0-8467.7 ppm). Iron concentration in agricultural soil was analyzed comparatively higher than other metals.

Iron's transfer factors were found very high from textile industrial effluent to groundwater (5.93-9.64), from textile industrial effluent to pond water up to 25.00, while iron was transferred from pond water to groundwater with highest 0.40 coefficient. However, transfer factor of iron was noticed low in comparison to other metals from shallow groundwater to agricultural soil. Metal enrichment factor in all seasons and in all sediment samples was calculated 1000.0 due to the comparative parameter for other metals for determination metals enrichment factor. In geo-accumulation index for drain sediment and pond sediment iron was not indexed at very high rank, it was found very low because the natural occurrence of iron in earth crust shale was very high, so iron was not a pollution causing agent in industrial area during the present study. Metal accumulation factor of iron was found from 2-4 for groundwater in Panipat region. Iron was detected in groundwater with 2-4 times high quantities in comparison to control site's groundwater. Iron in textile industrial effluent and pond water was found very high, while in groundwater samples it was also found very high for drinking purposes (WHO, 2006) Helena (1999), Yusus and Sonibare (2004), Umar and Ahmad (2000), Kumar *et al.* (2001) and Yadav *et al.* (2002) observed the similar trends of iron concentrations in groundwater and Singh *et al.* (1994) in agricultural soil of industrial area.

Manganese was found as a second abundant element in the effluent drain and pond sediment after iron. However, manganese was comparatively low in textile industrial effluent (0.11-0.31 ppm) and very low in groundwater samples (0.28-0.55 ppm). In groundwater samples manganese remains below 1.0 during the study even in shallow aquifer. In agricultural soil, manganese was comparatively low in concentrations (5.8-13.6 ppm) regarding other metals. Manganese was third most enriched element in groundwater samples. Mn was found slightly high at experimental sites in the case of soil samples. Manganese has positive relationship with all the physico-chemical parameters of groundwater as correlated with temperature (0.6892), total solids (0.8508) electric conductivity (0.8544), total alkalinity (0.7074) except dissolved oxygen.

Transfer factor of Mn from textile industrial effluent to pond water was noticed from 1 to 4 almost in all seasons, while below 1 in winter season. Transfer factor was calculated near 1 in all seasons from pond water to groundwater. Transfer factor from groundwater to agricultural soil was noticed very low at experimental sites as well as at control site. Enrichment factor of Mn was found around 100-200, highest after Fe in drain sediment, while enrichment factor was noticed higher in pond sediment. The value of Mn in geo-accumulation index was noticed always below 1 with negative rank even highest value in drain sediment as well as pond sediment due to the high natural shale value of Mn in earth crust. Metal accumulation factor of Mn in groundwater was high and ranged between 2 to 4, while in soil environment it was found below 2 always. Mn was found slightly high in groundwater for drinking purposes, because it should not be more than 0.5 mg/l for health views in drinking water (WHO, 2006).

Mathess (1974) reported that fresh groundwater constitutes 0.00-0.007 ppm cadmium, WHO (1984) recommended Cd concentration should not exceed maximum of 0.01 ppm in raw drinking water supply, 0.5 ppm in irrigation water. USEPA (1990) recommended maximum permissible Cd concentration as 0.04 and 0.02 ppm to protect aquatic life including fish. Hart (1982) reported a concentration range of Pb 0.0003-0.03 ppm in natural fresh water, while Frostner and Wittman (1979) the likely level as 0.002 ppm. Siddiqui and Pathania (2002) reported that the high value of Pb, Cr and other trace metals in groundwater in Jalandhar and Ludiana industrial area.

Manganese, Iron and potassium may deficient in alkaline soil (Miller and Turk, 2002), while manganese and iron is soluble in acidic soil (Brady, 1995). Manganese usually is present in sufficient quantity in most soils and need not be added for plant growth (Miller and Turk, 2002). Yusus and Sonibare (2004), Umar and Ahmad (2000), Kumar *et al.* (2001) and Yadav *et al.* (2002) gave a report of Mn status in groundwater, Fakayode and Oniyawa (2002) in soil system and Nasr *et al.* (2006) in sediments of aquatic system affected by industrial pollution.

Nickel is one of the second trace elements of the present study after cadmium. As it is very toxic to plants and animals in low concentrations and has a tendency to accumulate in body parts. However, Ni was presented in very minor concentrations in textile industrial effluent (0.007-0.067 ppm), and pond water (0.047-0.13 ppm), it was found comparatively high due to the regular input. Ni concentrations in pond were noticed higher during the summer season due to the loss of water by vaporization from pond system. About similar conditions were observed in groundwater samples. In deep groundwater, Ni was almost absent like cadmium. In agricultural soil all metals were found higher than groundwater so, nickel was also found high (5.3-10.5 ppm). Ni was found very high like other metals due to the settling of metals on bottom beds in pond sediment (49.0-63.7 ppm) and drain sediment (58.0-75.0 ppm). Nickel has no positive correlations with any physico-chemical parameter of soil and water except total alkalinity, Calcium, electric conductivity and water holding capacity. Highest transfer factor of Ni from textile industrial effluent to pond water and groundwater (14.14) was found in monsoon seasons. However, transfer factor from pond water to groundwater (0.83-1.29) was not found effective comparative to transfer factor of other metals from groundwater to agricultural soil at control site. Like cadmium, Ni was sometimes found infinitive in winter seasons.

Enrichment factor of Ni (7.35-11.7) was not effective in comparison to iron in drain sediment, but in pond sediment (8.38-14.85) it was slightly high due to the high retention time of wastewater in pond. The observations of present study reveals that Ni never reflected the pollution load in drain or pond sediment through the geo-accumulation index because it was found always below than 1. Metal accumulation factor in agricultural soil was recorded around 1-2, while in groundwater it was found slightly high or sometimes infinitive. Ni was found high in groundwater for drinking purposes, because it should not be more than 0.07 mg/l for health views, while it should be less than 0.02 mg/l generally in drinking water (WHO, 2006). Nickel may accumulate in aquatic life but its presence is not magnified along food chains (lenntech.com). So, there is not so big fear of nickel toxicity to higher animals including human beings, but agricultural plant health will be affected in contaminated agricultural soil. This trend of Ni concentrations in present study have been supported on the basis of recommendation of Kumar *et al.* (2001) and Yadav *et al.* (2002) in groundwater, Kasem and Singh (1999), Singh *et al.* (1994) and Fakayode and Oniyawa (2002) in soil and Nasr *et al.* (2006) in sediment. Jayabaskeran and Sree Ramulu (1996) observed a great heavy metal movement in sewage irrigating light textured soil than in heavy textured soil. It may be pointed out that the builds up of heavy metal like Cr, Pb, Ni warrant toward continuous monitoring and suitable measures are needed before these become toxic.

Lead was found highest in textile industrial effluent, but it accumulated in drain sediment in comparatively less quantity. Lead mostly remained in effluent (0.25-0.50 ppm) and fluctuated in pond (0.5-1.8 ppm). Rest quantity of lead settled down on bed sediment of drain and pond, so, the lead concentration was found low in pond sediment (44.0-69.0 ppm) and high in drain sediment (47.2-70.0 ppm). Lead was the second most abundant metal species after iron. Lead was found in high concentrations in shallow groundwater (0.85-1.84 ppm) in comparison to deep groundwater (0.28-0.41 ppm). Pb was also second most enriched element after iron in agricultural soil (25.1-62.4 ppm) due to the high distribution of these metals in earth crust. Lead is deposited mostly in bones and some soft tissues. High concentrations of lead may create toxicity in human (Kudesia, 1992).

Pb was found slightly high in groundwater (0.85-1.84 ppm) for drinking purposes and also in pond water (0.50-1.8 ppm), because it should not be more than 0.01 mg/l for health views, while it should be less than 0.1 mg/l generally in drinking water (WHO, 2006). Helena (1999), Umar and Ahmad (2000), Kumar *et al.* (2001) and Yadav *et al.* (2002) were observed the same trend of occurrence of Pb concentrations in groundwater, Fakayode and Oniyawa (2002), Krebs *et al.* (1999) and Singh *et al.* (1994) in soil environment and Bordas and Bourg (1998), Nasr *et al.* (2006) and Vazquez *et al.* (2007) in sediment.

However, zinc was not found so high in textile industrial effluent (0.11-0.34 ppm) in comparison to lead, but it was found in very high concentrations in drain sediment (143.5-260.0 ppm). Similarly, it was found in less quantity in pond water (0.11-0.2 ppm) and accumulated in very high quantity in bed sediment (116.2-240.0 ppm) of pond. Zinc concentration was fluctuated in pond sediment (116.2-240.0 ppm) season to season, while in drain sediment (143.5-260.0 ppm) the fluctuation of zinc concentrations was not observed in high concentration. Zinc concentration (0.13-0.22 ppm) was found at fifth position in groundwater samples among the all seven metals, while zinc concentration was found at fourth position in agricultural soil (7.7-19.4 ppm) among all metals studied. However, the concentrations of zinc in agricultural soil was found manifold than groundwater, pond water and textile industrial effluent due to the natural occurrence of zinc in soil profile. During the study, in all experimental soils zinc concentration was found slightly higher than those at control soil (7.3-16.5 ppm). Positive correlations were found between zinc and temperature (0.8704), electric conductivity (0.2569), Cl (0.7613), pH (0.4955) in agricultural soil. Positive relationship was noticed between heavy metals of groundwater with Zn. Transfer factor of zinc from textile industrial effluent to pond water was not found more than 1. However, transfer factor of Zn was found more than 1 from pond water to groundwater during almost all seasons.

Transfer factor of Zn from groundwater to agricultural soil was found slightly higher than 1 at experimental sites, while it was also found higher even more than 2 at control site. Enrichment factor of zinc in drain sediment was ranked at third position, while it was noticed slightly high in pond sediment due to the long retention time. In geo-accumulation index, the value of zinc was found more than 1.0 in all seasons, and ranked below 1 in geo-accumulation index in drain sediment.

Metal accumulation factor of Zn was calculated between 2 to 3 in groundwater samples, while metal accumulation factor was found near 1 in the case of agricultural soil and always remain below 2. Zn was found low in groundwater and in pond water for drinking purposes, because it should not be more than 3.0 mg/l for health views, while it should be less than 5.0 mg/l generally in drinking water (WHO, 2006).

Zinc is sometimes found in insufficient quantity in soil (Miller and Turk, 2002). Zinc salts are relatively non-toxic, but high concentrations may cause health problems like vomiting, renal damage, etc. (Kudesia, 1992). High amount of Zn may harm to living organism of that ecosystem (Aslam *et al.*, 2004). Helena (1999), Umar and Ahmad (2000), Kumar *et al.* (2001), Yadav *et al.* (2002), Yusus and Sonibare (2004) and Aslam *et al.* (2004) reported the same fluctuations of Zn concentrations in groundwater, Singh *et al.* (1994), Krebs *et al.* (1999) and Fakayode and Oniyawa (2002) in soil and Bordas and Bourg (1998), Nasr *et al.* (2006), Vazquez *et al.* (2007) and Korfali and Davies (2004) in sediment.

Heavy metals added to the soil in the field strongly absorbed by first component of upper part of undisturbed soil horizons. Subsequent redistribution to lower soil layers probably takes a very long time due to slow adsorption and heavy metals in upper layer tend to increase with time (Alumaa *et al.*, 2001). The trace heavy metal ions activities, depends on the solubility in the soil and often a much better indicator of heavy metal in soil solution reflected the soil metal fraction, that is most directly available for plants uptake, absorption by soil biota or leaching to groundwater (McBride *etal.*, 1997; Cances *et al.*, 2003 and Lofts *et al.*, 2004).

Minor or secondary elements such as magnesium, manganese, calcium, zinc, copper, boron, iron, and perhaps others are essential but except in certain specific cases either they are injurious to crop and soil in large quantity. If any element is lacking in soil or it is present in improper proportions normal plant growth will not occur (Miller and Turk, 2002). It is urgently required that industries should continuously monitor textile effluent and take necessary actions to properly treat wastewater prior their disposal to water bodies and save already depleting natural water resources (Aslam *et al.,* 2004).

Transfer of Heavy Metals

In the recent years there has been a growing concern with environmental protection. This can be achieved either by decreasing the afflux of pollutants to the environment or by their removal from contaminated media. The former is a feasible choice only for pollutants of anthropogenic origin, whereas, the latter is unavoidable for those of natural origin (Gomez-Serrano *et al.,* 1998). The presence of heavy metals in the aquatic environment has been of great concern to scientists and engineers because of their increased discharge, toxic nature, and other adverse effects on receiving waters (Singh, 2006).

In heavy metals analysis, the total concentrations of the metals are often determined (Okonwo *et al.,* 1999, 2001; Castro Dentros *et al.,* 2003). However, total concentration of trace metals provides no information concerning the fate of the metal in the terms of its interaction with sediments, its mobility, bioavailability, or resultant toxicity (Christie, 1995). It is now widely accepted that measuring total metal concentrations cannot fully assess the role of aquatic sediments as a sink or as a source of pollutants. In addition, determination of total elements does not given an accurate estimate of the likely environmental impact. Instead, it is desirable to have information on the potential availability of metals (whether toxic or essential) to biota under various environmental conditions. Since the mobility of heavy metals, as well as their bioavailability and related eco-toxicity to plants, critically depends upon chemical forms in which, a metal is present in the sediment, considerable interest exists in trace element speciation (Davidson *et al.,* 1994).

Heavy metals are introduced into aquatic system from various anthropogenic sources. Heavy metals constitute a special group of contaminants of aquatic systems and deserve special attention. Since metals are not removed by natural degradation processes, they may become enriched in sediments over time. Metal contamination of sediments is an issue of growing concerns worldwide (NRC, 1989). Both natural processes and anthropogenic activities are responsible for introducing metals in to the aquatic system (Nriagu, 1989). Many contaminants discharged into surface waters rapidly become associated with the particulate matter and incorporated in sediments. Metals in aquatic systems become part of the water-sediment system and their distribution is controlled by a dynamic set of physical-chemical interactions and equilibrium, largely governed by pH and type of legends and chelating agents, oxidation state of the mineral components and the redox conditions of the system. Metal contaminated sediments may release heavy metals back to the overlying water column and, thus, pose risk to aquatic life and ecosystems. Due to their particle reactivity, heavy metals tend to accumulate in sediment as a result may persist in the environment long after their primary sources have been removed (Forstner and Wittman, 1981).

Heavy metals travel from one level to another in the ecological system due to the accumulation in abiotic and biotic components. In the study area, basically anthropogenic source of heavy metals are the dye houses in textile industrial area. Heavy metals reach in effluent drain and also its bottom sediment and further into pond and bed sediment of pond near village Binjhole adjoining to textile industrial area of Panipat. Heavy metals percolate slowly from the bed sediment of Binjhole pond towards the groundwater table and there may be a threat to deep aquifers also by leaching process of metals. As the groundwater was mostly used in irrigation and drinking purposes in most rural parts of Haryana, which have affected the agricultural soil health after irrigation by contaminated groundwater and ultimately influence on the agricultural productivity, vegetation health and grain quality also. The use of groundwater for drinking purposes of human and cattle in contaminated conditions, a threat of various health hazards and diseases have occurred to public at present. Some of the diseases caused by toxic heavy metals have been lethal to living organisms.

Accumulation of Heavy Metals

The development and pollution are the two faces of the same coin having the forever positive relationship between each other. Due to the industrialization, a huge quantity of waste materials discarded outside from the industries, in which various toxic substances including heavy metals are present. These large quantities of discarded heavy metals have been created havoc in the environmental ecosystem and affected all biotic components of the surrounding water and agricultural system.

The accumulating processes of run off derived metals from the point and non-point pollution sources and their significant accumulation in the sediment of drain and pond with furthermore retention in agricultural soil has been observed by many investigators (Harper *et al.*, 1984; Nightangle, 1987 and Yusuf *et al.*, 1986). The extent of contamination due to heavy metal deposition in different industrial sites in relation to quite vulnerable metal toxicity among soil biota and plants have been investigated by Moon *et al.* (1991) and recorded Cd, Cu, Pb, Zn level in the industrial drain and top soil of agricultural fields.

Heavy metals have a tendency to accumulate and store in a component or trophic levels. The storage of heavy metals has created harmful effects for biotic components. The heavy metals accumulated in sediment and percolated down in to groundwater known as bio-accumulation of metal. From groundwater, heavy metals may turn into two ways; one is though irrigation and second is through drinking by human beings. The accumulation of heavy metals through the food chain is called the biomagnifications. The concentrations of metal increased at every next trophic level, whereas metals can cause various harmful effects on irrigated agricultural soil, human beings and livestock by alteration in some biochemical reaction in body cells (De, 2002).

In the present study, heavy metals in pond water were observed slightly lower than textile industrial effluents, whereas the heavy metal concentrations in effluent drain were higher than pond sediment. Shallow groundwater of about 40 ft depth was highly polluted in comparison to increasing depth of water table. Although the effect of spatial variation was also found in the study, when the heavy metals concentrations in groundwater and soils at control site were presented comparatively in less or very less quantities. Some metals like cadmium and nickel were not detected in groundwater at control site. After all, if the conditions will be same in textile industrial area then definitely there is a fear to contamination of groundwater into horizontal direction also and this will affect the groundwater of surroundings and deep aquifers with the time.

Conclusion

The present study reveals that there was heavy metals pollution in pond water and sediment, which affected the groundwater of the industrial area of Panipat region. Beside this, the bottom sediment of common effluent drain stored the heavy metals in high quantities, which have no secure or pre-managed future. Thus, textile industrial effluents discharging from dye houses contaminated the surface water, drain sediment, pond sediment and groundwater also. Furthermore, the impact of heavy metals reflected negative impacts on agricultural soil after irrigation by contaminated groundwater. Heavy metals remained in upper part of soil horizons mostly but moved in to lower soil layers probably after a long time due to slow adsorption (Alumaa *et al.*, 2001). Heavy metals strongly absorbed by first component to which they make contact and plants uptake them. So there is a fear to accumulate the heavy metals in different body parts of vegetation. High quantity of heavy metals accumulated in organisms may be harmful for health of organisms and create various diseases, which may magnify through food web of the water and agricultural ecosystems.

Heavy metals from textile industries affected the shallow groundwater quality in shallow badly, while deep groundwater affected slightly. Vertically, heavy metals contamination decreased with the increasing depth of water table. In the case of horizontal distribution, the heavy metal concentrations decrease with increasing distance from pollution source. In the circle of 1 Km from pollution source almost heavy metals level was found higher than the outer wide circle in the industrial area. Heavy metals contamination was found to be low in soil at control site and also in outer circle of textile industrial area. However, the area of around 1 Km from textile industrial area was found highly affected by heavy metals in Panipat region. Textile industrial effluent or contaminated groundwater has been altered the physico-chemical parameters of agricultural soil. It was indicated that the comparison of agricultural soil to control soil that some characteristics were influenced by contaminated irrigated groundwater. Heavy metals in agricultural soil were found slightly higher than control soil. Due to the regular discharging of metals residues in the textile industrial area, threat have been appear on plant vegetation, livestock and human beings also of the region. So, some appropriate measures for the amendment and control of textile pollution should be adopted by industrialists, municipality, government, NGOs and local people also at various levels for the protection of environmental components.

It is evident from the data that the physico-chemical characteristics of effluents exceed the prescribed Indian standards. It is therefore obvious that scientific treatment is necessary to minimize the pollution effects before textile industrial effluent is discharge on the land. But disposal of these effluents after proper dilution may be a favourable approach. After dilution, the effluent characteristics come within the prescribed disposal limits and the pollution load per unit effluent volume is decreases.

Effective steps should be taken for improving water use efficiency, as the integrated management of water resources should be based on the treatment of wastewater as an integral part of ecosystem and as a natural resource whose quantity and quality determine the nature of its utilization. Wastewater use in turn has its impact on the water quality and therefore utilization of water has to be managed as not to contribute to the deterioration of water quality. Secondly, conjunctive use of groundwater and surface water resources needs to be planned in the irrigation project from the beginning. There is a need to take effective steps for improving wastewater use efficiency through renovation and modernization of recycling treatments of wastewater.

Different guidelines and standards limits should be immediately applied in small scale textile industries for utilization of groundwater and open discharging the effluents into surface water or on land. A regular bio-monitoring should be applied for the assessment of effluent quality of every dye house outlet. An effluent treatment plant of appropriate capacity must be installed in the industrial area for the primary and biological treatment of for composite effluent from textile industries. Unlike most organic pollutants, heavy metals are generally refectory and cannot be degraded or readily detoxified biologically. Hence the safe and effective disposal of wastewater containing heavy metals is always a challenge to industrialists and environmentalists, since cost-effective treatment alternative are not available (Manahan, 1984). However, some researchers like Singh (2006) developed a low cost adsorbent of heavy metals from some cheap byproducts.

Beside this, water supply of the industrial sector should be managed and regularly analyzed the suitability of drinking water quality. Groundwater safety measures are immediately necessary in the industrial region. Restoration of affected agricultural soil is still an environmental aspect for Haryana state. Almost agricultural soils were noticed alkaline already, however, the soil in the industrial area of selected location was not so alkali, but its pH was more than 7. All the above observation and data indicated that the qualities of effluents sometimes do not meet the CPCB, BIS, MINAS and WHO specified norms. The effluents may also cause various health problems among the human being through contaminated groundwater supply by municipal tube wells water scheme. In order to achieve the standards it is suggested to install the effluent treatment plants for the treatment of composite effluents of the textile industries.

Chapter 8

Water and Health

The origin of water on earth is not clear so far. However, the current presumption is that the primordial earth had no oceans and perhaps very little atmosphere. It is believed that the volatile constituents bound in the earth crust, oozing to the surface through volcanoes, rock movements and hot springs, condensed to form the ocean, atmosphere and mountains. This way perhaps this remarkable combination of hydrogen and oxygen called water come into being and eventually became an indispensable component of earth's environments.

Water is the most Essential commodity for all living creatures. Man needs it for his physiological existence water is the elixir of life. It is the source of energy and govern the evaluation and function of the universe on the earth 97.3 per cent of the worlds water *i.e.* 1.4 billion cubic kilometer ocean salty water cannot be used for agriculture domestic and industries purposes. Out of it 2.7 per cent fresh water *i.e.* 73×10^6 cubic in the form of streams, lakes, wells and tube wells *i.e.* 0.6 per cent $8.5 \times 10^{15} m^3$ is groundwater which oceans at the depth of 80 meter below the ground surface (Bolen back 1983, Kataria 1994).

In Vedic Literature water is mainly called, as Apha we all know that waters is the elixir of life. All living organisms, plants and animals are structurally 30-98 per cent water. Besides being a structural component of body water is essential environmentally too. Water is the medium through which nutrients of soil underneath and above on the surface move horizontally as well as vertically water has great power of dissolving most of the inorganic components in mixture form and in the pure elemental and ionic form. Water provides a medium for interaction among different compounds water is a great nurture for plants as well as for animals water helps in conducting of energy the form of heat helps in maintaining a soothing climate, water acts as a rejuvenator and medicament itself (Rig 1.23.19).

Chemically, water is the mono-oxide of Hydrogen *i.e.* (H.O.H). It freely disassociated into hydrogen and hydroxyl ions. The ionization product (H^+) and (OH^-) is about 10^{-14}. Water is a colourless liquid and possesses a high dielectric constant (81), and there for salts gets highly ionized when dissolved in water but not so in the other solvents. Due to this property, water is known as universal solvent present on the earth that is why the natural water contains many salts in dissolved form. On the basis of amount these slats dissolved in the earth only 3 per cent of the total water (approximately 1.4 billion cubic kilometer) is fresh and suitable for human consumption of this, less than 0.01 per cent is associated with rivers and streams (Water Resources 1980).

8.1 Water and Health

In vedic literature, water is called as a medicine for human health. No body can imagine life without water. Water is our basic requirement even animal and plant can't survive without water. It is common statement that *"Life is not possible without water"*. Thus life and water are inseparable faces of the same coin. Water is not only essential for life but predominant inorganic constituent of living matter forming in general nearly three quarters of the weight of living cells. It makes some 5 per cent of the body weight of an adult human and can forms as 98 per cent of the mass of certain jellyfish. Organisms that contain relatively small amounts of water are generally in dormant state or show very slow development; seeds and certain invertebrates that live in arid environments are examples. On the other hand high rainfall over a landmass in variably means a large biomass per unit area.

Water is a resource circulated throughout all ecosystems and is one of the most important factors of them (Odum 1971). From the point of view of eco-hydrology, for the main elements of an ecosystem including inorganic and organic compound, producers and consumers both the quantity and the chemistry of water have a decisive impact on its biomass and biotic composition. Information obtained form current hydro-chemical analyses of water specimens is very important but it is sometimes difficult to Ganga the overall situation especially the occurrence of nutrient, which may be very important to an ecosystem.

Every thing originated in the water and every thing is sustained by water all life on earth depends on water. Water is very essential for the survival of all living organism there is a wide variety of beneficial uses of water. Man uses water for many purposes, drinking, irrigation, fisheries, industrial purposes, transportation and waste disposal. The origin of water on the earth is not clear so far. However, the current presumption is that the primordial earth had no oceans and perhaps volatile constituents bound in the earth crust oozing to the surface through volcanoes rock movements and hot springs, condensed to form the ocean and the atmosphere. This way come into being and eventually become an indispensable component of earth' environment.

Water quality includes all physical, chemical and biological factor that influence the beneficial use of water, Lotic habitats are those exiting in relatively fast running streams, springs, river and brooks. The lakes represent lentic habitats; marshes and swamps represent ponds and wetland. The above classification of fresh water environment is based on currents and ratio of the depth of surface area.

8.2 Phases of Surface Water

1. Permafrost
2. Hail
3. Fog
4. Water vapour
5. Dew
6. Ice
7. Aquifer
8. Sleet
9. Frost

8.3 Sources of Water

1. Rainfall
2. Groundwater
   - ☆ Dug wells
   - ☆ Bore wells
   - ☆ Tube wells
   - ☆ Hand pumps
3. Surface water
   a) Fresh water:
      i. Lentic (still water)
      - ☆ Ponds
      - ☆ Lakes
      - ☆ Wetlands

      ii. Lotic water (running water)
      - ☆ Rivers
      - ☆ Springs
      - ☆ Streams
      - ☆ Brooks

   b) Marine water
      - ☆ Ocean
      - ☆ Sea

Undergroundwater is one of the major water resources for domestic, agricultural and industrial uses. Pollutants are being added to the groundwater and soil system through various human activities and rapid growth of industrialization, which affects the human health directly or indirectly.

Environmental assessment in this context is concerned with groundwater concentrations of cadmium, zinc, lead and arsenic, the levels of which are often increased by the effects of human activity. Metals are spread via the atmosphere from smelters and other metal industrial plants. Significant increases in groundwater concentrations of metals from such sources may occur within a local radius of ca. ten kilometers; but their distribution through the atmosphere extends over far greater areas. Large quantities of lead have also been released into the atmosphere through the use of leaded petrol. Such releases have declined in recent years, but the greater portion of airborne lead pollution from previous years is still present in the surface layers of earth. Levels of cadmium, lead, zinc and arsenic in groundwater are often related to pollution or other human activities. Cadmium, lead and arsenic are dangerous to human health even in fairly low concentrations.

Metals have also entered land and water by more direct routes. In some areas, the use of cadmium-contaminated phosphate fertilizer has led to a doubling of cadmium levels in cropland during the past fifty years. Since zinc is often used for anti-corrosion purposes, it may enter the soil via galvanized steel products embedded in the ground. Both arsenic and zinc are included in many wood preservatives, and can be spread via waste spill at impregnating facilities or via the wood products themselves. Waste sites are another source of zinc and arsenic, which may leach out from discarded materials that include those metals. Still greater quantities of polluting metals may leach from dumps of mining waste.

Naturally high levels of cadmium and arsenic can occur in areas where the bedrock contains high concentrations of metals; this applies to many types of shale and slate. Lowering the water table, for example by pumping out water, may lead to increases in the amounts of naturally-occurring metals that enter the groundwater. This is because metals in bedrock are often bound in sulphide minerals. The addition of oxygen results in oxidation of the sulphide compounds and the release of their metallic components. In some cases, however, arsenic may also be released as the result of a rise in the water table. It is with redox class 3 that the risk of water-soluble arsenic is greatest.

The solubility of cadmium and zinc increases significantly with a decline in pH value, which means that acidification, can lead to increased levels of those metals in groundwater. Normally, lead is very tightly bound to humus and clay, but it may dissolve more readily if the pH value declines. In the case of arsenic, the reverse usually applies: Solubility increases concurrently with rising pH values. Zinc is an element that is essential to life for most organisms, and humans can tolerate high concentrations. Aquatic organisms, on the other hand, are much more sensitive to zinc. Increased levels in groundwater indicate that concentrations of other metals may be above normal. This is especially true of cadmium, which is often present concurrently with zinc. In most cases, cadmium levels are, at most, one percent higher than those of zinc; but cadmium can be harmful even at such low concentrations. Lead is another heavy metal, which is toxic at low concentrations. If pregnant women are exposed to lead concentrations above medically established threshold levels, there is an increased risk of birth defects. The toxicity of arsenic is comparatively moderate, but over longer periods of time it may produce changes in skin tissue, even at levels below medically established thresholds.

8.4 Threshold Levels of Heavy Metals

Table 8.1: Threshold Levels of Heavy Metals

Class	*Level*	*Cadmium µg/l*	*Zinc µg/l*	*Lead µg/l*	*Arsenic µg/l*
1	Very low	< 0.05	< 5	< 0.2	< 1
2	Low	0.05-0.1	5-20	0.2-1	1-5
3	Moderate	0.1-1	20-300	1-3	5-10
4	High	1-5	300-1000	3-10	10-50
5	Very high	> 5	> 1000	> 10	> 50

In several cases, the limits for the above five classes correspond with those of the National Food Administration's threshold levels (see table below). For all metals, the boundary between class 2 and class 3 is defined as the level at which effects on organisms in lakes and watercourses begin to occur.

Table 8.2: Different Types of Threshold Levels of Heavy Metals

Type of threshold level	*Cadmium µg/*	*Zinc µg/l*	*Lead µg/*	*Arsenic µg/*
Warning based on technical grounds; survival of aquatic organisms is affected even with brief exposure		> 300		
Warning based on technical and aesthetic grounds; risk of taste changes and turbidity		> 1000		
Medically-established threshold; use with caution	> 1			> 10
Medically-established; unsuitable for drinking	> 5		> 10	> 50

Source: National Food Administration.

8.5 Reference Value of Heavy Metals

Reference values have been set so high that uncontaminated groundwater has low concentrations of metals in the majority of cases. Such water may, however, include pollutants that have traveled lengthy distances through the atmosphere.

Table 8.3: Reference Values of Heavy Metals

Metal	*Reference Value µg/l*
Cadmium	0.1
Zinc	100
Lead	1
Arsenic	1

Table 8.4: Deviations from Reference Values of Heavy Metals

Class	*Extent of Deviation*	*Cadmium µg/l*	*Zinc µg/l*	*Lead µg/l*	*Arsenic µg/l*
1	None or insignificant	< 0.1	< 100	< 1	< 1
2	Moderate	0.1-0.5	100-200	1-1.5	1-2
3	Significant	0.5-2	200-400	1.5-2.5	2-5
4	Large	2-5	400-700	2.5-5	5-10
5	Very large	> 5	> 700	> 5	> 10

The five classes indicate the extent to which measured levels of metals in groundwater deviate from the reference values, which are relatively unaffected by non-natural causes. The upper limit of class 1, separating it from class 2, is the same as the reference value. The values of class 5 indicate the strong possibility of a local source of contamination. But in isolated cases, there may be another cause of the unusually high levels, for example: Class 5 levels of cadmium may occur naturally in extremely acid water (pH < 5); likewise, high arsenic levels occur naturally at a number of locations in the regions of Bergslagen and Skelleftea, as well as other parts of the country in certain deep-drilled wells with high levels of iron (redox class 3) and high pH values.

Chapter 9
Groundwater Modeling

In the last two decades, the rapid growth of industrialization and urbanization has created negative impacts on the environment. Due to industrial, municipal and agricultural wastes containing pesticides, insecticides, fertilizers residues and heavy metals with water; groundwater has been polluted (Sastry *et al.*, 1999). Groundwater is used for domestic supply, industries, and agriculture in most part of the world as it is a replenishable resource and has inherent advantage over surface water.

There has been a tremendous increase in the demand of fresh water due to growth in population, advanced agriculture practices and industrial usages (Bachmat, 1994). Industries are discharging their metal and hazardous substances containing effluents in to fresh water with out any adequate treatment (Canter, 1987). The seepage of effluents posed serious threat to the groundwater resources (Groff, 1993).

In water cycle, rainfall is the main source that provides water for all living organisms in different ecosystems. Water from rain, enters in to soil and then flow vertically and down slope through the soil profile. Water flow may vary at place to place according to the physical conditions of soil of a particular area. Flow of heavy metals and groundwater movement beneath the earth may be measured by using various forms of formulas according to the physico-chemical and geographical conditions of soil and soil profile of the particular area.

9.1 Movement of Groundwater

Contamination of groundwater is more complex than surface water pollution mainly because of difficulty in its timely detection and slow movement, which requires special expertise to predict the path and rate of groundwater movement.

Generally, it is believed that movement of groundwater flows according to ground elevation or the general topography of the area. Groundwater in its natural state is invariably moving and this movement is governed by established hydraulic principle and can be expressed by Darcy's Law. According to Darcy's Law "flow rate through porous media is proportional to the head loss and inversely proportional to the length of the flow path".

$$Q = \frac{K(h_1-h_2)}{L}$$

where,

Q: Specific discharge,

K: Hydraulic conductivity,

h_1–h_2: Head loss, and

L: Distance of travel by groundwater.

In addition the complex geo-chemical reactions taking place in the subsurface between myriad contaminants and earth materials are not always well-understood. Ideally speaking contamination should be prevented from occurring. After several contaminants are found in groundwater, a decision must be made on whether to rehabilitate the aquifer or find alternative groundwater resources. So, groundwater must be managed and protected carefully for the most beneficial use in the future.

9.2 Groundwater Flow Pattern

Geology offers control on the rate of groundwater movement. The size of the pores and their relative connectivity (permeability) determine the rate at which water moves in to, though and out of aquifer.

The flow of groundwater in saturated soil is described well by Darcy's law, which, for steady flow in one dimension (linear), may be given by (Wainwright and Mulligan, 2004).

$$q = -K\frac{dH}{dx}$$

Where, q is the discharge per unit area, K is the hydraulic conductivity; H is the hydraulic head and x is the distance in the direction of flow.

To make q positive, a minus sign appears before K, this correction is necessary because water flows down the hydraulic gradient, *i.e.* it flows along a negative gradient from areas of higher H to areas of lower H (Fetter, 1994).

Richards Equation

The water flow through unsaturated soil also be described by modified Darcy's law, where K is assumed to be a function of water content (θ), *i.e.*,

$$q = -K(\theta)\frac{dH}{dx}$$

L.A. Richards was the first person to realize that the Darcy's law could be applied to unsaturated flow, so the equation may be called as Richards equation.

Where, flow is unsteady such as in natural conditions, water content will change and so will K. Thus, in unsaturated flow, unlike saturated one, K is time-variant, except for some very simple cases.

The simplest of this class of model ignore or greatly simplify flow in the unsaturated zone and assume that flow in the soil is parallel with the soil surface, and that the hydraulic gradient at any point with in the saturated zone is equal to the slope (Beven, 2000). Thus, the Darcy's law becomes:

$$Q = KhSin\beta$$

Where, Q is the discharge per unit width of hill slope in the down slope direction, h is the depth below the water table measured orthogonally to the impermeable plane at the base of the soil and β is the slope of the plane and soil surface (Masters, 1998).

Groundwater moves more quickly in coarse sand sometimes as much as several meters per day while in case of clay which are generally impermeable the movement may seldom move not more than an inch per year (CPCB, 2003).

The general direction of groundwater flow can be established on the basis of the local topography using topographical maps or aerial photos and the presence of stream or river which act as groundwater discharge boundaries. The near surface groundwater flow generally follow surface drainage pattern. If the flow direction can not be established, three small diameter wells are temporarily installed in the aquifer (CPCB, 2003). Relative water level in these wells reveals the direction of flow. Other methods of delineating groundwater flow pattern include electromagnetic induction, earth resistivity survey, radioactivity and temperature logging.

9.3 Darcy and Linear Velocity/Flow Velocity

If we combine the usual relationship between flow rate, velocity and cross section area, than-

$$Q = Av$$

With Darcy's law we can calculate velocity,

$$v = \frac{Q}{A} = \frac{KA(dH/dL)}{A} = K\frac{dH}{dL}$$

Where, υ is known as the Darcy velocity, not a real velocity because much of the cross section area is made up of solids, the actual area through which all of the flow takes place is much smaller, so the real groundwater velocity is considerably faster than the Darcy velocity (Masters, 1998). A represent the cross section area, and A' being the area of voids filled with water, than υ' be the actual average linear velocity, as

$$Q = Av = Av'$$

$$v' = \frac{Q}{A'} = \frac{ALv}{A'L} = \frac{Total\ volume \times v}{Total\ volume}$$

The actual average linear velocity through the aquifer is the Darcy velocity divided by porosity, so-

$$v' = \frac{Darcy\ velocity}{Porosity} = \frac{v}{\eta}$$

$$v' = \frac{k(dH/dL)}{\eta}$$

Where, υ′ is the average linear velocity of groundwater, υ is Darcy velocity and η is the porosity of soil medium, which is the important physical factor may influence the groundwater movement.

Retardation Factor

If a contamination moves at the same speed as the groundwater, some contaminants are absorbed on the soil particles, which reduce the overall solute flow rate relative to that of the groundwater itself (Masters, 1998). This reduction in speed is called the retardation factor,

$$\text{Retardation tractor} = R = \frac{v'}{v_S} \geq 1$$

Where, υ′ is the average linear velocity of groundwater, $υ_s$ is the average linear velocity of retarded constituent.

9.4 Flow of Heavy Metals in Unsaturated Soils

Equations for the Flow Model

The flow model is formulated using Richards equation for unsaturated movement of water in one dimension with the inclusion of root water uptake. The equations governing vertical water movement are

$$q = -K\left(\frac{\partial \psi}{\partial z} - 1\right) \tag{1}$$

$$\frac{\partial \theta}{\partial t} = \left(\frac{\partial q}{\partial z} - S\right) \tag{2}$$

Where, q is water content by volume (cm^3/cm^3), q is the water flux (cm/h), S is the sink term for root water extraction (cm^{-1}), K is the hydraulic conductivity (cm/h), Y(<0) is the soil matrix potential (cm), z is the vertical coordinate direction taken positive downward, and t is time (h) (Nedunuri *et al.*, 1995). The two equations can be combined to yield the Richards equation in terms of q as

$$\frac{\partial \theta}{\partial t} = \frac{\partial}{\partial z}\left[K(\theta)\left(\frac{\partial \psi}{\partial z}-1\right)\right]-S(\psi) \qquad (3)$$

Where, the sink term depends on Y which controls the water availability to roots. The K(q) and Y(q) are non-linear functions, and have been described by Brooks and Corey. The pore water velocity at any depth is given by

$$v(z) = \frac{\partial(\theta)}{\theta} \qquad (4)$$

The equation describing the root water uptake is expressed as

$$S(\psi) = \alpha(\psi)S_{max} \qquad (5)$$

Where, $\alpha(\Psi)$ determines the proportion of the maximum possible root extraction rate S_{max} for a particular value Ψ. The expression for $\alpha(\Psi)$ is given as

$$\alpha(\psi) = \exp\left(-\delta\left|\psi - \psi_b\right|\right), \psi \leq \psi_b \qquad (6)$$

Where, Ψb is the air entry pressure head. Equation 6 indicates an exponential reduction in rate of availability of water to the roots as the soil becomes drier.

9.5 Equation for the Transport Model

The equation describing the movement of contaminants is-

$$\frac{\partial C}{\partial t} = \frac{\partial}{\partial z}\left[D\frac{\partial C_k}{\partial z}\right]-\frac{\partial}{\partial z}[v(z)C_k]+R_k \qquad (7)$$

$k = 1, 2, \ldots Nc$

Where, C_k is the concentration of the k-th component at any spatial location z at any time t, D is the dispersion coefficient, and R_k is the chemical source/sink term representing the changes in aqueous component concentrations (www.engg.ksu.edu).

9.6 Groundwater Monitoring Techniques

Groundwater monitoring involves various investigations prior to the actual assessment of groundwater quality. The topography, geology, soil profile, aquifer characteristics, source of pollution, land-use pattern and above all the extent of groundwater development are some of the important pre-requisites to decide the scheme of groundwater monitoring. For representative information on groundwater quality the first information to look for is the movement of groundwater.

Monitoring of groundwater quality involves collection of representative sample and simultaneous collection of hydro-geological information or data about the location of sampling station. The sampling and hence the groundwater quality data may adversely affect if one or more of the following factors are not appropriately looked into (CPCB, 2003).

1. Samples are not taken at appropriate time intervals, seasonally and at improper locations.
2. Sample was taken from stagnant water in the open dug well or suddenly in case of bore well without removing the casing storage effect. The error can be largely overcome if three to ten well bore volumes are removed, and in case of hand pumps sufficient water is pumped out before taking the sample.
3. Contamination due to entrained sediment.
4. Hydraulic characteristics of the soil formation near the screen, resulting in possible dilution of the contaminant.
5. Release of carbon dioxide during pumping and hence increase in pH which may cause many metallic ions to come out of solution.
6. The sample preservation is not properly done and analyzed without considering the parameter-specific prescribed period of retention time.

9.7 Concept of Groundwater Modeling

According to many instances of groundwater monitoring, the ability to predict how the contaminant plume will behave in future can be done by using the mathematical modeling techniques to estimate the spread of a contaminant and its strength at any point in the plume. A model is a device that represents an approximation of a field situation and provides a frame work for synthesizing field information and for testing of ideas about how the system works.

The concept of groundwater modeling is chiefly based on following factors:

1. Advection of the constituent with the water flowing through the aquifer.
2. Dispersion of the constituents.
3. Characteristics of recipient medium.
4. Sources of pollution.

The steps to build the model are shown in the structure given below:

Factors > Processes > Model > Prediction > Assessments

Table 9.1: The Basic Step of Groundwater Monitoring and Modeling

Step 1 *Factors*	*Step 2* *Processes*	*Step 3* *Model*	*Step 4* *Prediction*	*Step 5* *Assessments*
Hydrology/ Hydraulics	Dispersion/ Dilution	Attenuation and Convection	Contaminant concentrations in space and time	Comparison of model results with field data
Physical and chemical soil characteristics	Precipitation			
	Adsorption			
Waste leachate properties.	Biological activity etc.			

Chapter 10

Quality Improvement Measures and Standards

The increased contamination of aquatic environment by domestic sewage and industrial effluents has focused attention on the behavior and interaction of many rare and neglected elements in the biosphere. Bioaccumulation of heavy metals in living organisms occurs due to their property of persistence and non-degradability of the complex organic compounds. Aquatic plants in general are more sensitive to environmental dangers. Sensitive aquatic macrophytic species can serve as indicators and those resistances to pollution as accumulators. In the present study, *Eichhornia crassipes,* an aquatic weed has tremendous capacity to remove the pollutant from the water and also to grow and multiply rapidly in sewage containing aquatic environment.

Modern civilization is dependent upon the water for irrigation, domestic need, industrial and for power generation. So we can say that life and water are inseparable faces of the same coin. The intensity of human activities and the diversity and magnitude of wastes introduced into environment as a consequence are increasing at an accelerating rate with improvements in technology, advancing industrialization and create a significance impact on environment and effect all living beings including man himself. The increasing rate of population growth and progress of human beings in different fields is creating an alarming condition of environmental pollution.

According to Edwards (1972) "Pollution is the release of substances or energy into the environment by man in quantities that damages either his health or resources". Municipal sewage, domestic wastes, industrial wastes and effluents, synthetic chemicals like pesticides, detergents and toxic metals of fertilizer have polluted maximum water bodies. Water pollution by radioactivity and thermal pollution are also noteworthy. Sewage can be regarded as the water borne waste products of man. Sewage consists of discharge of water from bathrooms, kitchen and laboratories.

Raw sewage consists of 99.9 per cent water and 0.1 per cent solids, which include 70 per cent organic and 30 per cent inorganic compounds. Sewage is an excellent medium for growth of pathogenic bacteria, viruses and protozoa. Domestic sewage, which is primary compost of, spent water containing wine, faeces, soapy waste, food materials and paper makes the water extremely anesthetic. Water becomes totally unfit for drinking and domestic use.

Sewage containing oxidizable and fermentable matter causes depletion of dissolved oxygen in the receiving water bodies affecting the aquatic flora severely. Oxygen deficiency also leads to the production of objectionable odours in water. Highly organic suspended matters, which purify the water, get deposited on riverbeds, behind weir or sludge. Suspended matter present in sewage has a tendency to blenket the stream thereby interfering with the spawning of fish and reduction of aquatic biota. Accumulation of sewage and domestic waste in water bodies retards the self-regulatory capabilities of aquatic organisms. Discharge of nutrient rich effluents, sewage and domestic waste in combination with industrial wastes poses serious health problem in man.

Every community produces both liquid and solid wastes. The wastewater is essentially the water supply of the community after it has been fouled by a variety of uses. Wastewater is commonly classified as domestic, industrial or agricultural according to its source. Sewage water originates from domestic sources and contain varying amount of industrial wastewater of occasionally rainwater. The quantity of sewage water is about 80 per cent of the total consumption and varies between 100 to 300 capita per day depending upon the standard of living and season of the year. Most city jewelers are unable conceive that individual produces about 150 liter of sewage per day of it simply does not disappear underground.

The wastewater collected from cities and towns must ultimately be returned to the receiving water or on the land after the containments have been removed to protect the environment depending upon the local needs of the end use of the treated water.

10.1 Remediation

Bioremediation is the use of living organisms (primarily microorganisms) to degrade environmental pollution or to prevent pollution through waste treatment. Research projects are being modified to expand the range of microorganisms used for bioremediation. There is search for naturally occurring microbes that have better pollutant degradation kinetics, attack a wider range of microbial growth condition and could grow under extreme environmental condition, such as tolrance to organic solvents, growth under extremely alkaline substrate or high temperatures.

Phytoremediation is the process of recovery of hazardous substances from soil and wastewater by using plants. Aquatic macrophytes such as species of *Salvina, Lemna, Azolla, Eichhornia, Sedeges* and even tree species are also known to tolerate, uptake and even accumulate heavy metals and other toxicants in their cell.

In rural and small urban localities, the volume of sewage generated is low. It is very expensive to treat such low volumes by conventional technology because according to the economics of scale, lower the quantity of waste to have treated by any technology, higher is the cost of treatment per litre. This is one of the reasons why most villages and small town in developing countries don't have any sewage treatment facility. This leads to wide spread water and soil pollution and frequent outbreaks of water born diseases. Against the background, the emerging technology of using aquatic macrophytes especially water hyacinth, which enable cost effective treatment of sewage, assumes greater significance and promise.

To minimize the pollution load of water bodies various guidelines, quality standards and maximum permissible limits of pollutants have been set up in different countries. However, WHO standards are being applied in most countries of the world as they are considered as ideal. The Indian standards (BIS) or CPCB standards show relatively high deviation from the WHO standards for environmental quality. Maximum permissible limits for some water quality parameters are different than those of WHO. Limits for toxic metals (Cd, Cu, Hg, Mn, Ni, Pb, Zn) in drinking water are high in BIS or CPCB standards compared to WHO standards. Heavy metals are highly toxic and cause various health problems in humans and also affect cattle and vegetation. Thus, it is essential to control their migration to the unpolluted environment for the safety of human health and livestock (Bharti, 2007).

10.2 Water Quality Standards

The first sets of standards were published as Indian Standard, 2490 in 1963. These standards set the tolerance limits for Industrial Effluent discharged into inland surface water. The Indian Standard 2296 was also published in 1963, which set the tolerance limits for inland surface water subject to pollution.

The first attempts of water pollution control thus approached the problem using two independent concepts, one dealt with the control of pollution through specification of various parameters and their maximum concentration in the effluent itself. These could be referred to as 'Effluent Standards'. The other approach was to set standards for the receiving water body presumply after a reasonable opportunity for dilution. These standards could be called 'Receiving Water Body Standards' and were related to identified beneficial use to the receiving water body. It is extremely significant that the very first attempt of control of water pollution was through setting of standards simultaneously adopting the above two approaches.

Effluent Standards versus Receiving Water Body Standards

As the Indian Standards were evolved, a choice between Effluent Standards and Receiving Water Body Standards emerged, considering the fact that most of the Indian rivers are not perennial and hence carry only the effluents during major period of the year, the choice generally favoured the Effluent Standards. Effluent Standards are also relatively easy to enforce. This has generally led the various Pollution Control Boards to favoured Effluent Standards. In specific cases however, such as deep marine outfall disposal, it is necessary that the receiving water body standards be considered.

10.3 Minimal National Standards

Around 1976, the CPCB developed the concept of evolving industry specific effluent standards based on a comprehensive study of the problems of that particular industry relevant polluting parameters, their pollution potential, best pollution control technology available in the country and the cost of such a technology.

From these, comprehensive documentation evolved the idea of Minimal National Standards (MINAS). These standards were considered to be the minimum standards for specified industries. In the Environmental Protection Act (EPA), 1986, some of these standards were incorporated. Since these are the Minimal National Standards, the State Boards are only permitted to make them more stringent and in no case can relax them. It is essential that effluent standards satisfy three basic criteria namely, Necessity, Implement ability and Enforceability. The stipulation of standards for a specific parameter is necessary when the effect of the parameters on the environment were thoroughly studied and established. No compromise, therefore, can be made with regard to concentration of toxic pollutants. By implement ability it should mean that with best pollution control technology currently available in this country, these parameters could be brought down to the specified limits. Enforceability means either the industry or the concerned authorities have the expertise to analyze, to monitor the wastewater discharged from the industry. Each of the parameters included in the MINAS must satisfy each of these three fundamental criteria. At the same time, the set of parameters chosen must ensure the primary objective of protection of environment. Keeping in mind the basic criteria, major polluting parameters from the industries and the performance of the treatment plants, the standards are evolved.

Table 10.1: Standards for Effluent from Textile Industry by CPCB
MoEF Notification, New Delhi, 25th September, 2000
Bharat Ka Rajpatra- Bhaag–II, Khand 3 (i)
(For large scale industries)
Standards for effluents from textile industry

Parameter	*Concentration Not to Exceed, mg/l, except pH*
pH	5.5-9.0
Total suspended solids	100
BOD	30
COD	250
Total residual chlorine	1
Oil and grease	10
Total chromium as Cr	2
Sulphide as S	2
Phenolic compounds as C_6H_5OH	1

Notes:

1. Where the treated effluent is discharged into municipal sewer leading to terminal treatment plant, the BOD may be relaxed to 100 mg/l and COD to 400 mg/l.
2. The quantity of effluent (l/kg of product) shall not exceed 100, 250 and 80 in composite cotton textile industry, composite woolen textile industry and textile processing industry, respectively.

Table 10.2: Standard for the Textile Effluent According to Bureau of Indian Standard (BIS)

Parameters	*BIS (IS : 3307, 1965) (IS : 3306, 1965)*
pH	5.5-9.0
Temperature °C	40.0
Colour Co-pt (mg/l)	0.0
Total Solids (mg/l)	-
Total Dissolved Solids (mg/l)	-
Total Suspended Solids (mg/l)	100
Biochemical Oxygen Demand (mg/l)	30.0
Chemical Oxygen Demand (mg/l)	250
Oil and Grease (mg/l)	10
Ammonical Nitrogen (mg/l)	50

Table 10.3: Standards for Small Scale Industries (*i.e.* total discharge upto 25 KL/day) Notified Under the Environmental (Protection) Act, 1986

Parameter	*Cotton Textile Industry (Composite)*	*Woolen Mill (Composite and Processing)*
Common		
pH	5.5-9.0	5.5-9.0
SS	100	100
BOD	150	150
Oil and Grease	10	10
Bio-assay Test	90 per cent survival of fish after 96 hours	90 per cent survival of fish after 96 hours
Special		
Total Chromium (as Cr)	2	2
Sulphide (as S)	2	2
Phenolic Compounds (as C_6H_5OH)	5	5

Table 10.4: Environmental Standards (cpcb.nic.in, 2006)
Dye and Dye Intermediate Industry : Wastewater Discharge Standards

Parameter	*Concentration not to exceed, mg/l (except pH, colour and bio-assay test)*
pH	6.0 to 9.0
Colour, hazen unit	400
Suspended solids	100
BOD (27°C, 3 days)	100
Oil and Grease	10.0
Phenolics as C_6H_5OH	1.0
Cadmium as Cd	0.2
Copper as Cu	2.0
Manganese as Mn	2.0
Lead as Pb	0.1
Mercury as Hg	0.01
Nickel as Ni	2.0
Zinc as Zn	5.0
Chromium as Cr	
Hexavalent	0.1
Total	2.0
Bio-assay test	90 per cent survival in 96 hours

Table 10.5: WHO's Guidelines for Drinking Water Quality, 2006, are the International Reference Point for Standard Setting and Drinking Water Safety

Element/ Substance	*Symbol/ Formula*	*Normally Found in Freshwater/Surface Water/ Groundwater*	*Health Based Guideline by the WHO*
Aluminium	Al		0.2 mg/l
Ammonia	NH_4	< 0.2 mg/l (up to 0.3 mg/l in anaerobic waters)	No guideline
Antimony	Sb	< 4 μg/l	0.02 mg/l
Arsenic	As		0.01 mg/l
Asbestos			No guideline
Barium	Ba		0.7 mg/l
Berillium	Be	< 1 μg/l	No guideline
Boron	B	< 1 μg/l	0.5 mg/l
Cadmium	Cd	< 1 μg/l	0.003 mg/l
Chloride	Cl		No guideline
Chromium	Cr^{+3}. Cr^{+6}	< 2 μg/l	0.05 mg/l

Contd...

Table 10.5–Contd...

Element/ Substance	*Symbol/ Formula*	*Normally Found in Freshwater/Surface Water/ Groundwater*	*Health Based Guideline by the WHO*
Colour			Not mentioned
Copper	Cu		2 mg/l
Cyanide	CN		0.07 mg/l
Dissolved oxygen	O_2		No guideline
Fluoride	F	< 1.5 mg/l (up to 10)	1.5 mg/l
Hardness	mg/l $CaCO_3$		No guideline
Hydrogen sulfide	H_2S		No guideline
Iron	Fe	0.5–50 mg/l	No guideline
Lead	Pb		0.01 mg/l
Manganese	Mn		0.4 mg/l
Mercury	Hg	< 0.5 μg/l	0.006 mg/l
Molybdenum	Mb	< 0.01 mg/l	0.07 mg/l
Nickel	Ni	< 0.02 mg/l	0.07 mg/l
Nitrate and nitrite	NO_3. NO_2		50 mg/l and 3 mg/l
Turbidity			Not mentioned
pH			No guideline
Selenium	Se	< < 0.01 mg/l	0.01 mg/l
Silver	Ag	5 – 50 μg/l	No guideline
Sodium	Na	< 20 mg/l	No guideline
Sulfate	SO_4		No guideline
Inorganic tin	Sn		Not mentioned
TDS			No guideline
Uranium	U		0.015 mg/l
Zinc	Zn		No guideline

Source: www.who.in.

Table 10.6: Drinking Water Quality Standards

Parameters	*BIS*	*CPCB*	*UPPCB*	*WHO*
Temperature	40 °C	±5°C than receiving water	40 °C	-
SS	100 mg/l	100 mg/l	100 mg/l	5-25 mg/l
pH	5.5-9.0	5.5-9.0	5.5-9.0	7.0-8.5
BOD_5 at 20°C	30 mg/l	30 mg/l	30 mg/l	-
COD	250 mg/l	250 mg/l	250 mg/l	-
Cl	-	-	-	75-200 mg/l

Contd...

Table 10.6–Contd...

Parameters	*BIS*	*CPCB*	*UPPCB*	*WHO*
Ca	-	-	-	75-200 mg/l
Mg	-	-	-	30-150 mg/l
Cd	2.0 mg/l	2.0 mg/l	2.0 mg/l	0.1 mg/l
Cu	3.0 mg/l	3.0 mg/l	3.0 mg/l	0.05-1.5 mg/l
Fe	-	3.0 mg/l	-	0.1-1.0 mg/l
Pb	0.1 mg/l	0.1 mg/l	0.1 mg/l	0.1 mg/l
Mn	-	2.0 mg/l	-	0.05-0.5 mg/l
Ni	2.0 mg/l	3.0 mg/l	2.0 mg/l	-
Zn	5.0 mg/l	5.0 mg/l	5.0 mg/l	5-15.0 mg/l

Table 10.7: Recent Standards for Heavy Metals (in mg/l)

	Inland Surface Water		*Drinking water*		
	BIS /CPCB	*WHO*	*WHO, 1993*	*WHO, 2006*	
			Drinking Water	*Normal*	*Health Based*
Cd	2.0	0.1	0.003-0.005	<1µg/l	0.003
Cu	3.0	0.05-1.5	2.0	-	2.0
Fe	3.0	0.1-1.0	0.2	0.5-50	-
Mn	2.0	0.05-0.5	0.5-0.05	-	0.4
Ni	2.0	-	0.02	0.02	0.07
Pb	0.1	0.1	0.01	-	0.01
Zn	5.0	5.0-15.0	3.0	<20	-

Source: www.lenntech.com

Chapter 11

National Water Policy

Land, water and energy are the basic need for any development of the region. Out of this water has become the matter of concern in almost all the countries. The central theme in the "decade of 80" was water, because of the major droughts faced by many countries. This period has paradoxically coincided with the International Decade of drinking water. The rapidity with which the water crisis occupied the center stage of Indian planning warns that water resources should not be taken for granted any more.

Water is not automatically available because the ecological process that gives stability to the water cycle governs its renewability. The seasonal rains in most parts of India, lasting hardly for 90 days, have made the agriculture and social life adapt to seasonability. Frequent droughts have made it necessary to employ various water conservation methods, including traditional ones.

The water logging and subsequent salinity problems which have become common phenomenon in most of the surface irrigation project, the increasing concentration of nitrates, phosphates, fluoride and potash in intensely cultivated areas and major urban areas point out that groundwater management and conservation of groundwater as a precious national resources should be taken up seriously by the planners and by the general public. Otherwise, the unplanned exploitation coupled with contamination and salinity ingress would cause damage to the entire groundwater system. Such damage has already been noticed in certain coastal areas and in parts of the Lakshadweep and Andman-Nicobar Islands in the Arabian Sea and Bay of Bengal Islands respectively.

11.1 Water Demand

India is the wettest country in the world, should consider itself singularly blessed as it is one of the few countries which is self sufficient in water resources. It receives an average yearly rainfall of 1170 mm while central west America which is supposed to be the food bowl of the world, gets a more 200 mm annual rainfall.

It is estimated that in our country about 19 million kiloliters of fresh water is tappable but we utilize only fraction of this amount. It is estimated that by the turn of the 21st century, we will be able to utilize only about 10.9 million kiloliters of water, that is, just a little more than half the amount available. Out of this available supply, 79.59 per cent will be used for irrigation, 13.75 per cent for the production for electric power, 3.2 per cent in industries and 3.47 per cent for domestic purposes.

In our country, piped or organized water supply was given a thrust in 1970s and in the beginning three metropolitan cities "Calcutta, Bombay and Madras (now Kolkata, Mumbai, and Chennai) were covered". Still by 1991, about 44 per cent of the population had access to piped water supply. Out of the beneficiaries so covered 71 per cent live in the urban areas and 29 per cent in rural areas. If the current trend continues for another three years, India will suffer an acute drinking water crisis. The present crisis is due to mismanagement in the storage of rainwater. At present India utilizes only about 10 per cent of the available rainfall and, according to some projections at the current rate we will use a maximum of 25 per cent of available annual rainfall in the next 30 years. About one-third of the villages in the country have been declared water deficient, the main reason for within is rapid industrialization. In the last 10 years, the urban population has risen by 46.2 per cent.

Today, India is not the only country which is facing crisis but there are 26 another countries, collectively home of 230 million people, in this category. The lack of water becomes a severe constraint in food production, economic development and protection of natural systems. In many reasons as demand continue to rise and water supply projects get more difficult to built, water budgets are becoming badly imbalanced. Water that is reliably available to the human population is about 9000 cubic kilometers per year or about 1800 cubic meters per person. This may sound like a lot of water but the available resource are neither distributed evenly nor used efficiently. More than 600 million people live in the world arid and semi-arid areas where they face chronic water scarcity. It is a serious problem in these areas because about 47 per cent of rain fed crop land is affected by desertification.

Africa has the largest number of water scare countries *i.e.* eleven and the Middle East is the world's worst affected region, with nine of its 14 countries facing shortages. Chine with 2 per cent of the world's people and just 8 per cent of its freshwater faces severe constraints. Water labels beneath Beijing have been dropping one of two meters a year, and a third of its wells have gone dry. Some 100 Chinese cities and town, in the northern and coastal regions, have suffered severe shortage in the last few years. Paris of Mexico, Thailand and the western United States also suffer from a declining water table.

11.2 Need for Policy

Water is a prime natural resource, a basic human need and a precious national asset. Planning and development of water resources must take into account interalia, the national water resource position and financial and material aspects of explotation and sustenance of water resources under an integrated national prospective.

It has been estimated that out of the total precipitation of around 400 million hectare meters in the country, the surface water availability is about 178 million hectare meters. Out of this, about 50 per cent can be put to beneficial use because of topographical and other constraints. In addition, there is a groundwater potential of about 42 million hectare meters. The availability of water is highly uneven in terms of both space and time. Precipitation is confined to only about three or four months in the year and varies from 10 cm in the western parts of Rajasthan to over 1000 cm at Cherrapunji in Meghalaya. Further, water does not respect state boundaries. Not merely rivers but even underground aquifers often cut across state boundaries. Water as a resource is one and indivisible: rainfall, river waters, surface ponds and lakes and groundwater are all part of one system. Similarly, water is also a part of a large ecological system.

Flood sand droughts affect vast areas of the country, transcending state boundaries. Almost one third of the country is drought-prone. Floods affect an average area of around 9 million hectares per year. According to the National Commission on Floods, the area susceptible to floods is around 40 million hectares. The approach to the management of drought and floods has to be coordinated and guided at the national level. Even the planning and implementation of individual irrigation of multi-purpose projects, though done at the State level, involve a number of aspects and issues such as environmental protection, rehabilitation of project affected people and livestock, public health consequences of water impoundment, dam safety, etc.

On these matters integrated approaches and guidelines are necessary. Moreover certain problems and weaknesses have affected a large number of projects all over the country. There have been substantial time and cost overruns on projects. In some irrigation commands, problems of water-logging and soil salinity have emerged, leading to the degradation of good agriculture land. There are also complex problems of equity and social justice in regard to water distribution. The development and exploitation of the country's groundwater resources also give rise to questions of judicious and scientific resource management and conservation. All these question need to be tackled on the basis of common policies and strategies. The growth process and the expansion of economic activities inevitably lead to increasing demands for water for diverse purposes like domestic, industrial, agricultural, hydro-power, navigational and recreational use.

Domestic and industrial water needs have largely been concentrated in or near the principal cities, but the demand from rural society his expected to increase substantially as the development programme improve economic conditions in the rural areas. The demand for water for hydro and thermal power generation and for other industrial uses is also likely to increase substantially. As a result water which is already a scare resource will become even most scare in future.

The emphasized the need for the utmost efficiency in water utilization and a public awareness of the importance of its conservation. Another important aspect is water quality. Improvements in existing strategies and the innovation of new techniques based on strong research and developmental activities in science and technology will be needed to eliminate the pollution of surface and groundwater resources, to improve water quality and to set up the recycling and re-use of water. Science and technology and training have also important roles to play in water resources development in general.

Water is one of the most crucial elements in developmental planning. The need for a national water policy is thus abundantly clear that water is a source and precious national resource to be planned, developed and conserved on an integrated and environmentally sustainable basis, keeping in view the needs of the states concerned.

11.3 National Water Policy

As it is necessary to plan and develop this prime natural resource carefully under the given national perspectives, the Ministry of Water Resources drafted the National Water Policy in 1987. A brief outline of the chief contents of the policy is as follows:

1. A standard national information system should be established with a network of data banks to collect information about resource's availability and their utilization.
2. Comprehensive and reliable projections of future demands for water for diverse purposes.
3. Maximum utilization of water resources with the help of conservation augmentation and prevention of losses.
4. Establishment of organizational set up for the planned development and management of a river basin as a whole with the help of appropriate legislations.
5. Transfers of surplus water from one river basin in other having shortage of the same.
6. Recycling and reuse of water.
7. Water resources projects should be multipurpose projects with priority for drinking water use, the other uses being irrigation flood control, power generation, navigational, piscicultural, industrial and recreational.
8. Study of environmental socio-economic impacts of the projects during construction and later stages should be an essential component of project planning. The adverse impact if any should be minimized and off-set by adequate compensatory measures.
9. Integrated multi-disciplinary approach with more attention to needs of tribal and scheduled castes or weaker sections of the populations.
10. Appropriate provision should be made for periodic maintenance of the systems and structures created.
11. Continuous surveillance and monitoring of the system created.

12. Periodic reassessment of groundwater resources.
13. Periodic quality monitoring in case of surface water and groundwater.
14. Strict control on groundwater exploitation to ensure social equity and to avoid adverse effects like salinity ingress.
15. Integrated and co-ordinated development of surface and groundwater their conjunctive use.
16. Water rates should be such as to convey the scarcity value of the resources to the users and to foster the motivation for economy in water use.
17. Efforts should be made to involve people in various aspects of management especially distribution of water and collection water rates.
18. Water zoning: Industrial, agricultural and urban development should be planned on the basis of water availability in a particular zone.
19. Drought prone areas should be made less vulnerable to nature's vagaries through:
    a) Soil moisture conservation measures;
    b) Rain water harvesting;
    c) Minimization of evaporation losses; and
    d) Full development of groundwater and surface water potential.
20. Adequate training facilities for personnel engaged/involved in the discipline of water management.

Base on the National Water Policy document, the conference of "State Secretaries connected with groundwater exploitation work and concerned officials from the Government of India", gave recommendations as follows (June 1989, New Delhi):

1. Establishment of block wise and district wise computerized data storage and retrieval system.
2. Periodic assessment of groundwater potential.
3. Continuous monitoring of water quality and mapping of protection zones were considered to be of paramount importance for urban, industrial and agricultural areas with intensive chemical inputs. The States should arrange follow up action in areas identified as potentially hazardous from groundwater pollution angle.
4. The designs of groundwater structures should be on the basis of the recommendations of the District Development Plans prepared by Central Groundwater Board (which is the apex Government organization in the field of groundwater management commonly known as CGWB), so that they could withstand the vagaries of the monsoon.
5. Based on predicted level of the monsoon rainfall, management of the total resource would be worked out to ensure fulfillment of the needs of domestic water supply followed by needs of agricultural production.
6. Area specific methodologies would be developed for artificial recharge of groundwater. These projects would be prioritized and implemented in a phased manner.

Chapter 12

The Case Study of Panipat at a Glance

In the modern industrial era, rapidly growing of industrialization and urbanization has posed adverse impacts on the environment and its all components. Air, soil and water everything on earth surface may changed its quality due to the increasing pollution. The effluents have consisting high concentrations of heavy metals, organic pollutants and toxic colours, may alter the water quality of the region. Textile industries consumed substantial volume of water and chemicals in wet processing of textile.

Panipat district of Haryana was selected on the basis of existence of a large number of textile industries, those results in huge amount of effluents. Panipat city is well known for the textile industries. Many dye units have come up along the both side of a common seasonal drainage (*Nala*) nearby Binjhole village is located in the south-west of Panipat city. Due to the effluent being discharge by textile industries, water used to dry down in the *Nala* itself but with the dramatic rise in the dying units, huge quantities of the effluents have transported through this *Nala* that ultimately joins the village pond.

A pond (12 ha), attached with a seasonal drain is situated near Binjhole village and collected complete treated and non-treated effluents from the textile industries and being regularly fed by the effluents from the dyeing industries. Samples were collected monthly in morning hours (from 9.00 am to 11.00 am) from different sampling sites *viz.* tube wells and hand pumps of Binjhole village, pond, textile effluents drain and irrigated agricultural fields. Control site was selected in non-industrial area, 10 km. far from experimental sampling sites. Sampling sites were selected as follows:

(Site-A) Effluent Drain on Jatal Road, (EXP)

(Site-B) Effluent Drain End with Pond, (EXP)

(Site-C) Binjhole Village, (EXP)

(Site-D) Agricultural fields, (EXP) and

(Site-E) Control 10 Km Jatal Road, (C)

The following routine sampling and analysis programme was devised to assess the effect of textile industrial effluent on the quality of the groundwater and soil characteristics:

Collection of samples	:	Monthly [Between 20th -22nd of each month]
Time of sample collection	:	Between 9.00 A.M. to 11.00 A.M.
On site analysis	:	Temperature, pH, DO and free CO_2
Analysis within 24 hours	:	COD and physico-chemical parameters
Analysis within 15 days	:	Heavy metals by AAS
Calculation of data	:	After completion of analytical work

In the present study, water, soil and sediment samples were collected and analyzed for determination of physico-chemical parameters and heavy metals from April 2005 to March 2007 regularly. The results are depicted in the form of table and graphs in previous section of this report. The main aspects of the present study are discussed here for arising any response about the study and its significance. During the study, physico-chemical characteristics and heavy metals in textile effluents, surface (pond) water, groundwater, soils and sediments were found with some fluctuation according to seasons, place and other factors. The present work is supported with total 63 tables of data, 120 figures, 12 reference tables, in total 191 pages.

The summarized findings are given here on for experimental and control sites during year 2005-2007.

12.1 Physico-chemical Parameters of Textile Effluents

Physico-chemical parameters of an environment component provide a basic idea about the general quality and nature of the medium. Physico-chemical parameters of textile industrial effluents were monitored for the quality assessment at preliminary level in which, the temperature of textile industrial effluents was recorded maximum 32.83 ± 2.65 in monsoon 2006 and minimum 25.74 ± 1.04 in winter 2005-06 ranged between 24.8 to 35.7 °C.

The maximum pH value of textile industrial effluents were recorded 10.72 ± 0.75 in monsoon 2006 and minimum 9.65 ± 0.07 in summer 2007 ranged between 9.5 and 11.5 on pH scale. The maximum total solids in textile industrial effluents were recorded 2643.67 ± 220.73 in monsoon 2006 and minimum 2210.40 ± 39.20 in winter 2005-06 ranged between 2183 to 2780 mg/l. Total dissolved solids in textile industrial effluents were recorded maximum 2219.75 ± 55.18 in summer 2006 and minimum 2045.67 ± 184.03 in monsoon 2005 ranged between 1842 to 2298 mg/l. Total suspended solids in textile industrial effluents were recorded maximum 570.67 ± 94.87 in monsoon 2006 and minimum 141.80 ± 86.18 in winter 2005-06 ranged between 49 to 680 mg/l and maximum electric conductivity in textile industrial effluents was recorded 3.45 ± .096 in summer 2006 and minimum 3.196 ± 0.29 in monsoon 2005 ranged between 2.88 to 3.58 mho/cm.

Dissolved oxygen in textile industrial effluents was recorded maximum 3.52 ± 0.48 in winter 2006-07 and minimum 0.87 ± 0.77 in summer 2005 ranged between 0.00 to 4.5 mg/l. Free carbon dioxide in textile industrial effluents was recorded maximum 2.8 ± 0.30 in summer 2005 and minimum 0.38 ± 0.53 in winter 2005-06 ranged between 0.00 to 3.1 mg/l. The biochemical oxygen demand in textile industrial effluents was recorded maximum 170.0 ± 6.25 in summer 2005 and minimum 138.6 ± 16.52 in winter 2005-06 ranged between 122.0 to 188.0 mg/l.

The chemical oxygen demand in textile industrial effluents was recorded maximum 445.0 ± 40.93 in monsoon 2005 and minimum 351.5 ± 14.85 in summer 2007 ranged between 310 to 490 mg/l. Total alkalinity in textile industrial effluents was recorded maximum 889.0 ± 47.88 in monsoon 2006 and minimum 633.60 ± 69.55 in winter 2005-06 ranged between 535 to 946 mg/l.

The concentration of calcium in textile industrial effluents was recorded maximum 27.37 ± 6.78 in monsoon 2006 and minimum 10.58 ± 3.2 in winter 2005-06 ranged between 5.6 to 34.5 mg/l. The concentration of magnesium in textile industrial effluents was recorded maximum 16.07 ± 3.46 in summer 2005 and minimum 4.70 ± 2.12 in summer 2007 ranged between 2.4 to 18.9 mg/l.

The concentration of potassium in textile industrial effluents was recorded maximum 8.83 ± 0.58 in monsoon 2006 and minimum 7.40 ± 0.14 in summer 2007 ranged between 7.0 to 9.5 mg/l. Sodium in textile industrial effluents was recorded maximum 177.0 ± 7.21 in monsoon 2006 and minimum 145.0 ± 4.24 in summer 2007 ranged between 142.0 to 186.0 mg/l. Chloride in textile industrial effluent was recorded maximum 825.67 ± 29.94 in monsoon 2006 and minimum 588.50 ± 9.19 in summer 2007 ranged between 565.0 to 860 mg/l. These textile industrial effluents drained regularly in to a large pond near village Binjhole throughout the year.

12.2 Physico-chemical Parameters of Pond Water

Water quality of Binjhole pond had altered by the input of textile industrial effluent in to pond in many ways. To assess the water quality of surface water of the region village pond was selected. Temperature of pond water was recorded maximum 32.4 ± 2.5 in monsoon 2006 and minimum 25.3 ± 1.2 in winter 2005-06 ranged between 24.0 to 35.2 °C. The maximum pH value of pond water was recorded 8.63 ± 0.15 in summer 2005 and minimum 8.17 ± 0.12 in monsoon 2005 ranged between 8.1 and 8.8 on pH scale. The maximum total solids in pond water were recorded 2365.67 ± 104.18 in monsoon 2006 and minimum 2033.20 ± 30.38 in winter 2005-06 ranged between 2012.0 to 2476.0 mg/l. Total dissolved solids in pond water were recorded maximum 2037.50 ± 44.46 in summer 2006 and minimum 1861.00 ± 102.62 in monsoon 2005 ranged between 1776.0 to 2093.0 mg/l. Total suspended solids in pond water were recorded maximum 487.67 ± 23.0 in monsoon 2006 and minimum 96.80 ± 76.81

in winter 2005-06 ranged between 20.0 to 525.0 mg/l. Maximum electric conductivity in pond water was recorded 3.18 ± 0.07 in summer 2006 and minimum 2.91 ± 0.16 in monsoon 2005 ranged between 2.78 to 3.27 mho/cm.

Dissolved oxygen in pond water was recorded maximum 3.43 ± 0.81 in monsoon 2005 and minimum 1.23 ± 0.25 in summer 2005 ranged between 0.9 to 4.2 mg/l. Free carbon dioxide in pond water was recorded maximum 2.77 ± 0.25 in summer 2005 and minimum 0.50 ± 0.46 in winter 2005-06 ranged between 0.00 to 3.1 mg/l. The biochemical oxygen demand in pond water was recorded maximum 136.0 ± 4.58 in summer 2005 and minimum 99.0 ± 4.24 in summer 2007 ranged between 90.0 to 161.0 mg/l. The chemical oxygen demand in pond water was recorded maximum 359.67 ± 19.65 in summer 2005 and minimum 274.50 ± 4.95 in summer 2007 ranged between 252.0 to 405.0 mg/l. Total alkalinity in pond water was recorded maximum 732.33 ± 50.21 in monsoon 2006 and minimum 563.40 ± 45.11 in winter 2005-06 ranged between 505.0 to 785.0 mg/l.

The concentration of calcium in pond water was recorded maximum 236.7 ± 51.3 in monsoon 2006 and minimum 104.0 ± 11.4 in winter 2005-06 ranged between 80.0 to 280.0 mg/l. The concentration of magnesium in pond water was recorded maximum 141.7 ± 41.0 in summer 2005 and minimum 42.5 ± 16.3 in summer 2007 ranged between 25.0 to 178.0 mg/l.

The concentration of potassium in pond water was recorded maximum 8.43 ± 0.21 in monsoon 2006 and minimum 7.3 ± 0.85 in summer 2005 ranged between 6.5 to 8.6 mg/l. Sodium in pond water was recorded maximum 158.67 ± 8.62 in monsoon 2006 and minimum 128.5 ± 8.62 in summer 2007 ranged between 120.0 to 168.0 mg/l. Chloride in pond water was recorded maximum 770.67 ± 21.13 in monsoon 2006 and minimum 576.50 ± 17.68 in summer 2007 ranged between 512.0 to 793.0 mg/l. Many alterations were found in pond water due to the input of textile industrial effluent in pond.

12.3 Physico-chemical Parameters of Groundwater

Groundwater samples were collected from hand pumps of Binjhole village and tube well of agricultural fields of the region. The water quality of groundwater samples were collected from various depths of water table for regular bio-monitoring. Temperature of groundwater was recorded maximum 31.63 ± 2.5 in monsoon 2006 and minimum 25.5 ± 0.71 in summer 2007 ranged between 25.0 to 34.5 °C. The maximum pH value of groundwater was recorded 7.7 ± 0.1 in summer 2005 and minimum 7.43 ± 0.06 in monsoon 2005 ranged between 7.3 to 7.8 and pH scale. The maximum total solids in groundwater were recorded 1464.67 ± 191.20 in summer 2005 and minimum 782.75 ± 121.72 in winter 2006-07 ranged between 641.0 to 1688.0 mg/l. Total dissolved solids in groundwater were recorded maximum 1331.67 ± 169.74 in summer 2005 and minimum 710.25 ± 110.09 in winter 2006-07 ranged between 612.0 to 1537.0 mg/l. Total suspended solids in groundwater were recorded maximum 133.0 ± 21.52 in summer 2005 and minimum 72.50 ± 11.73 in winter 2006-07 ranged between 62.0 to 155.0 mg/l. Maximum electric conductivity in groundwater was recorded 2.08 ± 0.265 in summer 2005 and minimum 1.11 ± 0.17 in winter 2006-07 ranged between 0.90 to 2.40 mho/cm.

Dissolved oxygen in groundwater was recorded maximum 4.0 ± 0.45 in monsoon 2006 and minimum 2.57 ± 0.21 in summer 2005 ranged between 2.4 to 4.8 mg/l. Free carbon dioxide in groundwater was recorded maximum 0.93 ± 0.15 in summer 2005 and minimum 0.10 ± 0.20 in winter 2006-07 ranged between 0.00 to 1.1 mg/l. The biochemical oxygen demand in groundwater was recorded maximum 35.67 ± 8.15 in summer 2005 and minimum 11.00 ± 2.82 in summer 2007 ranged between 5.0 to 57.0 mg/l. The chemical oxygen demand in groundwater was recorded maximum 93.00 ± 16.52 in summer 2005 and minimum 27.50 ± 7.78 in summer 2007 ranged between 12.0 to 130.0 mg/l. Total alkalinity in groundwater was recorded maximum 596.67 ± 48.91 in monsoon 2006 and minimum 493.60 ± 31.1 in winter 2005-06 ranged between 448.0 to 653.0 mg/l.

The concentration of calcium in groundwater was recorded maximum 143.3 ± 47.3 in monsoon 2006 and minimum 46.0 ± 16.7 in winter 2005-06 ranged between 20.0 to 180.0 mg/l. The concentration of magnesium in groundwater was recorded maximum 97.7 ± 35.2 in monsoon 2006 and minimum 14.02 ± 22.0 in summer 2007 ranged between 15.0 to 141.0 mg/l. The concentrations of potassium in groundwater was recorded maximum 6.10 ± 0.10 in monsoon 2005 and minimum 4.90 ± 0.14 in summer 2007 ranged between 4.8 to 6.2 mg/l. Sodium in groundwater was recorded maximum 97.67 ± 11.24 in monsoon 2006 and minimum 51.00 ± 13.45 in winter 2005-06 ranged between 36.0 to 110.0 mg/l. Chloride in groundwater was recorded maximum 151.33 ± 17.95 in monsoon 2006 and minimum 47.50 ± 3.54 in summer 2007 ranged between 24.0 to 172.0 mg/l. Potable water quality monitoring is essential for prevention and control of drinking water pollution problems and health hazards in human beings and livestock in the region.

12.4 Physico-chemical Parameters of Drain Sediment

Drain sediment samples were collected from bottom sediment of textile effluent drain, for the evaluation of sediment soil quality and contamination potential. Temperature of drain sediment was recorded maximum 24.93 ± 3.49 in summer 2005 and minimum 15.46 ± 3.13 in winter 2005-06 ranged between 10.5 to 28.1 °C. Water holding capacity of drain sediment was recorded maximum 31.71 ± 0.51 in summer 2005 and minimum 24.50 ± 5.11 in monsoon 2005 ranged between 20.23 to 32.25 per cent. The bulk density of drain sediment was recorded maximum 1.45 ± 0.07 in summer 2005 and minimum 1.33 ± 0.03 in winter 2005-06 ranged between 1.29 to 1.51 g/cm^3. The soil moisture of drain sediment was recorded maximum 20.85 ± 2.25 in monsoon 2006 and minimum 13.33 ± 1.15 in summer 2005 ranged between 12.5 to 22.3 per cent .

pH of drain sediment on pH scale was recorded maximum 8.32 ± 0.04 in monsoon 2006 and minimum 7.84 ± 0.12 in summer 2005 ranged between 7.7 to 8.5. Electric conductivity of drain sediment was recorded maximum 32.50 ± 0.89 in summer 2005 and minimum 27.86 ± 0.92 in winter 2005-06 ranged between 26.8 to 33.7 mho/cm. Chloride concentration in drain sediment was recorded maximum 836.67 ± 145.1 in summer 2005 and minimum 609.0 ± 33.94 in summer 2007 ranged between 580.0 to 998.0 µg/g. Potassium concentration in drain sediment was recorded maximum 33.50 ± 0.71 in summer 2007 and minimum 27.33 ± 1.53 in monsoon 2005 ranged between 26.0 to 34.0 µg/g. Sodium concentration in drain sediment was recorded maximum 587.50 ± 41.72 in summer 2007 and minimum 157.0 ± 98.85 in monsoon 2005 ranged between 95.0 to 617.0 µg/g. Drain sediment samples were found highly affected by textile industrial effluent and its pollution load.

12.5 Physico-chemical Parameters of Pond Sediment

Pond sediment samples were collected from bottom sediment of pond near Binjhole village, for the evaluation of sediment soil quality and contamination potential. Temperature of pond sediment was recorded maximum 24.83 ± 3.48 in summer 2005 and minimum 15.35 ± 3.19 in winter 2005-06 ranged between 10.25 to 28.0 °C. Water holding capacity of pond sediment was recorded maximum 36.36 ± 0.77 in summer 2005 and minimum 29.52 ± 4.99 in monsoon 2005 ranged between 25.33 to 37.75 per cent . The bulk density of pond sediment was recorded maximum 1.49 ± 0.05 in monsoon 2006 and minimum 1.36 ± 0.03 in winter 2005-06 ranged between 1.32 to 1.58 g/cm^3. The soil moisture of pond sediment was recorded maximum 20.94 ± 2.37 in monsoon 2006 and minimum 13.54 ± 1.14 in summer 2005 ranged between 12.8 to 22.5 per cent .

pH of pond sediment on pH scale was recorded maximum 8.37 ± 0.08 in winter 2005-06 and minimum 7.92 ± 0.09 in summer 2005 ranged between 7.82 to 8.45 on pH scale. Electric conductivity of pond sediment was recorded maximum 31.93 ± 0.81 in summer 2005 and minimum 27.30 ± 0.91 in winter 2005-06 ranged between 26.3 to 33.0 mho/cm. Chloride concentration in pond sediment was recorded maximum 827.0 ± 147.55 in summer 2005 and minimum 588.50 ± 10.61 in summer 2007 ranged between 561.0 to 991.0 µg/g. Potassium concentration in pond sediment was recorded maximum 33.50 ± 0.71 in summer 2007 and minimum 28.33 ± 1.53 in monsoon 2005 ranged between 27.0 to 34.0 µg/g. Sodium concentration in pond sediment was recorded maximum 556.00 ± 32.53 in summer 2007 and minimum 143.67 ± 93.0 in monsoon 2005 ranged between 87.0 to 589.0 µg/g. Pond sediment samples were found highly affected by textile industrial effluent and its pollution load.

12.6 Physico-chemical Parameters of Agriculture Soils

Agricultural soil samples were collected from agricultural fields of the region for the assessment of soil quality and impact of textile industrial effluent and contaminated groundwater. Temperature of agricultural soil was recorded maximum 27.87 ± 3.56 in summer 2005 and minimum 17.40 ± 3.92 in winter 2005-06 ranged between 11.3 to 31.5 °C. Water holding capacity of agricultural soil was recorded maximum 43.75 ± 0.49 in summer 2005 and minimum 36.13 ± 4.92 in monsoon 2005 ranged between 31.58 to 44.35 per cent . The bulk density of agricultural soil was recorded maximum 1.24 ± 0.04 in monsoon 2006 and minimum 1.15 ± 0.03 in summer 2005 ranged between 1.08 to 1.31 g/cm^3. The soil moisture of agricultural soil was recorded maximum 10.76 ± 2.26 in monsoon 2006 and minimum 3.52 ± 1.04 in summer 2005 ranged between 2.75 to 12.2 per cent .

pH of agricultural soil on pH scale was recorded maximum 7.63 ± 0.15 in monsoon 2006 and minimum 7.11 ± 0.01 in summer 2007 ranged between 7.1 to 7.9.

Electrical conductivity of agricultural soil was recorded maximum 27.73 ± 1.66 in monsoon 2006 and minimum 24.18 ± 1.12 in winter 2005-06 ranged between 22.3 to 30.0 mho/cm. Chloride concentration in agricultural soil was recorded maximum 734.33 ± 112.01 in summer 2005 and minimum 499.50 ± 37.47 in summer 2007 ranged between 416.0 to 871.0 µg/g. Potassium concentrations in agricultural soil was recorded maximum 36.50 ± 0.71 in summer 2007 and minimum 31.00 ± 1.00 in monsoon 2005 ranged between 30.0 to 37.0 µg/g. Sodium concentrations in agricultural soil was recorded maximum 599.0 ± 36.77 in summer 2007 and minimum 192.0 ± 91.99 in monsoon 2005 ranged between 133.0 to 665.0 µg/g. some alterations were found in agricultural soil quality during the study.

12.7 Heavy Metals in Textile Effluent

Generally, textile industries used synthetic dyes, which contain toxic metals and the dye residue remains in outlet wastewater of dye houses. So, for the assessment of heavy metals concentrations in the textile industrial effluent, effluent samples were collected from common effluent drain of textile industrial area. The cadmium concentration in textile effluent was recorded maximum 0.052 ± 0.042 in summer 2005 and minimum 0.005 ± 0.004 in winter 2006-07 ranged between 0.001 to 0.1 ppm. The concentration of copper in textile effluent was recorded maximum 0.44 ± 0.085 in summer 2005 and minimum 0.165 ± 0.066 in winter 2005-06 ranged between 0.1 to 0.49 ppm. The concentration of iron in textile effluent was recorded maximum 0.543 ± 0.17 in summer 2006 and minimum 0.240 ± 0.10 in winter 2006-07 ranged between 0.12 to 0.78 ppm. The concentration of manganese in textile effluent was recorded maximum 0.310 ± 0.014 in summer 2006 and minimum 0.110 ± 0.02 in monsoon 2006 ranged between 0.09 to 0.32 ppm.

The concentration of nickel in textile effluent was recorded maximum 0.067 ± 0.037 in summer 2006 and minimum 0.007 ± 0.003 in monsoon 2006 ranged between 0.0012 to 0.061 ppm. The concentration of lead in textile effluent was recorded maximum 0.500 ± 0.21 in summer 2005 and minimum 0.250 ± 0.085 in monsoon 2005 ranged between 0.21 to 0.81 ppm. The concentration of zinc in textile effluent was recorded maximum 0.338 ± 0.146 in winter 2006-07 and minimum 0.110 ± 0.012 in monsoon 2005 ranged between 0.09 to 0.51 ppm. Some part of heavy metals present in textile industrial effluent reach in pond water and some quantity of heavy metals may settle down on bottom sediment.

12.8 Heavy Metals in Pond Water

Heavy metals reached in to pond water due to the direct link of effluent drain with pond. So, the pond water quality was assessed for the evaluation of heavy metals concentrations in pond water. The cadmium concentration in pond water was recorded maximum 0.012 ± 0.006 in summer 2006 and minimum 0.001 ± 0.002 in monsoon 2005 ranged between 0.00 to 0.02 ppm. The concentration of copper in pond water was recorded maximum 0.427 ± 0.045 in summer 2005 and minimum 0.245 ± 0.07 in winter 2006-07 ranged between 1.9 to 13.2 ppm. The concentration of iron in pond water was recorded maximum 9.70 ± 3.04 in summer 2005 and minimum 4.23 ± 1.45 in monsoon 2005 ranged between 0.09 to 0.6 ppm. The concentration of manganese in pond water was recorded maximum 0.530 ± 0.061 in summer 2005 and minimum 0.170 ± 0.059 in winter 2006-07 ranged between 0.09 to 0.6 ppm.

The concentration of nickel in pond water was recorded maximum 0.130 ± 0.014 in summer 2006 and minimum 0.047 ± 0.038 in winter 2006-07 ranged between 0.008 to 0.015 ppm. The concentration of lead in pond water was recorded maximum 1.80 ± 0.56 in summer 2005 and minimum 0.503 ± 0.346 in winter 2006-07 ranged between 0.05 to 2.3 ppm. The concentration of zinc in pond water was recorded maximum 195.0 ± 0.007 in summer 2007 and minimum 0.110 ± 0.02 in monsoon 2005 ranged between 0.05 to 0.21 ppm. Heavy metals leached down in to groundwater table through soil profile after some time, however some quantity remain suspended on bottom sediment of pond.

12.9 Heavy Metals in Groundwater

Heavy metals leached down in to groundwater table through soil profile with the time. So, groundwater samples from various depths were collected for the evaluation of groundwater quality and concentrations of heavy metals. The cadmium concentration in groundwater was recorded maximum 0.01 ± 0.002 in summer 2006 and minimum 0.002 ± 0.002 in monsoon 2006 ranged between 0.00 to 0.01 ppm. The concentration of copper in groundwater was recorded maximum 0.443 ± 0.067 in summer 2005 and minimum 0.288 ± 0.093 in winter 2006-07 ranged between 0.137 to 0.58 ppm. The concentration of iron in groundwater was recorded maximum 3.82 ± 0.59 in summer 2006 and minimum 1.50 ± 0.24 in monsoon 2005 ranged between 0.46 to 4.6 ppm. The concentration of manganese in groundwater was recorded maximum 0.550 ± 0.08 in summer 2005 and minimum 0.284 ± 0.078 in winter 2005-06 ranged between 0.19 to 0.63 ppm.

The concentration of nickel in groundwater was recorded maximum 0.153 ± 0.01 in summer 2006 and minimum 0.044 ± 0.035 in winter 2006-07 ranged between 0.009 to 0.16 ppm. The concentration of lead in groundwater was recorded maximum 1.84 ± 0.31 in summer 2005 and minimum 0.850 ± 0.24 in winter 2005-06 ranged between 0.55 to 2.12 ppm. The concentration of zinc in groundwater was recorded maximum 0.215 ± 0.03 in summer 2006 and minimum 0.126 ± 0.051 in winter 2006-07 ranged between 0.06 to 0.25 ppm. Heavy metals were found in high quantity in shallow groundwater of hand pump and remain low in deep groundwater of tube well.

12.10 Heavy Metals in Effluent Drain Sediment

Heavy metals have a tendency to settle down on bottom of water body due to the heavy atomic weight and some quantity of heavy metals settle down on sediment of drain. So, effluent drain sediment samples were collected for the assessment of sediment quality and heavy metals load. The cadmium concentration in drain sediment was recorded maximum 5.78 ± 0.231 in summer 2005 and minimum 4.25 ± 0.544 in winter 2005-06 ranged between 3.54 to 6.12 ppm. The concentration of copper in drain sediment was recorded maximum 54.83 ± 7.33 in summer 2006 and minimum 45.87 ± 5.82 in summer 2007 ranged between 37.85 to 55.36 ppm. The concentration of iron in drain sediment was recorded maximum 8467.67 ± 2448.18 in summer 2005 and minimum 5656.0 ± 1216.72 in monsoon 2005 ranged between 4256.0 to 11235.0 ppm. The concentration of manganese in drain sediment was recorded maximum 1105.00 ± 107.08 in summer 2006 and minimum 791.33 ± 51.73 in monsoon 2005 ranged between 568.0 to 1105.0 ppm.

The concentration of nickel in drain sediment was recorded maximum 75.00 ± 3.0 in summer 2005 and minimum 58.0 ± 3.0 in monsoon 2005 ranged between 49.0 to 79.0 ppm. The concentration of lead in drain sediment was recorded maximum 70.00 ± 4.36 in summer 2005 and minimum 47.20 ± 3.83 in winter 2005-06 ranged between 42.0 to 78.0 ppm. The concentration of zinc in drain sediment was recorded maximum 260.00 ± 83.54 in summer 2005 and minimum 143.50 ± 32.29 in winter 2006-07 ranged between 95.0 to 385.0 ppm. Heavy metals were found in very high quantity in textile industrial effluent drain sediment due to the dye contamination and natural concentrations of heavy metals in earth crust also.

12.11 Heavy Metals in Pond Sediment

With a tendency to settle down on bottom of stagnant water body due to the heavy atomic weight, some quantity of heavy metals settle down on sediment of pond. So, pond sediment samples were collected for the assessment of sediment quality and heavy metals load. The cadmium concentration in pond sediment was recorded maximum 5.72 ± 0.45 in summer 2005 and minimum 4.05 ± 0.59 in winter 2005-06 ranged between 3.26 to 6.32 ppm. The concentration of copper in pond sediment was recorded maximum 52.46 ± 6.51 in summer 2006 and minimum 38.65 ± 9.23 in summer 2007 ranged between 24.45 to 57.21 ppm. The concentration of iron in pond sediment was recorded maximum 6013.40 ± 856.58 in winter 2005-06 and minimum 3803.50 ± 113.84 in summer 2007 ranged between 3542.0 to 8514.0 ppm. The concentration of manganese in pond sediment was recorded maximum 942.00 ± 48.14 in summer 2005 and minimum 605.00 ± 161.71 in monsoon 2006 ranged between 465.0 to 998.0 ppm.

The concentration of nickel in pond sediment was recorded maximum 63.75 ± 11.12 in summer 2006 and minimum 49.0 ± 5.94 in winter 2006-07 ranged between 41.0 to 79.0 ppm. The concentration of lead in pond sediment was recorded maximum 69.00 ± 6.24 in summer 2005 and minimum 44.00 ± 5.39 in winter 2005-06 ranged between 37.0 to 75.0 ppm. The concentration of zinc in pond sediment was recorded maximum 240.00 ± 92.60 in summer 2005 and minimum 116.25 ± 32.76 in winter 2006-07 ranged between 75.0 to 321.0 ppm. Heavy metals were found in very high quantity in pond sediment due to the dye contamination and natural concentrations of heavy metals in earth crust also.

12.12 Heavy Metals in Agriculture Soil

Heavy metals were assessed in agricultural soil for quality monitoring and evaluation of heavy metals input in to agricultural soil system in the vicinity of textile industrial area. The cadmium concentration in agricultural soil was recorded maximum 3.4 ± 0.7 in summer 2006 and minimum 0.70 ± 0.60 in monsoon 2005 ranged between 0.1 to 4.0 ppm. The concentration of copper in agricultural soil was recorded maximum 36.3 ± 7.4 in summer 2005 and minimum 13.4 ± 6.0 in winter

2005-06 ranged between 5.0 to 42.5 ppm. The concentration of iron in agricultural soil was recorded maximum 82.8 ± 18.1 in summer 2006 and minimum 26.8 ± 10.8 in monsoon 2005 ranged between 14.40 to 100.5 ppm. The concentration of manganese in agricultural soil was recorded maximum 13.6 ± 1.4 in summer 2005 and minimum 5.8 ± 1.7 in winter 2006-07 ranged between 2.0 to 15.0 ppm.

The concentration of nickel in agricultural soil was recorded maximum 10.5 ± 0.8 in summer 2005 and minimum 5.3 ± 1.1 in winter 2006-07 ranged between 3.8 to 12.1 ppm. The concentration of lead in agricultural soil was recorded maximum 62.4 ± 10.9 in summer 2005 and minimum 25.1 ± 6.7 in winter 2006-07 ranged between 16.0 to 73.5 ppm. The concentration of zinc in agricultural soil was recorded maximum 19.4 ± 4.7 in monsoon 2006 and minimum 7.7 ± 2.8 in winter 2005-06 ranged between 4.3 to 23.5 ppm. Agricultural soil was found slightly affected by heavy metals due to the direct contact or due to the irrigation by contaminated groundwater.

12.13 Physico-chemical Parameters of Groundwater at Control Site

Groundwater samples were collected from control site from the depth of 80 ft for the comparison of groundwater quality at experimental sites during the study. Temperature of groundwater on control site was recorded maximum 30.60 ± 2.15 in monsoon 2005 and minimum 26.64 ± 0.47 in winter 2005-06 ranged between 26.0 to 32.8 °C. The maximum pH value of groundwater on control site was recorded 7.45 ± 0.07 in summer 2007 and minimum 7.13 ± 0.058 in summer 2005 ranged between 7.1 and 7.6 on pH scale. The maximum total solids in groundwater on control site were recorded 780.50 ± 110.30 in winter 2006-07 and minimum 446.75 ± 76.16 in summer 2006 ranged between 363.0 to 907.0 mg/l. Total dissolved solids in groundwater on control site were recorded maximum 709.50 ± 100.14 in winter 2006-07 and minimum 404.50 ± 69.02 in summer 2006 ranged between 328.0 to 822.0 mg/l. Total suspended solids in groundwater on control site were recorded maximum 71.00 ± 10.17 in winter 2006-07 and minimum 42.25 ± 7.18 in summer 2006 ranged between 35.0 to 85.0 mg/l. Maximum electric conductivity in groundwater on control site was recorded 1.11 ± 0.16 in winter 2006-07 and minimum 0.632 ± 0.11 in summer 2006 ranged between 0.513 to 1.284 mho/cm.

Dissolved oxygen in groundwater on control site was recorded maximum 3.57 ± 0.50 in monsoon 2005 and minimum 2.37 ± 0.51 in summer 2005 ranged between 1.8 to 4.1 mg/l. Free carbon dioxide in groundwater on control site was recorded maximum 0.93 ± 0.15 in summer 2005 and minimum 0.17 ± 0.29 in monsoon 2005 ranged between 0.00 to 1.1 mg/l. The biochemical oxygen demand in groundwater on control site was recorded maximum 15.67 ± 6.11 in monsoon 2006 and minimum 6.60 ± 5.69 in winter 2005-06 ranged between 0.00 to 22.0 mg/l. The chemical oxygen demand in groundwater on control site was recorded maximum 40.67 ± 15.04 in monsoon 2006 and minimum 16.00 ± 13.17 in winter 2005-06 ranged between 0.00 to 57.00 mg/l. Total alkalinity in groundwater on control site was recorded maximum 303.00 ± 22.91 in summer 2005 and minimum 216.40 ± 41.62 in winter 2005-06 ranged between 156.0 to 331.0 mg/l.

The concentration of calcium in groundwater on control site was recorded maximum 110.0 ± 10.0 in monsoon 2005 and minimum 40.0 ± 14.1 in summer 2005 ranged between 30.0 to 120.0 mg/l. The concentration of magnesium in groundwater on control site was recorded maximum 101.0 ± 23.3 in monsoon 2006 and minimum 21.0 ± 4.2 in summer 2007 ranged between 15.0 to 127.0 mg/l. The concentration of potassium in groundwater on control site was recorded maximum 5.80 ± 0.10 in monsoon 2005 and minimum 4.65 ± 0.21 in summer 2007 ranged between 4.4 to 5.9 mg/l. Sodium in groundwater on control site was recorded maximum 58.33 ± 10.26 in monsoon 2005 and minimum 18.50 ± 3.54 in summer 2007 ranged between 9.0 to 67.0 mg/l. Chloride in groundwater on control site was recorded maximum 13.00 ± 3.78 in monsoon 2005 and minimum 4.90 ± 1.11 in summer 2006 ranged between 3.9 to 21.0 mg/l. Physico-chemical parameter of groundwater at control site was found having less pollution load comparatively to groundwater samples from hand pump and tube well of Panipat region even at different depth of water table.

12.14 Physico-chemical Parameter of Soil at Control Site

Soil samples were collected from control site for the comparison of agricultural soil quality at experimental sites during the study. Temperature of soil at control site was recorded maximum 27.83 ± 2.70 in summer 2005 and minimum 17.50 ± 3.48 in winter 2005-06 ranged between 11.8 to 31.0 °C. Water holding capacity of soil at control site was recorded maximum 44.18 ± 0.74 in summer 2005 and minimum 36.81 ± 1.89 in winter 2006-07 ranged between 32.17 to 45.0 per cent. The bulk density of soil at control site was recorded maximum 1.22 ± 0.01 in monsoon 2006 and minimum 1.11 ± 0.051 in summer 2006 ranged between 1.05 to 1.25 g/cm^3. The soil moisture of soil at control site was recorded maximum 11.06 ± 1.86 in monsoon 2006 and minimum 3.73 ± 1.29 in summer 2005 ranged between 2.8 to 12.82 per cent.

pH of soil at control site was recorded maximum 7.47 ± 0.058 in monsoon 2005 and minimum 7.23 ± 0.15 in summer 2005 ranged between 7.1 to 7.6. Electric conductivity of soil at control site was recorded maximum 23.75 ± 0.27 in summer 2006 and minimum 23.00 ± 0.163 in winter 2006-07 ranged between 22.5 to 24.1 mho/cm. Chloride concentration in soil at control site was recorded maximum 529.33 ± 114.47 in summer 2005 and minimum 313.80 ± 64.32 in winter 2005-06 ranged between 225.0 to 671.0 µg/g. Potassium concentration in soil at control site was recorded maximum 34.50 ± 0.71 in monsoon 2005 and minimum 30.33 ± 2.52 in winter 2006-07 ranged between 28.0 to 36.0 µg/g. Sodium concentration in soil at control site was recorded maximum 380.00 ± 12.73 in monsoon 2005 and minimum 95.35 ± 37.01 in winter 2006-07 ranged between 11.0 to 468.0 µg/g. Soil quality of control site was slightly differ from the agricultural soil of Binjhole region.

12.15 Heavy Metals in Groundwater at Control Site

Heavy metals assessment at control site was also necessary to compare with the heavy metals concentrations in groundwater samples collected from hand pump and tube well of Binjhole village from various depth of groundwater table. The cadmium concentration in groundwater on control site was recorded maximum 0.00 ± 0.00 and minimum 0.00 ± 0.00 ranged between 0.00 to 0.0005 ppm. The concentration of copper in groundwater on control site was recorded maximum 0.684 ± 0.80 in summer 2007 and minimum 0.90 ± 0.061 in winter 2005-06 ranged between 0.00 to

0.18 ppm. The concentration of iron in groundwater on control site was recorded maximum 0.868 ± 0.07 in summer 2006 and minimum 0.363 ± 0.128 in winter 2006-07 ranged between 0.18 to 0.99 ppm. The concentration of manganese in groundwater on control site was recorded maximum 0.204 ± 0.03 in summer 2005 and minimum 0.075 ± 0.037 in winter 2006-07 ranged between 0.02 to 0.245 ppm.

The concentration of nickel in groundwater on control site was recorded maximum 0.044 ± 0.011 in summer 2005 and minimum 0.00 ± 0.00 in winter 2005-06 ranged between 0.00 to 0.055 ppm. The concentration of lead in groundwater on control site was recorded maximum 0.282 ± 0.013 in summer 2007 and minimum 0.178 ± 0.127 in winter 2006-07 ranged between 0.00 to 0.291 ppm. The concentration of zinc in groundwater on control site was recorded maximum 0.099 ± 0.015 in summer 2005 and minimum 0.043 ± 0.013 in winter 2006-07 ranged between 0.02 to 0.112 ppm. Heavy metals load in groundwater at control site was found very low in comparison to groundwater of experimental sites in Binjhole village's territory.

12.16 Heavy Metals in Soil at Control Site

Estimation of the occurrence of heavy metals in control soil was essential to compare with the agricultural soil of Binjhole village region. The cadmium concentration in soil on control site was recorded maximum 2.23 ± 0.31 in summer 2005 and minimum 0.225 ± 0.21 in winter 2006-07 ranged between 0.00 to 2.80 ppm. The concentration of copper in soil on control site was recorded maximum 27.4 ± 5.7 in summer 2005 and minimum 10.0 ± 4.1 in winter 2005-06 ranged between 3.5 to 32.0 ppm. The concentration of iron in soil on control site was recorded maximum 68.6 ± 17.5 in summer 2005 and minimum 18.4 ± 9.0 in winter 2006-07 ranged between 6.90 to 84.50 ppm. The concentration of manganese in soil on control site was recorded maximum 11.5 ± 1.1 in summer 2005 and minimum 4.8 ± 1.5 in winter 2006-07 ranged between 3.2 to 12.5 ppm.

The concentration of nickel in soil on control site was recorded maximum 8.43 ± 1.2 in summer 2005 and minimum 2.6 ± 1.87 in winter 2006-07 ranged between 0.00 to 9.20 ppm. The concentration of lead in soil on control site was recorded maximum 42.3 ± 3.1 in summer 2005 and minimum 18.25 ± 5.9 in winter 2006-07 ranged between 11.3 to 46.8 ppm. The concentration of zinc in soil on control site was recorded maximum 16.5 ± 3.3 in monsoon 2006 and minimum 7.3 ± 0.8 in summer 2007 ranged between 5.1 to 19.1 ppm. Heavy metals in soil at control site were found slightly low in comparison to agricultural soil of Panipat region.

Transfer Factor (Coefficient)

Transfer factor was measured for evaluation of migration rate of heavy metals from a component to another. Transfer coefficients for cadmium, copper, iron, manganese, nickel, lead and zinc from textile industrial effluent to groundwater, were found maximum 1.50, 1.79, 9.64, 4.12, 14.14, 4.65, 1.21 and minimum 0.12, 0.99, 5.93, 1.27, 1.45, 2.01 and 0.38 respectively. Transfer coefficients for cadmium, copper, iron, manganese, nickel, lead and zinc were found maximum 0.83, 1.53, 25.0, 4.27, 11.76, 3.63, 1.00 and minimum 0.05, 0.97, 15.93, 0.77, 1.64, 1.19 and 0.54 respectively from textile industrial effluent to pond water. From pond water to groundwater of 40 ft depth, transfer coefficients for cadmium, copper, iron, manganese, nickel, lead and zinc were found maximum 2.20, 1.17, 0.40, 1.84, 1.29, 1.91, 1.35 and minimum 0.87, 0.97, 0.34, 0.96, 0.83, 1.02 and 0.70 respectively. From pond water to groundwater of 80 ft depth, transfer coefficients for cadmium, copper, iron, manganese, nickel, lead and zinc were found maximum 1.11, 0.68, 0.39, 0.88, 0.82, 1.05, 1.21 and minimum 0.18, 0.54, 0.12, 0.46, 0.27, 0.48 and 0.43 respectively. From pond water to deep groundwater of 120 ft depth, transfer coefficients for cadmium, copper, iron, manganese, nickel, lead and zinc were found maximum 0.13, 0.52, 0.16, 0.75, 0.61, 0.57, 0.74 and minimum 0.00, 0.41, 0.10, 0.40, 0.19, 0.22 and 0.25 respectively.

Transfer coefficients for cadmium, copper, iron, manganese, nickel, lead and zinc were found maximum 7.20, 0.82, 0.22, 0.27, 1.28, 0.36, 1.39 and minimum 2.27, 0.45, 0.17, 0.18, 0.69, 0.26 and 0.40 respectively from groundwater of 40 ft depth to agricultural soil. From groundwater of 80 ft depth to agricultural soil, transfer coefficients for cadmium, copper, iron, manganese, nickel, lead and zinc were found maximum 40.00, 1.48, 0.58, 0.48, 4.18, 0.72, 1.53 and minimum 4.45, 0.88, 0.16, 0.38, 1.07, 0.48 and 0.59 respectively. From deep groundwater of more than 120 ft depth to agricultural soil, transfer coefficients for cadmium, copper, iron, manganese, nickel, lead and zinc were found maximum 100.00, 1.91, 0.79, 0.63, 6.24, 1.55, 2.39 and minimum 22.33, 1.08, 0.40, 0.46, 1.54, 0.88 and 0.78 respectively. From groundwater to soil in control site, transfer coefficients for cadmium, copper, iron, manganese, nickel, lead and zinc were found maximum 670.00, 1.75, 0.80, 0.68, 2.51, 1.59, 2.37 and minimum 114.29, 0.23, 0.33, 0.54, 1.89, 1.00 and 0.84 respectively.

Enrichment Factor

Enrichment factor for heavy metals were measured for the availability of heavy metals in sediment samples in comparison to Iron. In the drain sediment, metal enrichment factor for cadmium, copper, iron, manganese, nickel, lead and zinc were found maximum 0.90, 8.47, 1000, 175.63, 11.70, 10.73, 38.80 and minimum 0.53, 5.74, 1000, 109.92, 7.35, 5.83 and 18.12 respectively.

Metal enrichment factor for cadmium, copper, iron, manganese, nickel, lead and zinc were found maximum 1.26, 12.04, 1000, 243.85, 14.85, 14.23, 47.57 and minimum 0.67, 7.61, 1000, 139.09, 8.38, 7.32 and 19.89 respectively in the pond sediment.

Geo-accumulation Index

Geo-accumulation index for heavy metals was calculated for the evaluation of accumulation level of heavy metals in the sediment samples during the study. Geo-accumulation index and values for highest values of cadmium, copper, iron, manganese, nickel, lead and zinc in geo-accumulation index were found maximum 12.85, 0.81, 0.12, 0.87, 0.74, 2.33, 1.82 and minimum 3.24, 0.68, 0.08, 0.62, 0.57, 1.57 and 1.01 respectively in drain sediment.

Rank of cadmium in geo-accumulation index was found maximum 3.68 in summer 2005 and minimum 3.24 in winter 2005-06. For copper, iron, manganese and nickel geo-accumulation index rank was found in negative values below 0.00. For lead geo-accumulation index rank was found maximum 1.22 in summer 2005 and minimum 0.65 in winter 2005-06. Geo-accumulation index rank for zinc was maximum 0.87 in summer 2005 and minimum 0.01 in winter 2006-07. While for highest concentrations of heavy metals, highest values for cadmium, copper, iron, manganese, nickel, lead and zinc was found 13.60, 0.92, 0.16, 0.98, 0.77, 2.60 and 2.70 respectively. Geo-accumulation index rank was found 3.77, 1.38 and 1.43 for cadmium, lead and zinc respectively.

The values of cadmium, copper, iron, manganese, nickel, lead and zinc in geo-accumulation index were found maximum 12.70, 0.78, 0.08, 0.74, 0.63, 2.30, 1.68 and minimum 9.01, 0.57, 0.05, 0.47, 0.48, 1.47, 0.82 respectively for highest values in pond sediment.

Rank of cadmium in geo-accumulation index was found maximum 3.67 in summer 2005 and minimum 3.17 in winter 2005-06. For copper, iron, manganese and nickel geo-accumulation index rank was found in negative values below 0.00 For lead geo-accumulation index rank was found maximum 1.20 in summer 2005 and minimum 0.55 in winter 2005-06. Geo-accumulation index rank for zinc was maximum 0.75 in summer 2005 and minimum 0.00 in monsoon 2005. While for highest concentrations of heavy metals, highest values for cadmium, copper, iron, manganese, nickel, lead and zinc was found 14.04, 0.85, 0.11, 0.78, 0.77, 2.50 and 2.25 respectively. Geo-accumulation index rank was found 3.81, 1.32 and 1.17 for cadmium, lead and zinc respectively. Mostly heavy metals were found in the safe zone of geo-accumulation index.

Metal Accumulation Factor

Metal accumulation factor was calculated for the evaluation of heavy metal accumulation in groundwater and soil environment. In the present study of groundwater, metal accumulation factors for cadmium, copper, iron, manganese, nickel, lead and zinc were found maximum 250.00, 3.28, 4.41, 4.17, 7.14, 6.84, 3.00 and minimum 57.14, 0.53, 2.64, 2.65, 3.36, 4.45 and 2.02 respectively.

Metal accumulation factor for cadmium, copper, iron, manganese, nickel, lead and zinc found maximum 8.56, 3.15, 1.49, 1.31, 2.05, 1.48, 1.20 and minimum 0.91, 1.31, 1.12, 1.11, 1.17, 1.28 and 1.02 respectively in soil system.

In the present study, heavy metals in pond water were observed slightly lower than textile industrial effluents, whereas the heavy metals concentrations in effluent drain were higher than pond sediment. Shallow groundwater of about 40 ft depth was highly polluted in comparison to increasing depth of water table. Although the effect of spatial variation was also observed in the study, when the heavy metals concentrations in groundwater and soils at control site were presented comparatively in less or very less quantities. Some metals like cadmium and nickel were not detected in groundwater at control site. It is concluded that in the textile industrial area of Panipat region, there was heavy metals pollution in pond water and sediment, which affected the groundwater of the region. Beside this, the bottom sediment of common effluent drain stored the heavy metals in high quantities, which have no secure or pre-managed future.

Thus, textile industrial effluents discharging from dye houses contaminated the surface water, drain sediment, pond sediment and groundwater also. Furthermore,

Table 12.1: Summarized Data for Heavy Metals

Metal (ppm)	*Textile Effluent (Min-Max.)*	*Pond Water (Min-Max.)*	*Groundwater (Min-Max.)*	*Drain Sediment (Min-Max.)*	*Pond Sediment (Min-Max.)*	*Agriculture Soil (Min-Max.)*	*Control Groundwater (Min-Max.)*	*Control Soil (Min-Max.)*
Cu	0.001-0.1	0.0-0.02	0.0-0.01	3.54-6.1	3.26-6.32	0.1-4.0	0.0-0.0005	0.0-2.8
Cd	0.1-0.49	0.24-4.3	0.14-0.58	37.8-55.4	24.5-57.2	5.0-42.5	0.0-0.18	3.5-32.0
Fe	0.12-0.78	1.9-13.2	0.46-4.6	4256-11235	3542-8574	14.4-100.5	0.18-0.99	6.9-84.5
Mn	0.09-0.32	0.09-0.6	0.19-0.63	568-1105	465-998	2.0-15.0	0.02-0.25	3.2-12.5
Ni	0.001-0.06	0.008-0.015	0.009-0.16	49.0-79.0	41.0-79.0	3.8-12.1	0.0-0.055	0.0-9.2
Pb	0.21-0.81	0.05-2.3	0.55-2.1	42.0-78.0	37.0-75.0	16.0-73.5	0.0-0.29	11.3-46.8
Zn	0.09-0.51	0.05-0.21	0.06-0.25	95-385	75-321	4.3-23.5	0.02-0.11	5.1-19.1

the impact of heavy metals reflected negative impacts on agricultural soil after irrigation by contaminated groundwater. Heavy metals from textile industries affected the shallow groundwater quality in shallow badly, while deep groundwater affected slightly. Vertically, heavy metals contamination decreased with the increasing depth of water table. In the case of horizontal distribution, the heavy metal concentrations decrease with increasing distance from pollution source. In the circle of 1 Km from pollution source almost heavy metals level was found higher than the outer wide circle in the industrial area. Heavy metals contamination was found to be low in soil at control site and also in outer circle of textile industrial area. However, the area of around 1 Km from textile industrial area was found highly affected by heavy metals in Panipat region.

Textile industrial effluent or contaminated groundwater has been altered the physico-chemical parameters of agricultural soil. It was indicated that the comparison of agricultural soil to control soil that some characteristics were influenced by contaminated irrigated groundwater. Heavy metals in agricultural soil were found slightly higher than control soil. Due to the regular discharging of metals residues in the textile industrial area, a threat have been appear on plant vegetation, livestock and human beings also of the region. So, industrialists, municipality, government, NGOs and local people also at various levels for the protection of environmental components should adopt some appropriate measures for the amendment and control of textile pollution.

Bibliography

Aboulhassan, M.A.; Souabi, S.; Yaacoubi, A. and Baudu, M. (2005): Treatment of textile wastewater using a natural flocculants, *Environmental Technology*. 26: 705-711.

Abson, J.N.; Furness, C.D. and Howe, C. (1967): Development of the simicar resipirometer and its application to waste treatment, *Waste Poll. Control*, 66: 607.

Aggarwal, R.R. and Mehrotra, R.R. (1952): Soil survey and soil water in Uttar Pradesh, *Suptd. Printing and Stationery*, Aallahabad, pp: 361-372.

Ajmal, M. and Khan, A.U. (1985): Effects of a textile factory effluent on soil and crop plants, *J. Env. Poll.*, 37(2): 131-137.

Al-Degs, Y.; Khraisheh, M.A.M.; Allen, S.J. and Ahmed, M.N. (2000): Effect of carbon surface chemistry on the removal of reactive dyes from textile effluent, *Water Research*, 34(3): 927-935.

Alumaa, P.; Steinnes, E.; Kirso, U. and Petersell, V. (2001): Heavy metal sorption by different Estonian soil types at low equilibrium solution concentrations, *Proc. Estonian Acad. Sci. Chem.*, 50(2): 104–115.

Ameta Suresh C., Punjabi Pinki Bala, Kothari Shilpa, Sancheti Anjali (2003): Effect of untreated and photocatalytically treated dyeing industry effluent on growth and biochemical parameters of *Allium cepa* (Onion), *Poll. Res.*, 22(3): 389-392.

Annadurai, G.; Juang, R.S. and Lee, D.J. (2002): Use of cellulose based wastes for adsorption of dyes from aqueous solutions, *J. Haz. Mat.*, B92: 263-274.

APHA (1998): Standard methods for the examination of water and wastewater, *American Public Health Association*, 19th edition, 1015, Fifteenth Street, NW, pp: (1-1) – (10-150).

Aslam, M.M.; Baig, M.A.; Hassan, I.; Qazi, I.A.; Malik, M. and Saeed, H. (2004): Textile wastewater characterization and reduction of its COD and BOD by oxidation, *Electron. J. Environ. Agri. Food Chem.*, 3(6): 804-811.

Ather, M. and Vohra, S.B. (1985): Heavy metals and environment, *New Age International Publisher Limited, Wiley Eastern Limited, New Delhi*, pp: 635.

Aurangabadkar, K.; Swaminathan, K.; Sandhya, S; Uma, T.S.; Jothikumar, N. and Paramasivam, R. (2001): Groundwater quality around a municipality solid waste dumping site at Chennai, *IJEP*, 21(4): 323-327.

Bachmat, Y. (1994): Groundwater as part of the water system. In: Groundwater contamination and control, ed. V. Zoller. Marcel Dekker, Inc., New York, pp: 1-560.

Bachmat, Y. (1994): Groundwater as part of the water system. In: Groundwater contamination and control, Ed. V. Zoller. *Marcel Dekker, Inc.*, New York, pp: 560.

Bae, W.; S.H. Lee; and G.B. Ko (2004): Evaluation of predominant reaction mechanisms for the Fenton process in textile dyeing waste. *Water Sc. and Tech.* 49 (4): 91-96.

Baruah N.K.; Kotoki, P.; Bhattacharya, K.G. and Borah, G.C. (1996): Metal speciation in Jhanjhi river sediments. *Sci. Total Environ.*, 193: 1-12.

Basu, A.K.; Rao, C.G.S.; Dhaneshwar, G; Mukharjee, M.K.A.; Kutty, D. Ray and Trivedi, R.C. (1973): Study on Hooghly estuaries pollution in the vicinity of tissue mill, *Proc. Symp. on Environ. Poll.*, pp: 24-26.

Beven, K. (2000): On the future of distributed modeling in hydrology, *Hydrological Processes*, 14: 3183-3184.

Bhakuni, T.S. and Bopardikar, M.V. (1967): Method of recovery Zinc from spinning bath waste of Viscos reyon factory by ion exchange process, *Env. Health*, 9(4): 327-338.

Bhargava, G.P.; Singhla, S.K. and Abrol, I.P. (1972): Characteristics of some typical saline sodic soils occurring in Karnal district Haryana state, Report No. 2, Division of soils and agronomy; *CSSRI, Karnal*, pp: 35.

Bharti, P.K. (2007): Effect of textile industrial effluents on groundwater and soil quality in Panipat region (Haryana), Thesis submitted to Gurukula Kangri University, Hardwar, pp. 191.

Bhat, R. and Kulkarni, R. (2005): COD reduction of dye industry effluent, *J. Ind. Poll. Control*, 21 (1): 147-150.

Bordas, S. and Bourg, C.M. (1998): A critical evaluation of sample pretreatment of contaminated sediments to be investigated for the potential mobility of their heavy metal load, *Water, Air and Soil Pollution*, 103: 137-149.

Bousselmi, L.; Gelsson S.V. and H. Schroeder (2004): Textile wastewater treatment and reuse by solar catalysis: results from a pilot plant in Tunisia, *Water Sc. and Tech.*, 49(4): 331-337.

Brady, N.C. (1995): The nature and properties of soils, Prentice Hall of India Pvt. Ltd., New Delhi, pp: 621.

Cances, B.; Ponthieu, M.; Castrec-Rouelle, M.; Aubry, E. and Benedatty, M.F. (2003): Metal ion speciation in a soil and its solution: Experimental data and modal results, *Geoderma*, 113: 341-355.

Canter, L.W. (1987): Groundwater quality protection, *Lewis publications*. Inc., Chelsea, Ml, pp: 1-650.

Castro Dentros, T.S. Dentos Neto, A.A. Moura, M.C., P.A. Barros Neto, E$.I., Forte K.R., Leite, R.H.L. (2003): Heavy metals extraction by micromulsion, *Wat. Res.*, 37: 2709-2717.

Chao, Keh-Ping; Tsai, Ching-Tsan; Wang, John H.C.; Lin, Shaw-Tao and Chiang, Chow-Feng (2005) Health risk assessment of a heavy metal contaminated site in Taiwan. *Jr. Practice Periodical of Hazardous, Toxic, and Radioactive Waste Management*, pp: 167-171.

Christie, S.L. (1995): Speciation, In: Fitfield, F.W., Haines. P.J. (Eds). *Environmental Analytical Chemistry*, Chapman and Hall, UK, pp: 245-273.

Chung, K.T. (1983): The significance of azo-reduction in the mutagenesis and carcinogenesis of azo dyes, *Mutation Research*, 114: 269-279.

Considine, D.M. (1974): Chemical and process technology encyclopedia, *McGraw Hill Book Co.*, New York, pp: 655.

CPCB (2003): Parivesh: Groundwater, *Central Pollution Control Board*, Delhi-32, July, pp: 40.

Darra, B.L.; Mehta, K.M. and Pareek, B.L. (1964): Quality of irrigation water in Rajasthan, *J. Indian Soc, Soil Sc.*, 12: 121-130.

Dash, S.K. and Sahoo, H.K. (1998): Quality assessment of groundwater in a part of Sundargarh district, *Indian J. Environmental Protection*, 19(4): 273-278.

Davidson, C.M., Thomas R.P., McVey, S.E., Perala, R., Littlejohn, D. and Ure, A.M. (1994): Evaluation of a sequential extraction procedure for the speciation of heavy metals in sediments. *Anal.Chim. Acta.*, 291: 277-286.

Dayama, O.P. (1987): Influence of dyeing and textile water pollution on nodulation and germination of gram, *J. Acta. Ecol.*, 9(1-2): 34-37.

De, A.K. (2002): Environmental Chemistry, *New Age International (P) Limited Publishers*, pp: 392.

De, A.K. (2005): Adsorption of cadmium and zinc on coal fly ash, *J. Ind. Poll. Control*, 21 (1): 27-30.

Desh, D. (1981): Activated sludge treatment of liquid waste- A review, *J. IAWPC Tech.*, Vol-VII, pp: 69-72.

Dhar, N.R.; Khoda, A.K.M.B.; Khan, A.H.; Bala, P. and Karim, M.F. (2005) A study of effect of acid activated saw dust on the removal of different dissolved tannery dyes (Acid dye) from aqueous solutions. *Journal of Environ. Sci. and Engg*. 47 (2): 103-108.

Doneen, L.D. (1954): Stalinization of soils of salts in irrigation waters, *Trans. Geophys. Union*, 35: 943-952.

Dowedy, R.H. and Larson, W.E. (1975): The availability of sludge borne metals in various vegetable crops, *J. Env. Qual.*, 4: 278-282.

Dulka, J.J. and Risby, T.H. (1976): Ultra trace metals in some environmental and biological systems, *Analytical Chemistry*, 48(8): 640-653.

Dutta, P.S. (1999): Groundwater situation in Delhi- Red alert, Nuclear research laboratory, *IARI, New Delhi*, pp: 39.

Dutta, P.S. (2005): Groundwater ethics for its sustainability, *Current Science*, 89 (5): 812-817.

Dwivedi, A.K.; Shashi and Singh, J. (2005): Water pollution and groundwater recharging, *Current Science*, 91 (4): 407-408.

Eaton, F.M. (1950): Significance of carbonates in irrigation waters, *Soil Sc.*, 39: 123-133.

ECIL methods manual (2004): Methods Manual, Atomic Absorption Spectrophotometer, AAS 4129, *Electronic Corporation of India Limited*, Hyderabad-500062, pp: 85.

Eckenfelder, W.W. and Bornad, J.L. (1971): Treatment cost relationship for industrial waste, *Chemical Engg. Progress*, 67(9): 76.

Fakayode, S.O. and Onianwa, P.C. (2002): Heavy metal contamination of soil, and bioaccumulation in Guinea grass (*Panicum maximum*) around Ikeja industrial Estate, Lagos, Nigeria, *Environmental Geology*, 43: 145-150.

Fetter, C.W. (1994): Applied Hydrology, 3rd edition, Macmillan, New York.

Florence, T.M. (1982): The speciation of trace elements in waters, *Talanta*, 29: 345-369.

Forstner, U. (1985): chemical forms and reactivities of metals in sediments in chemical methods for assessing bioavailable metals in sludge and soils, In: Leschber, R., Davies, R.D. and Hermite, L.P. (Eds), *Elsevier, London*, pp: 1-30.

Forstner, U. and Wittman, G.T.W. (1981): Metal pollution in the aquatic environment, second ed. *Springer, Berlin*, pp: 486.

Forstner, U.; Ahlf, W.; Calmano, W.; Kersten, M. and Schoer, J. (1990): Assessment of metal mobility in sludge and soild wastes, In: Metal speciation in the environment (Eds.- Broeckaert, J.A.C; Gucer, S. and Adams, F.) Springer, Berlin, pp: 1-41.

Fransson, Asa (2007): A case study to verify methods for estimating transmissivity distributions along boreholes, *Hydrogeology Journal*, 15: 307–313.

Fraser, B.G. and Williams, D.D. (1998): Seasonal boundary dynamics of a groundwater/ surface-water ecotone, *Ecology*, 79(6): 2019-2031.

Frostner, U., and Wittman, G.T.W. (1979) Metal pollution in the aquatic environment, *Berlin Germany, Springer- Verlag.*, pp: 545.

Gardner, W.R. (1957): Some steady state solutions of the unsaturated moisture flow equation with application to evaporation from a water table, *Soil Sci.*, 85: 228-232.

Garg, V. K; Sharma, I.S. and Bishnoi, M.S. (1998): Fluoride in under groundwater of Uklana town, District Hisar, Haryana, *Poll. Res.* 17(2): 149-152.

Garg, V.K. and Kaushik, P. (2006): Influence of short-term irrigation of textile mill wastewater on the growth of chickpea cultivars, *Chemistry and Ecology*, 22(3): 193-200.

Garg, V.K.; Chaudhary, A.; Deepshikha and Dahiya, S. (1998): An appraisal of groundwater quality in some villages of district Jind, *Indian J. Environmental Protection*, 19(4): 267-272.

Gharaibeh, M.A.; Eltaif, N.I. and Bayan Al-Abdullah (2007): Impact of Field Application of Treated Wastewater on Hydraulic Properties of Vertisols, *Water Air Soil Pollut*, DOI- 10.1007/s11270-007-9423-z.

Golia, E.E.; Dimirkou, A. and Mitsios, I.K. (2007): Accumulation of Metals on Tobacco Leaves (Primings) Grown in an Agricultural Area in Relation to Soil, *Bul.l Environ. Contam. Toxicol.*, DOI 10.1007/s00128-007-9111-0.

Goltenboth, F. (1994): Impact of textile factory in Salatiga on Ledok river system, Central Java, *J. Environ. Toxicol.*, South East Asia, pp: 301-317.

Gomez-Serrano, Garcia-Macias, A., Espinosa-Mansilla, A. and Valenzuela-Calahorro (1998): Adsorption of mercury, cadmium and lead from aqueous solution on heat treated and sulphurized activated carbon. *Wat. Res.*, 32:1-4.

GoogleEarth (2006): Software for satellite imagery, available at http://www.google.com.

Gopal, B. (1994): Conservation of inland water in India: An overview. Verh. Internet. Verein. *Hydrobiologia*, 384: 267.

Govindan, V.S. and Sundarlingam, V.S. (1979) Studies on treatment of textile mill wastewater by stabilization pond method. *J. Pollution Control*, 5 (2):137-145.

Gray, C. W. and Mclaren, R. G. (2006): Soil factors affecting heavy metal solubility in some New Zealand soils, *Water, Air and Soil Pollution*, 175: 3-14.

Groff, K.A. (1993): Textile waste, *Environmental Research*. 65: 421.

Gupta, R.K., Singh, R.R. and Abrol, I.P. (1989): Influence of simultaneous changes in sodicity and pH on hydraulic conditions of an alkali soil, *Soil Science*, 147: 28-33.

Harper, H.H.; Yusuf, Y.A. and Wanielista, M.P. (1984): Fate of heavy metals in stromwater management system, In: *Lake and Reservoir Management, USEPA, Washington*, pp: 329-334.

Hart, B.T. (1982): Trace metals in natural waters, *Speciation Chem. Australian*, 49: 260-265.

Hedge, V.S.; Shalini, G. and Kanchanagouri, D.G. (2006): Provenance of heavy minerals with special reference to Ilmenite of the Honnevar beach, central west coast of India, *Current Science*, 91 (5): 644-648.

Helena, B.A.; Vega, M.; Barrado, E.; Parado, R. and Fernandez, L. (1999): A case of hydro-chemical characterization of an alluvial aquifer influenced by human activities, *Water, Air and Soil Pollution*, 112: 365-387.

Hesse, P.R. (1994): A textbook of soil chemical analysis, *CBS Publishers and distributors*, 485, Bholanathnagar, Shahdhra, Delhi-110032, pp: 520.

Hitz, H.R.; Huber, W and Rud, R.H. (1978): The absorption of dyes on activated sludge, *Journal of society of dyers and Colorists*, February: 71-76.

Horing, R.H. (1976): Characterization and treatment of textile dyeing wastewaters, In: Proceedings of national technical conference, *American association of textile chemists and colorist*, pp: 100-104.

Hoston, A.K. (1965): An index number system for rating water quality, *J. Wat. Poll. Cont. Fed.*, 37(3): 300-306.

http//:www.engg.ksu.edu as access on 16 November, 2006.

Huang, J.; Huang, R.; Jiao, J.J. and Chen, K. (2007): Speciation and mobility of heavy metals in mud in coastal reclamation areas in Shenzhen, China, *Environ. Geol.*, DOI 10.1007/s00254-007-0636-7.

Hynes, H.B.N. (1970): The ecology of running waters, *Liverpool University Press, Liverpool, 4th impression*: 1-555.

Ireland, M.P. (1983): Heavy metals uptake and tissue distribution, In: Earthworm ecology- from Darwin to vermiculture (Ed.- Satchell, J.E., *Chapman and Hall*), pp: 135-147.

IUPAC (1994): Atomic weights of the elements, 1993, International Union of Pure and Applied Chemistry, *Pure Appl. Chem.*, 66: 2423.

Jain, C.K. (2004): Metal fractionation study on bed sediments of river Yamuna, India, *Wat. Res.* 38: 569-578.

Jayabaskeran, K. J. Sree Ramulu U. S. (1996): Distribution of heavy metals in soil of various sewage farms in Tamil Nadu, *J. Indian Soc. Soil. Sci.*, 44: 401-404.

Jha, M.N. and Pandey, P. (1984): Impact of growing eucalyptus and soil monocultures on soil in natural sal area of Doon valley, *Ind. For.*, 110: 16-18.

Joshi, B.D.; Sharma, S.D.; Joshi, K.D. and Sengupta, S. (1998): Some physico-chemical properties of soil from different regions in relation to Eucalyptusization in Himalayan foothills, *Him.J. Env. Zool.*, 2(10): 58-60.

Judlins, J.F. and Hornsby, J.S. (1978): Color removal from textile dye waste using magnesium carbonate, *Journal of Water Pollution Control Federation*, 50: 2446-2456.

Kalra, Y.P. and Maynarol, D.G. (1991): Methods manual for forest soil and plant analysis. *For. Can., Northwest Reg. North. For. Cent. Edmonton, Alberta. Inf. Rep. Nor. X-319*, pp: 1-65.

Kanwar, J.S. (1961): Quality of irrigation water as an index of suitability for irrigation purposes, *Potash Review*, 13: 1-13.

Karle, K.K.; Bhusal, S.S.; Gunjal, P.S. and Kuchekar, S.R. (1992): Studied on groundwater quality in Pravaranagar area of Ahmednagar district, *Poll. Res.* 11(2): 65-68.

Kashem, M.A. and Singh, B.R. (1999): Heavy metal contamination of soil and vegetation in the vicinity of industries in Bangladesh, *Water, Air and Soil Pollution*, 115: 347-361.

Kataria, H.C. (1996): BOD and COD contents in bore well water of Bhopal (M.P.), *Natcon*, 8(1): 69-72.

Kataria, H.C.; Gupta, S.S. and Jain, O.P. (1995): Water quality of bore wells in B.H.E.L. area of Bhopal, *Poll. Res.* 14 (4): 455-462.

Katz, M. (1975): The effect of heavy metals on fish and aquatic environment. (Ed.- Krenkel, P.A.), *Pregamon Press, New York*, pp: 637-645.

Khan T.I.; Marwari Richa and Singh N. (2003): Impact of textile wastewater on *Solanum melongena* var-FI- Hybrid Kanhaiya in pot experiment with special emphasis on analysis of heavy metals. *Dimensions Polln*, 2: 108-116.

Khanna, D.R. and Bhutiani, R. (2004): Water analysis-at a glance, ASEA, Rishikesh, pp: 108.

Khurshid, Zaheeruddin and Mohd. Usman Shabeer (1997): Degradation of water quality due to heavy metals pollution in Faridabad district, Haryana, India, *Poll. Res.* 16(1): 41-43.

Korfali, S.I. and Davies, B.D.E. (2004): The relationship of metals in river sediments (Nahr-Ibrahim, Lebanon) and adjacent floodplain soil, *the CIGR J. of scientific research and development*, manuscript LW 04 010, 6: 1-22.

Krebs, R.; Gupta, S.K.; Furrer, G. and Schulin, R. (1999): Gravel sludge as an immobilizing agent in soils contaminated by heavy metals: A field study, *Water, Air and Soil Pollution*, 115: 465-479.

Krishna, A. K. and Govil, P. K. (2004): Heavy metal contamination of soil around Pali industrial area, Rajasthan, India, *Environmental Geology*, 47: 38-44.

Krull, R. and E. Dopkens (2004): Recycling of dye house effluents by biological and chemical treatment, *Water Sc. and Tech.* 49(4): 311-317.

Kudesia, V.P. (1992): Water Pollution, *Pragati Prakashan, Meerut*, pp: 407.

Kumar, Rakesh; Singh, R.D. and Sharma, K.D. (2005): Water resources of India, *Current Science*, 89 (5): 794-811.

Kumar, S.; Bhatacharjee, J.W. and Sharma, R.K. (1992): Relationship between fluoride, total hardness and total alkalinity in the groundwater of Barmer Dist. (Rajsthan), *Poll. Res.* 11(2): 111-116.

Kumar, S; Kushwaha, R.; Sapra, S. Gupta, A.B. and Bhargava, A. (2001): Impact of textile industry on groundwater quality of Sanagar, Jaipur, *J. Indian Water Works Association*, 33(4): 321-326.

Lagas, P.; Verdam, B and Loch, J.P.G. (1989): Threat to Groundwater quality by pesticides in the Netherlands, Groundwater management: quantity and quality, IAHS Publ. no. 188, *IAHS Press Walling ford, United Kingdom*, pp.171-180.

Lee, K.K.; A.M. Kassim and H.K. Lee (2004): The object of nitrogen supplementation on the efficiency of colour and COD removal by Malaysian white rot fungi in textile dying effluent, *Water Sc. and Tech.* 50 (50): 73-77.

Lofts, S.; Spurgeon, D.J.; Svendsen, C. and Tipping, E. (2004): Deriving soil critical limits for Cu, Zn, Cd, and pH: A method based on free ion concentration, *Environ. Sci. Tech.*, 38: 3623-3631.

Lokeshwari, H. and Chandrappa, G.T. (2006): Impact of heavy metal contamination of Bellandur Lake on soil and cultivated vegetation, *Current Science*, 91(9): 622-627.

Loska, K. and Wiechula, D. (2002): Speciation of cadmium in the bottom sediment of rybnik Reservoir, *Water, Air Soil Pollut.*, 141: 73-89.

Lung, W. (1990): Speciation analysis – why and how? *Fresenius J. Anal. Chem.* 337: 557-564.

Mahesha, N.N. and Prasad, N.R.R. (2004): Physico-chemical characteristics of bore well water in Arsikara taluk, Hassan, *IJEP*, 24(12): 897-904.

Malik, D.S. and Bharti, P. K. (2007): Soil quality of irrigated agricultural fields in textile industrial area of Panipat city, *Asian Journal of Experimental sciences*, 21 (2): 445-451.

Malik, D.S.; Bharti, P.K. and Sumit Grover (2006): Alteration in surface water quality near textile industries at Panipat (Haryana), *Environment Conservation J.*, 7(2): 65-68.

Malik, D.S.; Bharti, P.K; Kamboj, N. and Yadav, R. (2007): Quantification of heavy metals migration in groundwater regime due to discharge of textile industrial effluents in Panipat area, Haryana, *Pollution Research*, 26 (4): 123-125.

Malik, D.S.; Yadav, Rashmi and Bharti, P.K. (2004): Accumulation of heavy metals in crop plants through irrigation of contaminated groundwater in Panipat region. *Environ. Conserv. J.*, 5 (3): 101-104.

Manahan, S. (1984): Environmental Chemistry, Brooks/Colei, CA, USA, pp: 638.

Manchanda, H.R. (1990): Significance of the type of salinity for growing pulses under saline conditions, *Indo-Pak workshop on Soil Salinity and Water Management*, Feb, 10-14, pp: 2-3.

Mani, V.; Kaur, H. and Mohini, M. (2005): Toxic metals and environmental pollution, *J. Ind. Poll. Cont.*, 21(1): 101-107.

Martin, J.M. and Whitfield, M. (1983): The significance of the river input of chemical elements to the ocean, In: Wong, C.S.; Boyle, E.; Bruland, K.W.; Burton, J.D. and Goldburg, E.D. (Eds.), Trace metals in sea water, *Plenum Press, New York*, pp: 324-331.

Martin, M.H. and Cougherty, P.J. (1982): Biological monitoring of heavy metals pollution, *Applied Science Publishers*, pp: 463.

Masters, G.M. (1998): Introduction to environmental engineering and science, *Prentice-Hall India*, pp: 460.

Mathess, G. (1974): Heavy metals as trace constituents in natural and polluted groundwater, *Goel mijnbouw, New Jersy, USA*, 53: 149-155.

Mathur, R.P. (1993): Water and wastewater testing (A lab manual). *Nemchand and Bros. Pub.*, Roorkee, pp: 45.

McBride, M., Sauve, S. and Hendershot, W. (1997): Solubility control of Cu, Zn, Cd, and Pb, in contaminated soils, *Euro. J. Soil Science*, 48: 337-346.

McKay, G. (1979): Waste color removal from textile effluents, *American Dyestuff Reporter*, 68: 29-36.

Meier, W., and Mull, R. (1989), Nitrate in Groundwater, in Groundwater management: Quantity and quality, IAHS, Publ. no.188, *IAHS Press, Wallingford, United Kingdom*, pp: 181-190.

Michaels, G.B. and Lewis, E.L. (1985): Sorption and toxicity of azo and triphenyl methane dues to aquatic microbial populations, *Environmental Toxicology and Chemistry*, 4: 45-50.

Mico, C; Recatal, L.; Peris, M. and Sanchez, J. (2007): A comparison of two digestion methods for the analysis of heavy metals by flame atomic absorption spectroscopy, *Spectroscopy Europe*, 19 (1): 23-26.

Mido, Y. and Satake, M, (1995): Chemicals in the environment, *Discovery Publishing House, New Delhi*, pp: 491.

Millar, C.E. and Turk, L.M. (2002): Fundamentals of soil science, Biotech Books, Delhi-35, pp: 462.

Minhas, P.S. and Gupta, R.K. (1992): Quality of irrigation water – assessment and management, Publication and Information Division, ICAR, *Krishi Anusandhan Bhawan, Pusa, New Delhi*, pp: 123.

Minhas, P.S. and Khosla, B.K. (1986): Solute displacement in a silt loam soil as affected by the method of water application under different evaporation rates, *Agric. Wat. Mgmt*, 12: 63-75.

Mishra, E; Sharma, D. and Malkania, U. (1990): Fish mortality as affected by liquid effluents from Raja Textile Ltd., Rampur (U.P.), *Int. J. Ecol. Env. Sc.*, 16 (2-3): 119-123.

Mishra, N.K. and Sahoo, H.K. (2003): Evaluation of groundwater quality in and around Deogarh, *IJEP*, 23 (6): 667-672.

Mohan, Devendra and Yogesh Sharma (2002): Trace metals in a few water bodies of Pali, Rajsthan, *J. Natcon.* 14 (2): 363-366.

Mohan, R.; Chopra, N. and Chowdhary, G.C. (1998): Heavy metals (Fe, Pb, Cd, Zn) in the groundwater of Naini industrial area, Dist. Allahabad, U.P., *Poll. Res.* 17(2): 167-168.

Mohanasundaram, S. (2003): Affordable and effective textile processing effluent treatment plants. *South India's Premier Magazine on Environment Sc. and Engg.*, pp 76-82.

Moon, C. H.; Lee Y. S. and Yoon, T. H. (1991): *Environmental Technology*; 12: 413.

Mor, S.; Bishnoi, M.S. and Bishnoi, N.R. (2003): Assessment of groundwater quality of Jind city, *IJEP*, 23(6): 673-679.

MPCA (2001): Effects of Land Use on Groundwater Quality, St. Cloud Area, Minnesota – Short Report, *Minnesota Pollution Control Agency*, Groundwater Monitoring and Assessment Program, pp: 1-29.

Muller, G. (1979): Schwermetalle in den sedimenten des Rheinsveranderungen seit, *Umschau*, 79: 778-783.

Mustafa, W.; Somasundaram, S.S.N.; Shahul, P.H. and Palaniappan, R. (1998): Evaluation of groundwater quality in Tiruchirapalli city, *Indian J. Environmental Protection*, 19(4): 284-289.

Narayan, A.C. and Suresh, G.C. (1989): Chemical quality of groundwater of Mangalore city, Karnataka, *Indian J. Env. Hlth.*, 131: 220-226.

Nasr, S.M.; Okbah, M.A. and Kasem, S.M. (2006): Environmental assessment of heavy metal pollution in bottom sediments of Aden port, Yemen, *International Journal of Oceans and Oceanography*, 1(1): 99-109.

National Research Council (NRC) (1989): Contaminated marine systems- assessment and remediation, *National Academy Press, Washington*, D.C. pp: 185.

Navarro, A. and Font, X. (1993): Discriminating different sources of groundwater contamination caused by industrial waste in the Besos River Basin, Barcelona, Spain, *J. Applied Biochemistry*, APPGEY, 8: 277.

Nawar, S.S. and Doma, H.S. (1989): Removal of dyes from effluents using low-cost agricultural byproducts, *The Science of Total Environment*, 79: 271-279.

Nedunuri, K.V.; Govindaraju, R.S.; Erickson, L.E. and Schwab, A.P. (1995): Modeling of heavy metal movement in vegetated, unsaturated soils with emphasis on geochemistry, In: Proceedings of the 10th Annual Conference on Hazardous Waste Research, pp: 57-66.

Neelima Rajvaidya and Dilpi Kumar Markandey (1998). Advanced in Environmental Sciences and Technology. *A.P.H. Publishimg Corporation,* New Delhi, 6: 231-249.

Nemerow, N.L. (1971): Industrial water pollution: origins, characteristics and treatment, Addision, *Wesley Publishing* Co., pp: 532.

Nightingale, H. I. (1987): *Wat. Res. Bull.*, 23: 663.

Nongkynrih, P.; Dkhar, P.S. and Khathing, D.T. (1996): Micronutrients elements in acid alfisols of Meghalaya under rice cultivation, *J. Soil Si.*, 44(3): 455-457.

Nriagu, J.O. (1989): A global assessment of natural sources of atmosphere trace metals, *Nature*, 338: 47-49.

Nriagu, J.O. and Pacyna, J.M. (1988): Quantitative assessment of worldwide contamination of air, water and soil with trace metals, *Nature*, 333: 134-139.

Numberg, H.W. (1984): The volumetric approach in trace metal chemistry of natural waters and atmospheric precipitation, *Analyst. Chim. Acta*, 164: 1-21.

Obiri, Samuel (2007): Determination of Heavy Metals in Water from Boreholes in Dumasi in the Wassa West District of Western Region of Republic of Ghana, *Environ Monit Assess*, 130: 455–463.

Okonwo, J.O.; Lwenje, S.M.; Mtetwa, V.S.B.; and Mkhweli, T.B. (1999): Measurement of lead concentrations in the hair of school children in Manzini region, Swaziland, *Int. J. Environ. Stud.* 56: 419-428.

Okonwo, J.O.; Lwenje, S.M.; Mtetwa, V.S.B.; Gumedge, P.N. and Shilongonyane, M.M. (2001): Determination of urinary lead in school children in Manzini region, Swaziland, South Africa, *Environmentalist*, 21: 205-209.

Padmavathiamma, P.K. and Li, L.Y. (2007): Phytoremediation Technology: Hyper-accumulation Metals in Plants, *Water Air Soil Pollut.*, DOI- 10.1007/s11270-007-9401-5.

Palanivelu, K.; Priya, M.N.; Selvan, A.M. and Natesan, U. (2006): Water quality assessment in the tsunami affected coastal areas of Chennai, *Current Science*, 91 (5): 583-584.

Pathe, P.P.; Biswas, A.K.; Rao, N.N. and Kaul, S.N. (2005): Physico-chemical treatment of wastewater from clusters of small-scale cotton textile units, *Environmental Technology*, 26: 313-327.

Poots, V.J.; Mckay, G. and Heakt, J.J. (1976): The removal of acid dye from effluent using natural absorbents-II, *Water Research*, 10: 1067-1070.

Porter, J.J. (1970): The changing nature of textile processing and waste treatment technology, *Textile chemist and Colourists*, 2: 336.

Porter, J.J. (1972): Treatment of textile waste with activated carbon, *American Dyestuff Reporter*, 8: 2427.

PPCB (1993): A report on groundwater quality of Ludhiana city, Punjab Pollution control Board, pp: 50.

Prasad, B.V. and Chandra, P.R. (1997): Groundwater quality in an industrial zone, *Poll. Res.* 16(2): 105-107.

Prasad, S. (2003): Fundamentals of biostatistics (biometry), *Emkey Publication, Delhi-51*, pp: 168.

Pujari, G.K. and Sinha, B.K. (1999): Studies on the water and soil quality of some villages of Attabira area irrigated by Bargarg main canal originated from Hirakund reservoir of Orissa, *J. of Env. and Poll.*, 6(1): 71-76.

Ramani, K.V. and Srivastava, M.S. (1989): Effect of flyash from thermal power station on soil, *BHEL J.* 10: 1-4.

Rao, K.V.G.K.; Gupta R.K. and Kamra, S.K. (1987): Reclamation of waterlogged high SAR saline soil, CSSRI, Karnal, A feasibility report for *CIRB, Hissar, CSSRI, Karnal*, pp: 21.

Rao, N.S. and Rao, V.V.S.G. (1999): Pathlines of pollutants migration in groundwater of the Visakhpatanam urban area, India, *Hydrological Process*, 13: 1381-1389.

Reddy, K.J. and Jianping Lin (2000): Nitrate removal from groundwater using catalytic reduction, *Water Res.* 34 (3): 995-1001.

Reshma, S. and V.R. Prakasam (2003): Potability of dug wells of Mayyand Panchayat, Kerala, *Indian J. Environ. and Ecoplan.* 7(3): 471-476.

Rhoades, J.D. (1987): Use of saline water for irrigation, *Wat. Qual. Bull.*, 12: 14-20.

Richards, L.A. (1954): Diagnosis and improvement of saline and alkali soils, US Deptt. Agri., edited Handbook, No. 60: 160.

Rizk, Z.S.; Alsharhan, A.S. and Wood, W.W. (2007): Sources of dissolved solids and water in Wadi Al Bih aquifer, Ras Al Khaimah Emirate, United Arab Emirates, *Hydrogeology Journal,* DOI 10.1007/s10040-007-0188-7.

Royee, M.K.P. and Prakasham (2003): Water characteristics of dug and tube wells of Kollam municipality, *IJEP*, 23(6): 607-612.

Saed, K.; Noor, M.M.M. and Yusuf, B. (2005): Sugarcane bagasse as an adsorbent for dye removal, *J. Ind. Poll. Control*, 21 (1): 1-10.

Saksena, R.K.; Sharma, M.L. and Jodha H.R. (1966): Quality of groundwaters for irrigation in Ahor developmental block, Jalore, *Ann. Arid Zone*, 5: 204-218.

Samy, D.I.A. and Gananarethinam, J.L. (1980): Effect of distillery effluents on the growth of three aquatic macrophytes, *J. Comp. Physiol. Ecol.*, 5(4): 290-295.

Sastry, K.V. and Pratima Rathee (1999): Groundwater quality in three villages of Rohtak district. *J. Natcon*, 11(2): 175-182.

Sastry, K.V. and Rathee, P. (1999): Groundwater quality in three villages of Rohtak district, *J. Natcon*, 11(2): 175-182.

Saxena, M.M. (1994): In: Environmental analysis water, soil and air, *Agro Botanical Publishers*. pp: 1-180.

Schroeder, W.H. (1989): Development in the speciation of mercury in natural waters, *Trends Anal. Chem.* 8: 339-342.

Seeber, P.R. (1962): Cation hydrological facies of groundwater in the English town Formation, New Jersey, In: Short paper in geology, hydrology and topography, USGC, Profes. Paper, 450B: 124-126.

Senthinathan, S. and Azeez, P.A. (1999): Influence of dyeing and bleaching industries on groundwater of Tripur, Tamilnadu, India, *Bull. Environ. Contam. Toxicol.*, 62: 330-335.

Seshadri, S.; Bishop, P.L. and Agha, A.M. (1994): Anaerobic and aerobic treatment of selected azo dyes in wastewater, *Waste Management*, 15: 127-137.

Sharma, D.R. (1978): Testing of a model for prediction sodium hazards of irrigation waters, *J. Indian Soc. Soil Sci.*, 27: 204-208.

Sharma, D.R. and Parihar, S.S. (1973): Effect of depth and salinity of groundwater on evaporation and soil salinization, *Indian J. Agric. Sci.*, 43: 582-586.

Sharma, K. and K.S. Patel (1996): Spectrophotometric determination of Copper in water, *Journal of Indian Chemical Society*. 73, pp: 443-444.

Sharma, K.P.; Sharma, K.; Bhardwaj, S.M. and Chaturvedi, R.K. (1999): Environment impact assessment of textile printing industries in Sanganer, Jaipur: A case study, *J. Ind. Bot. Soc.* 78:71.

Sharma, R.C.; Abrol, I.P. and Bhumbla, D.R. (1975): Quality of groundwaters of Jind district, Haryana, Bulletin No. 3 (ICAR), *Central Soil Salinity Research Institute (BKAP), Karnal*, pp: 63.

Sharma, R.C.; Abrol, I.P. and Bhumbla, D.R. (1976): Quality of groundwaters of Karnal district, Haryana, Report No. 3 (ICAR), *Central Soil Salinity Research Institute (BKAP), Karnal*, pp: 47.

Sharma, S.; Jain, P.C. and Mathur, R. (1995): Quality assessment of groundwater in municipal and fringe areas near Gwalior, *IJEP*, 15(7): 534-538.

Sheth, K.N. and Desai, A.P. (2005): Study of bio-kinetics of textile wastewater by designing a Bench Scale Completely Mixed Activated Sludge Reactor Model, *J. Ind. Poll. Control*, 21 (1): 11-16.

Sheth, K.N. and Patel, M.R. (2001): Characterization, treatment and comparative cost analysis studies of textile processing wastewater of Vatva industrial complex, *J. Engg. and Tech., S.P. University*, 14: 235-241.

Shivakumar, K.; Pandey, A.K. and Bhiksham, G. (1977): Toxic trace element pollution in groundwater around Patancheru and Bollaram Industrial area, A.P., India: a graphical approach, *Env. Monitor. Assess.*, 45 (1): 57-80.

Shrivastava V.S. and Patil B.H. (2003): Metallic and some physico-chemical studies of soil and aquatic sediments, *Eco Env Conserv*, 9(1): 75-77.

Sial, R.A.; Chaudhary, M.F.; Abbas, S.T.; Latif, M.I. and Khan, A.G. (2006): Quality of effluent from Hatter Industrial Estate, *Journal Zhejang Univ. Science B*, pp: 974-980.

Siddiqui, Z.M. and Pathania, D. (2002): Studies on heavy metals in surface and groundwater of Jalandhar and Ludhiana districts, *IJET*, 22(2): 201-206.

Singh, A.K. (2006): Chemistry of arsenic in groundwater of Genges-Brahmaputra river basin, *Current Science*, 91 (5): 599-606.

Singh, A.K.; Jaiswal, R.K. and Mukharjee, S. (1996): Groundwater quality assessment of Varanasi city, *Hydrology Journal*, 19(4): 23-31.

Singh, B. and Bhumbla, D.R. (1968): Effect of quality of irrigation water on soil properties, *J. Res.* (Punjab Agri. Univ.), 5: 166-171.

Singh, B. Rana, D.S. and Bajwa, M.S. (1977): Salinity and sodium hazards of underground irrigation waters of Bhatinda district (Punjab), *Indian J. Ecol.*, 4: 32-41.

Singh, K.N.; Bains, S.S. and Dayanand (1969): Salinity problem in high water table areas- An appraisal, *Indian J, Agron.*, 14: 31-34.

Singh, K.P.; Takroo, R. and P.K. Roy (1987): Pesticides and heavy metal residue analysis in water, ITRC Manual. *ITRC*, Lucknow, pp: 1-160.

Singh, K.S. and Sharma, R.P. (1971): Studies on the effects of saline irrigation waters on physico-chemical properties of some soils of Rajasthan, *J. Indian Soc. Soil Sci.*, 18: 345-356.

Singh, T.B.; Jadon, S.P.S. and Mishra, G.J. (1994): Degradation of water and soil quality of Parwanoo area with respect to heavy metals, *IJEP*, 14 (4): 282-287.

Singh, V.K. (2006): Studies on distribution of heavy metals in Gomati river and development of low cost adsorbents for their removal, *Ph.D. Thesis, Gurukula Kangri University, Hardwar*, pp: 296.

Soltan, M.E. (1999): Evaluation of groundwater quality in Dakhla oasis (Egyptian Western Desert), *Environmental Monitoring and Assessment*, 57: 157-168.

Souther, R.H. and Alsapaugh, T.A. (1957): Biological treatment of mixture of textile wastes and domestic sewage. *Textile*, 2 (1): 135-139.

Srinivas, C.; Piska, R.S. and Reddy, R. (2002): Groundwater pollution due to the industrial effluents in Kothur industrial area, Mahaboonagar, A.P., *Asian J. Micobiol. Biotech. Env. Sci.*, 4(3): 39-42.

Srinivasan, M. and Murali, M. (1996): Some studies on effect pollutants on properties of soils, *IJEP*, 16(7): 522-523.

Stetzenbach; Klaws, J.; Irene M. Farnham; Vernon F. Hodge and Kevin H. Johannesson (1999): Using multivariate stastical analysis of groundwater major cation and trace element concentrations to evaluate groundwater flow in a regional aquifer, *Hydrol. Process*. 13: 2655-2673.

Tan, K.H. and Nopamornbodi, O. (1981): Electron microbeam analysis and scanning electron microscopy of soil-root interfaces, *Soil Sci.*, 131: 100-106.

Taylor, S.R. (1964): Abundance of chemical elements in the continental crust – A new table, *Geochim. Cosmochim. Acta*. 28: 1273-1275.

Thorat, P.R. and Pathade, G.R. (2001): Textile waste: Characterization and treatment, In: Environmental Pollution and management of wastewaters by microbial techniques (Eds.- Pathade, G.R. and Goel, P.K.), ABD Publishers, Jaipur, India, pp: 316-325.

Todd, D.K. (1980): Groundwater hydrology, John Wiley, New York: 554.

Trevors, J.T. and Saier Jr., M.H. (2007): Regulation of Pollution, *Water Air Soil Pollut*, DOI 10.1007/s11270-007-9344-x.

Trivedi, R.K. and Goel, P.K. (1984): Chemical and biological methods for water pollution studies Karad, *Environmental publication*, pp: 1-251.

Tyagi, P. and Budhi (2000): Degradation of groundwater quality due to heavy metals in industrial area of India- A review, *Indian J. Env. Prot.*, 20 (3): 174-181.

Umar, R. and Ahmad, M.S. (1999): Groundwater quality in parts of Central Ganga Basin, India, *Environmental Geology*, 39 (6): 673-678.

USEPA (1990): Operation and quality control manual, Environmental Protection Agency (EPA), Athens, GA 30613.

Va'zquez, N.N.; Gil, M.A.; Esteves, J. L. and Narvarte, M. A. (2007): Monitoring Heavy Metal Pollution in San Antonio Bay, Rio Negro, Argentina, *Bull. Environ. Contam. Toxicol.*, DOI 10.1007/s00128-007-9084-z.

Vaidge, A.A. and Datye, K.V. (1982): Environment pollution during chemical processing of synthetic fibers, *Colourage*, 14: 3- 10.

Vates, M.V. (1986): Septic tank density and groundwater contamination, *Groundwater*, 23(5): 586-590.

Vazquez, N.N.; Gil, J.L.; Esteves, and Narvarte, M.A. (2007): Monitoring heavy metal pollution in San Antonio Bay, Rio Negro, *Argentina, Bull. Environ. Contam. Toxicol.*, DOI- 10.1007/s001 28-007-9084-z.

Verma, N.K. and Kharat, D.S. (2000): Textile Industry, *Central Pollution Control Board*, New Delhi, pp: 82.

Verma, S.R. and Mathur, R.P. (1974): Studies on the toxicity of industrial wastes to macrobranchium deyanum, *Indian J. Environ. Hlth.*, 16(1): 1-11.

Vijith, H. and Satheesh, R. (2007): Geographical Information System based assessment of spatiotemporal characteristics of groundwater quality of upland sub-watersheds of Meenachil River, parts of Western Ghats, Kottayam District, Kerala, India, *Environ Geol*, DOI 10.1007/s00254-006-0612-7.

Vishwanath, G. and Ananthmurthy, K.S. (2002): Assessment of groundwater quality around a solid waste dumpsite, *Nat. Env. Poll. Tech.*, 1(4): 411- 413.

Wainwright, John and Mulligan, Mark (2004): Environmental modeling, *John Wiley and Sons Ltd.* pp: 408.

Walker, D.J.; Bernal, M.P. and Correal, E. (2007): The Influence of Heavy Metals and Mineral Nutrient Supply on *Bituminaria bituminosa*, *Water Air Soil Pollut*, DOI 10.1007/s11270-007-9422-0.

Walker, R. (1970): The metabolism of azo compounds: a review of the literature, *Food Cosmetics Toxicology*, 8: 659-661.

Watler, J.K. and Wint, M. (1981): Industrial effluent treatment, *Applied Science, UK*, pp: 548.

Wetzel, R.G. (1975): Limnology, *W.B. Saunders Co. Philadelphia*, pp: 743.

WHO (1984): Guideline for drinking water quality, Recommendations, *World Health Organization, Geneva*, pp: 130.

WHO (1988): International standards for drinking water. Word Health Organization, Geneva. pp: 142.

WHO (2006): Standards for drinking water, available at http//:www.lenntech.com

Widyanto, L.S. (1975): The effect of industrial pollutants on the growth of water hyacinth (Eicchornia crassipes Mart Solm.), In: *Proceedings of the third Indonesian Weed Science Conference*, Bandung, pp: 328.

Williums, D.R. (1972): Metals, Ligands and Cancer, *Chemical Review*, 72 (3): 203-213.

Wint, A. (1981): The disposal of toxic wastes, In: Industrial Effluent Treatment Vol-I (Eds.- Watler, J.K. and Wint, M.), pp: 1-19.

Wong, P.K. (2000): Application of enhanced photocatalytic oxidation in the treatment of azo dyes in textile wastewater, In: Advances in Wastewater Treatment Technologies (II), (Eds.- Trivedi, R.K. and Kaul, S.N., *Global Science Publications, Aligarh*, pp: 302.

www.http//mapsofindia.com, as access on 12 October 2006.

www.http://boulder.com as access on 07 June 2007.

www.http://cpcb.nic.in as accessed on 22 February 2007.

www.http://Earth&table.com as accessed on August 2006.

www.http://en.wikipedia.org as access on 18 October 2006.

www.http://epa.gov/safewater/mcl.html.

www.http://haryanaonline.com as access on 21 March, 2006.

www.http://holar.is as access on 29 June, 2007.

www.http://indiatravelogue.com as access on 21 March, 2006.

www.http://lef.org as accessed on August 2006.

www.http://lenntech.com/aquatic/metal.htm as accessed on May 2007.

www.http://tuberose.com as accessed on August 2006.

www.http://vistamagonline.com as accessed on August 2006.

www.http://who.in as accessed on 25 May 2007.

Yadav, A.K; P.K. Jain and Jyoti Sharma (2003): Assessment of groundwater quality of Behror textile of Alwar Dist., Rajasthan, *Aquacult*, 4(2): 265-270.

Yadav, J. P. and Lata, S. (2003): Pollution of fluoride in groundwater in Bahadurgarh block of district Jhajjar, Haryana, *Indian J. Environmental Protection*, 23(6): 680-686.

Yadav, R.K.; Goyal, B.; Sharma, R.K.; Dubey, S.K. and Minhas, P.S. (2002): Post-irrigation impact of domestic sewage effluent on composition of soils, crops and groundwater- a case study, *Environment International*, 28: 481-486.

Yu, F-Y; Li, C-W and Kang, S-F. (2005): Color, dye and doc removal, and acid generation during fenton oxidation of dyes, *Environmental Technology*, Vol-26: 537-544.

Yuandong, Z.; Shirong, L. and Jiangming, M. (2006): Water-holding capacity of ground covers and soil in alpine and sub-alpine shrubs in western Sichuan, China, *ACTA Ecologica Sinica*, 26(9): 2775-2782.

Yusuf, Y.A.; Wanielista, M.P.; Harper, H.H. and Hvitved-Jacobsen, T. (1986): Best Management practices- effectiveness of Retention and Detention Pond for control of contamination in Highway Runoff, *Final Report, University of Florida, Orlendo*, pp: 151.

Yusuff, R.O. and Sonibare, J.A. (2004): Characterization of textile industrial effluents in Kaduna, Nigeria and pollution implications, *Global nest: Int. J.*, 6(3): 212-221.

Zhang, J; Crittenen, J.C. and Hand, D.W. (1994): The solar photocatlytic decontamination of water, *Chemistry and Industry*, 19(1): 714-717.

Zheng, G.; Yue, L.; Li, Z and Chen, C. (2006): Assessment on heavy metals pollution of agricultural soil in Guanzhong District, *J Geographical Sciences*, 16, (1): 105-113.

Zilliox, Lothaire (1989): Industrial impact on the quality of groundwater in a large Basin: Case of the Rhine aquifer, *Water International*, 14(2): 62-68.

Zollinger, H. (1987): Colour chemistry synthesis properties and application of organic dyes and pigment, *New York VCH*, pp: 275.

Index

www.ingramcontent.com/pod-product-compliance
Ingram Content Group UK Ltd.
Pitfield, Milton Keynes, MK11 3LW, UK
UKHW022215051026
15624UKWH00001B/177

9 789351 242086